D1053317

THE
AGE
OF THE
HORSE

THE
AGE
OF THE
HORSE

**An Equine Journey Through
Human History**

SUSANNA
FORREST

Atlantic Monthly Press
New York

First published in Great Britain in 2016 by Atlantic Books,
an imprint of Atlantic Books Ltd.

First Grove Atlantic hardcover edition: May 2017

Published simultaneously in Canada
Printed in the United States of America

ISBN 978-0-8021-2651-1
eISBN 978-0-8021-8951-6

Atlantic Monthly Press
an imprint of Grove Atlantic
154 West 14th Street
New York, NY 10011

Distributed by Publishers Group West

groveatlantic.com

17 18 19 20   10 9 7 6 5 4 3 2 1

For Henry

Contents

Illustrations

BLACK & WHITE ILLUSTRATIONS

THE
AGE
OF THE
HORSE

Introduction

'Bridle Road to ... ' When I see this notice in England it has the same effect on me as Mescalin does on Mr Aldous Huxley. Here there are no such notices but you can see bridle roads leading over the plains and the sierras in every direction and to an addict the sight is intoxicating. Everyone has his weaknesses: some people run after women, others after Dukes; I run after priests and along *carriles* which, with their alluringly sinuous ways, are gravely tempting me to throw all my family duties to the wind and to go on riding along them forever.

Two Middle-Aged Ladies in Andalusia
by Penelope Chetwode (1963)

This is not a history of the horse. It is a wander down six bridle roads that in turn branch out into more pathways, whether on the Copper Age Kazakh steppe, in Mauryan India, industrial England, an eco settlement in rural Massachusetts or smoggy twenty-first-century Beijing. Sometimes we pass along them at a brisk canter, at other times we slow, dismount and survey the pebbles and the sand of the track. I could have followed them forever and happily wandered away my life – maybe I will – but for now here are itineraries that will guide you through six ways in which we have used the horse, and the routes that ideas, people and horses took across an ever-changing territory. On the way you may spot tempting *carriles* to explore on your own.

Horses are so common in history that we glance over them without seeing them when they fill royal parade grounds, frame battle scenes, clog teeming Victorian streets, or amble under kings and queens or before peasants' ploughs. My aim is to draw your focus to the horses in those images and in the margins of texts. The men and women of the past knew the value of horses, they lived alongside them, their scent, their character, their strength, the cost of their shoes and the grain they lipped up. Empires like the Mongolians' have both risen and fallen on the availability of fodder for their cavalries. Many in the present know the value of horses too, and in this book these contemporaries are not confined to a brief end chapter but woven into a tapestry in which the warps are open-ended, never completed. Every time the age of the horse is said to be over, a new use is found for *Equus caballus*.

For we use the horse in more ways than any other animal: we ride on its back, attach it to wagons and ploughs, strap packs to it, drink its milk, eat its meat, go to war on it, cherish it as a pet and have turned it into a symbol of everything from wealth to political power, purity, lasciviousness and human suffering. In 5,500 years of domestication, humans have transformed horses' bodies into everything from buttons to thrones.

The maps in *The Age of the Horse* are paired: horses valued for their wildness and horses valued for their sophisticated training; horses as both industrial machines and industrial products; horses as luxury status symbols and as anarchists. They also interact with one another, despite their geographical range from China to New England, and across time from 56 million years ago to the present day. In each case, there's a site I visited between the summers of 2013 and 2015: a steppe in Mongolia, a chandelier-lit manège in Versailles, a farm in New England, a sale barn in rural America, a polo field outside Beijing, a bullring in Lisbon and a fort in Virginia.

These maps chart the metamorphosis of the wild horse into a Nazi symbol. They recall the dancing horses that lent discipline to the court of Louis XIV and held the citizens of London, Paris and New

York spellbound, and the dray horses that were recast as royal mounts and solar engines. They show how horsemeat influenced presidential elections and inspired terrorist attacks, and they examine the complicated relationship between the value and worth of steeds coveted by Chinese emperors at the expense of their human subjects. Lastly, they give form to the political and emotional lives of equines themselves, and the way that their own nature enabled them to go to war in our service.

History is generally not told from the perspective of the horse – all attempts are obviously mere ventriloquy. Its acceptance of us and our actions never ceases to baffle, but it's that cooperation, whether driven by neuroendocrinology, the need to eat, fear, behavioural or genetic manipulation, that has enabled mankind to conquer tremendous territories, feed and clothe nations and calm the damaged. Whether the horse likes it or not, it has been given a place of privilege among other animals and called a comrade. It has also been grossly misused. Such are the costs and losses of being co-opted by a fellow mammal who has both saved the horse from extinction and destroyed it in numbers never tallied.

Here, then, are a handful of the stories we have told ourselves about horses over thousands of years, and six tales of the creature that the Comte de Buffon, a French naturalist, called in a telling paradox 'the most noble conquest of mankind'.

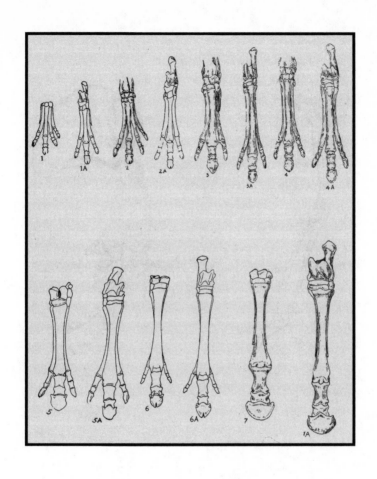

EVOLUTION

An Onion Can Turn into a Lily

The history of the horse family is still one of the clearest and most convincing for showing that organisms really have evolved, for demonstrating that, so to speak, an onion can turn into a lily.

Horses: The Story of the Horse Family in the Modern World and through Sixty Million Years of History
by George Gaylord Simpson (1951)

Any beginning in nature is arbitrary. So let it be 56 million years ago, when *Sifrhippus* or 'zero horse' was a 12-lb runt browsing on fruits and low branches in the Bighorn Basin in Wyoming, its back flat like a deer's, hind legs crouching. This is the onion that the palaeontologist George Gaylord Simpson said would eventually morph into the lily which is *Equus caballus*. It had four toes balancing that puny weight on its feet (a fifth went unused), for stability on unsure ground. By its looks, it could have been a proto-camel, a proto-deer, a proto-tapir, a giraffe, a moose-to-be. A rabbit, as creationists have claimed. Give us another ten years and we will have found new fossils, and there will be a new name, a few million years skipped back and our notion of the earliest 'horse' will shapeshift once again, even as the bones of a still earlier proto-horse lie stony and undiscovered under a Wyoming cliff. But it will do as our onion that turns into a lily, although it is perfectly happy and functional as an onion.

The evolution that follows is not the smooth, Russian-doll progression of old-fashioned natural history museums and biology diagrams in which those extraneous toes are sloughed away, the neck lengthens and cannon bones elongate until – shazam! – the family *Equidae* is fit to meet its human master, but instead a daily effort to live in climates and landscapes that shift by the millions of years. These proto-horse onions are not working at progress, but survival and reproduction. They are, in our own terms, very successful – they survived longer by far than we have managed to be human.

They are not a single-file parade but a broad band of animals running swiftly towards us, flowing, changing, some dropping back or vanishing, others running on. They may lose one feature only to regain it later, or find that they have compromised themselves in one way to make another way of, say, feeding or running easier. They appear and disappear around the globe, coexisting with one another. They are numberless – new species are found in old swamps and dry basins all the time, and rearticulated with computer graphics or artists' impressions, given stripes and crested manes.

The tiny 'dawn horses' became extinct on the Eurasian continent during the Oligocene 23 million years ago, but in North America they prospered for stretches of time unimaginable to us, until America was overrun by grasslands, and the proto-horses that come to the fore – *Mesohippus*, *Miohippus*, *Parahippus* – developed high-crowned teeth that ran deep into their jaws, for grinding down the lignin and silica in the grass. Their heads were long so that they could graze with their eyes above the level of that vast expanse of herbage and see what was approaching them. They were perhaps twice the size of *Sifrhippus* and his fellow dawn horses, and growing. Two of the toes had gone, the central toe become larger, fusing bones to make a percussive instrument to beat against the firmer ground and propel the horse away from predators no longer impeded by forest foliage. The sugars in the grass caused inflammation in these stiffer feet, but the added speed and ready forage was worth it.

To either side of that hoof were smaller digits that still provided balance, but they would become vestigial nubs of tough, horny material behind the fetlock or inside the knee as *Mesohippus* and his peers fell back and the 'true horses' – *Pliohippus, Dinohippus, Hippidion, Hipparion* – took up the baton over slow millennia. The central, divot-like toe spread and rounded, and the joints above it lost much of their ability to rotate. These *Equidae* were the size of small ponies, with heavy jaws and profiles that tended towards convexity, like the faces of rams. They were almost lilies. The first *hominin* put in an appearance in the fossil record during their era.

Four million years ago, and a wavering distinction could be drawn between horse, donkey and zebra. Two and a half million years ago, and the early horses passed over the land bridge of grassland steppe from eastern America to western Eurasia, and began to have to deal with a new predator. On fossil zebra bones in Ethiopia slashing grooves appear, made, it seems, while some creature with a stone tool was removing as much flesh and sinew as possible – more than he could manage with his teeth.

Then, over 1.8 million years ago, *Homo erectus* makes it out of Africa. Man has travelled at least as far as Schöningen in Germany 1.5 million years later, and he is hunting horses with spears. The cleaned bones of twenty *Equus mosbachensis* lie scattered by a lake shore, with the eight javelins of spruce and pine used to bring them down. At this time, there are horses all across the globe from the Americas to the Eurasian landmass. What they look like is unclear, but their bones indicate that they are still a tribe of multiple different species, albeit with frames that do not greatly differ from one another. There is no recognizable ancestor of the Shire horse or the Arabian, just small, swift, heavy-headed equines sought after as food by early men. They refuse to divide themselves into convenient wilder, smaller versions of the hot-house lilies we keep today, but they are becoming *Equus caballus*.

But with *Homo sapiens* comes a sudden flare of torchlight across Eurasia, and we can at last see what our horse looks like through human eyes: full-bellied and fine-legged, he flees across Lascaux cave walls, his dun sides shaded in with ochre and haematite and black hoofs marked in charcoal. At Niaux his leg joints have been accurately rendered by someone who has scraped them clean of flesh and sinew, and he has a shaggy beard and near smile on his pale muzzle. At Peche Merle he has a narrow black head and startling spots that dance across his sides and surround him. A side chamber at Lascaux reveals a small, fat, black and white horse running with duns and bays. You can hold them in your hand, too: mammoth ivory carved into an arched neck with notched crosses to indicate a mane; a lump of bright orange amber shaped into a horse with piggy eyes and ears pinned back; and on a fragment of bone, a wild-eyed, bristly *Equus caballus* is etched in detail down to the grooves along his nose and the irregular shape of his nostril. They are known, measured, sought.

Horses survived the extinctions that removed the woolly mammoths and rhinos, the short-faced bears, the sloths the size of oak trees, and the dire wolves and sabre-tooths that once pursued them, but in the Americas, as some of the grasslands gave way to forest and tundra, the horses began to shrink, and then, unable to cross the now-sunken land bridge, were hunted or simply dwindled away until there were none left on the continent somewhere in the eighth millennium BC. And then much of Europe also became more hostile to horses: they were hunted in greater numbers as more and more men appeared, and the continent began to be shaded over with woodland that ate up the grasslands for which they'd so slowly adapted.

Horses escaped south and west to Iberia, and east to the grasslands that reached from the Carpathians to Siberia. This landscape remained open but for scant forested areas, with an arid if harsh climate. Horse hoofs could find solid purchase on the dry ground. The animals could graze, keeping a watchful eye on their surroundings and to the far horizon, where the land became the sky. Here they began to grow larger than their poorer relations left behind in pockets of western and central Europe. What happened to the horses on the steppe would shape the future of *Equus caballus* and ensure both its extermination and its preservation.

DOMESTICATION

A Tooth, a Grave and Mare's Milk

When the Man and the Dog came back from hunting, the
Man said, 'What is Wild Horse doing here?' And the Woman
said, 'His name is not Wild Horse any more, but the First
Servant, because he will carry us from place to place for
always and always and always. Ride on his back when you
go hunting.'

The Cat That Walked by Himself
by Rudyard Kipling in *Just So Stories* (1902)

The villages of the Botai culture lay east of the Urals in the Copper
Age, by the banks of the Iman-Burluk river where the steppe was partly
interthreaded with sparse forests of pine and birch. After a Stone Age
of roaming hunter-gathering, the Botai had taken root in these roughly rectangular sunken houses with walls made from clay packed with
quartz, and thatched-over wooden roofs. The houses stood in rows or
gathered around shared spaces. There were over a hundred in one settlement alone.

What abundance of food made these people believe that perma-
nence was better than the restless pursuit of game? They seldom fished
or hunted for the local wild cattle or aurochs. They grew no grain or
other crop. They had dogs that resembled modern Samoyed hounds,
and some copperware, but their true economy was one of horsemeat,
horse bones and horse milk: over 90 per cent of the bones found at
their scattered sites were equine. The thatched roofs were insulated

with horse dung. Horse jawbones were used to scrape leather thongs made from horsehide. Horse cannon bones were turned into hooked spear heads, and tendons used for thread. Ribs and shoulder blades shaped and smoothed the Botai's clay, round-bottomed pots.

The hourglass-shaped short pastern bone, taken from just above the hoof, was stippled with dots, dashes and geometric patterns and left in caches in hollows under the houses – headless, feminine shapes that may be women in embroidered dresses. Only two Botai graves have been found, but one, which contains two men, a woman and a child, was scattered with the skulls and a few other bones of fourteen horses. Small pits contain equine skulls pared into masks or dishes, and slices of neck still on the vertebra left on altar-like stones.

Horsemeat protein was found in the Botai's pots. In the grasslands the meat and grease from equid joints provided fatty acids that men in a more diverse biosphere would have found in seeds and nuts. The pots yielded up something else, too: mare's milk, perhaps fermented into cheese, yoghurt or a mildly alcoholic, sour drink that made it easier to digest. It would have given the Botai the vitamin C they lacked, and more of those fatty acids. But how could the Botai have obtained mare's milk from wild horses?

Perhaps the rawhide thongs were used to make lassos or hobbles, and the area of earth in one village that is enriched by old horse dung and urine was a corral. Some of the horse skulls had been poleaxed – again, not an action that can be performed on a wild horse. They seem to have been butchered among the houses, which makes it hard to believe that wild horses could have been repeatedly driven in from the steppe by men on foot, or dragged as fresh kills by manpower alone. The butcher work was not especially thorough, suggesting that these were not hard-won meals stalked for days across the steppe, but acquired easily and processed rapidly.

The vertebrae of the Botai horses do not show any obvious signs of riding, but some believe that their teeth are worn by horsehair, leather or bone bits. The Botai may have ridden them in order to hunt local wild horses with those spears made from cannon bones, or used them to haul wild horse carcasses back to the village. Perhaps their dogs were enough to herd the wild horses. But these Copper Age Kazakhs were the earliest proven horsekeepers, and at the time they were making their horse leather and smoothing their pots, large horses began to head back east towards Europe again, and in the grasslands around them, more horse-centric peoples appear.

These horses also developed new coat colours unrecorded on Stone Age cave walls: horses that were bay, black, dun or spotted like snow leopards were now joined by chestnuts, smokey blacks, horses with cloud-like white sabino roaning, or dark bodies with pale manes – the negative of those Lascaux horses. The gene that made cream, blue-eyed horses or turned chestnut into palomino and bay into buckskin appeared. More mares were captured or lured in from the wild to add to the lineages. Horses began to alter in muscle and bone, and in their heart anatomy but also their mentality, which became more tolerant of humans, more able to learn and less fearful.

As the Copper Age hardened into the Bronze, sheep and cattle bones took the place of horses' in the middens, and horses were now placed in the graves of warriors and the wealthy. The horse people began to cover greater distances, either on the backs of their horses or with wagons drawn by them or oxen. They took with them languages, religious ideas, inventions and goods they had acquired or traded from both the Far East and the West. Most of all, they took horses.

The Tarpan, or Wild Horse.

WILDNESS

A Swift and Savage Breed

The bones of one's horse are not to be discarded
in a foreign land.

Mongolian saying

I

It's a thousand miles as the crow flies from the open steppe of Askania-Nova north of the Crimea to Berlin. In the summer of 1942, the dirt roads were dry and dusty but passable, and German troops, guns and horses flowed east and south, heading across the grasslands north of the Black Sea towards the Don River and onwards to Stalingrad.

Nearly 750,000 horses had been moved to the Eastern Front to haul the artillery of Operation Barbarossa, and by that summer, a third of them were dead or useless. Farm horses were requisitioned from Germany, from Poland and all the new *Lebensraum* the Wehrmacht had passed over on its way to the steppe. Up to 4,000 horses a week were taken from Poland alone for the Western Front. That summer, hundreds of thousands of horses flowed from the new, expanding heartlands of Greater Germany towards its raw edges. But two went the other way – not because they were being given temporary care or respite behind the lines, but because they were being taken back to the very centre of Nazi Germany, Berlin.

The dun stallion and mare were small, the size of ponies. They were not broken to harness nor to saddle. Unlike the Lipizzaner horses that the Nazis were also transferring from Piber in Austria to Czechoslovakia, they were not highly trained or even trained at all. It would have taken days to drive them in off the steppe of the Askania-Nova estate, where they grazed with five herdmates. They were probably packed into wooden crates for their train ride rather than standing in a wagon with six others like the farm horses off to the front, so that when they kicked and fought, they would not harm each other or fall badly. Scared, they battered against the boards with their heels, calling to one another. They would be the last of their herd to leave Askania-Nova, and the horses they left behind would not survive the war.

When the Germans had arrived in Askania-Nova on the heels of the retreating Red Army in September 1941, they had found the threshed wheat on fire in the fields, the farm machinery wrecked and the buildings tumbledown. Field Marshal Erich von Manstein set up his headquarters there, ordering his men to check for booby traps and take an inventory of the contents of the estate. They had counted the seven dun horses, and duly notified Berlin.

But why were these wild horses so important that they were allocated the resources and the manpower needed to get them back to Germany, and at whose command did they travel? What did the Nazis want with wild horses, and why were any on a German-Russian estate famous for its sheep herds in the first place? The story of the wild horse begins when it came into being – at the creation of the domestic horse.

II

When the Botai took a lasso to the horse, there were still wild horses that continued to evade, or resist, domestication. More mares were captured from different wild herds and brought into the corrals, but few stallions after the first one were pressed into service. As the human population grew and began to disperse and settle in the wild horses' grazing lands and by their watering holes, the horses slipped away from the flocks and the crops, heading for the deep steppe or desert in the east, or into the forests and mountains of Europe in the west. As decades and then centuries of domestication took effect, the species split into two: the useful source of meat, milk, skins and transportation – a man's wealth – and the wild remnant, arduous to hunt, a competitor for resources and territory.

The role of wild horses in the human record is, for a long time, minimal, both because they were elusive and because they were not a plentiful enough food source to meet man's interest. Sometimes domestic horses escaped and became feral, even for generations, but they were not the true wild horse, which was identifiable not just by its appearance but also by its almost telepathic alertness – it wanted to stay wild, and it must have been this temperament that meant it was not tamed.

Accounts of them veer from the credible to the semi-mythical. Herodotus mentioned wild, pale horses by the River Bug in Scythia at the eastern end of the steppe, and Pliny listed 'equiferi' when he tallied the world's species. Julius Capitolinus records that thirty wild horses, along with stags, ostriches, wild sheep, goat, boar and asses were presented by Gordian the Elder to the people of Rome for a wild beast hunt in the amphitheatre, but does not mention where they came from. Varro noted wild horses grazing in 'Hither Spain' and Strabo's *Geography* claimed they roamed the Alps alongside wild cattle.

The search for wild horses in these texts should be undertaken with caution – 'wild' could simply mean a domestic runaway. For thousands of years – and indeed, still now in places like the Camargue and Exmoor – horses were bred free-range, living in large groups to which good stallions would occasionally be added to improve their blood. Captured and brought into work, these animals too had had little experience of man. There are also episodes of mistaken identity – Oppian said there was a 'dread overweening tribe' of 'Wild Horses' in Ethiopia, with two poisonous tusks, cloven hoofs and a shaggy crest that ran from their necks along their spines to a bristly tail. Less credulous commentators take this beast to be a gnu, or nylghau. Others may have been zebras, which were still referred to as wild horses in the nineteenth century.

The territories the last holdouts roamed and their location – in different groups or even species – are impossible to gauge. They are scattered over a wide geographical area, and are often reported second-hand; or even third-hand, making them shaggy horse stories.

Gerolamo Cardano, the sixteenth-century Italian mathematician and physician, maintained that northern Gaul's boggy forests had been the haunt of woolly black horses with small, pale eyes and coats that gave off sparks if rubbed at night. In the sixteenth century, Leo Africanus claimed to have seen a wild, white, curly-tailed colt in North Africa, although he admitted that 'The wilde horse is one of those beasts that come seldome in sight.' Local Arabs covered snares with sand at the watering holes the horses frequented, and trapped them by the leg. These captives were eaten: 'The younger the horse be, the sweeter is his flesh.' Vladimir II Monomakh, eleventh-century grand prince of Kievan Rus', reminded his sons of his deeds in a testament he wrote for them before his death: 'Near Chernigov I have with my own hand caught ten or twenty wild horses in the forests, and I have besides caught elsewhere many wild horses with my hands, as I used to travel through Russia.'

Pope Gregory had admonished the pagan Germans for eating wild horses in 732 AD, but this Christian taboo took centuries to come into full force in wild horse country, and even then it was perhaps more a shortage of horses that eventually brought consumption to a close. Domestic horses were seldom eaten, but although the wild horses could be a source of fresh genes for the home herd – Duke Sobeslaus gathered 'herds of wild mares' from Pomerania and Silesia in 1132 – Christian Europeans knew as well as Leo Africanus's Arabs that they also made for excellent sport and eating.

Here they are in a Westphalian document of 1316, parcelled with the fish and game inventory of a forest handed to a noble called Herman

– a hunting target for the privileged, who, though they did not wrestle them to the ground with their bare hands like Vladimir, tested their own mounts against them as they crashed through the trees. They could run as swiftly as deer, and for longer, because like humans and unlike any other game, they could sweat to cool themselves.

Teutonic Knights were admonished by their Grand Master, Duke Albrecht, for hunting wild horses for their hides in his forest near Lyck in 1534. That year the German artist Hans Baldung made a series of three sinister woodcuts of horses in some dense, draped forest: all have their mouths open to bite – one another – and they writhe like souls in hell. Baldung associated horses with brute instinct and witchcraft. One of the horses, foregrounded, is ejaculating on the forest floor, as a furious mare kicks at him. A stallion peels back its top lip, its eyes blank and black.

Poachers in the ancient forest of Bialowiecza, then in Lithuania, were fined 360 grosze for butchering a wild horse in 1588. Erasmus Stella gave one of the first detailed accounts of these Prussian horses in *De Origine Borussorum* (*On the Origin of the Prussians*), saying they were close in conformation to domestic horses, but 'with soft backs, unfit to be ridden, shy and difficult to capture, but very good venison'. Andrias Schneebergius said they were 'mouse-coloured, with a dark streak on the spine, and the mane and tail dark' but 'inexpressibly violent if any person attempted to mount them'.

East on the steppe they were also hunted, partly as pests and partly for sport, and as the Enlightenment made its influence felt, natural historians and travellers began to make more close observations of the animals in both the forests of Europe and on the steppe. In this age of science and exploration, as the Russian Empire conquered more and more territory, gentlemen from the academies of St Petersburg and Europe undertook long expeditions deep into what Swift's Gulliver had called 'the great continent Tartary', which stretched from the edge of European civilization eastwards all the way to the Pacific. Carrying notebooks and gathering specimens, they endured long months of plodding through bleak landscapes, suffering bandits, disease, harsh winters, and summers buzzing with insects that could drain the blood of their mounts. All was hungrily recorded: the tough grasses, the habits of the local people, the topography and, of course, the wildlife.

From these accounts emerged what appeared to be two distinct species of wild horse. West, on the steppe of the Ukraine, southern Russia and the Pontic Caspian, grazing on feather grass, sheep's fescue and cherry shrubs, was the mouse-coloured Tarpan – the Turkic term for wild horse. The German scientist Samuel Gottlieb Gmelin described the horse's large head, foxy ears and 'fiery eyes' in 1771, saying it had a dense pelt and could run twice as swiftly as a tame horse.

Wild horses were not mentioned in Linnaeus's original taxonomy, but Linnaeus's Dutch colleague, Pieter Boddaert, was the first to coin the formal *Equus ferus* after observations of Tarpans in the Voronezh district of Russia in 1785. Julius von Brincken, the German-born forester who catalogued Bialowiecza for its new conqueror, the Tsar, used the name *Equus sylvestris* in 1828 for the small, robust ponies that had once lived in the forest's clearings, though he explained that they had not been seen there for a hundred years. Brincken's vanished ponies are often referred to as 'forest Tarpans', and Boddaert's as 'steppe Tarpans'.

East along the steppe in deep Central Asia, towards the Gobi and Taklamakan Deserts and the Altai mountains, lived the second wild horse, a sandy dun beast that the locals called the 'Takhi'. It had a brush-like, upright black mane, a burst of short hairs at the top of its tail and ears so large that it was sometimes confused with the wild ass. In 1719–1722 a Scottish doctor called John Bell, who had served Peter the Great of Russia, travelled from St Petersburg to Peking and mentioned these curious wild horses, seen near the River Tom, which flowed from the Abakan mountains into the River Ob:

> There is, besides, a number of wild horses, of a chestnut-colour, which cannot be tamed, though they are catched when foals. These horses ... are the most watchful creatures alive. One of

them waits always on the heights to give warning to the rest, and, upon the least approach of danger, runs to the herd, making all the noise it can; upon which all of them fly away, like so many deer. The stallion drives up the rear, neighing, biting, and kicking those who do not run fast enough. Notwithstanding this wonderful sagacity, these animals are often surprised by the Kalmucks, who ride in among them, well mounted on swift horses, and kill them with broad lances. Their flesh they esteem excellent food, and use their skins to sleep upon instead of couches.

These Takhi, had, of course, been coexisting with humans for long before Europeans made their observations: like the residual wild horses in Europe, they were game for the nobility and crafty lesser locals. In 1630, a Mongolian aristocrat called Chechen-Khansóloj-Chalkaskyden gave one to a Manchurian emperor as a gift, and in 1750, another Manchurian emperor, Qianlong, organized a grand hunt of this game of games, employing 3,000 beaters to drive 300 horses to their death at the guns. The Manchurians of this period believed the Takhi were the ancestors of the domestic horses on which their empire relied, although that was no protection from slaughter. Other Chinese were reportedly succeeding in training younger horses but could not tame the older, so killed them for meat.

Sometimes the word 'Tarpan' was used interchangeably with 'Takhi'. In his monograph on the origins of the domestic horse in 1841, Charles Hamilton Smith quoted an early nineteenth-century account from a Cossack attached to a Tartar chief in the service of the Tsar, who said that both Cossacks and Tartars distinguished between feral domestic horses (takja or muzin) and wild. He calls them Tarpan, and although some of them were 'mouse' in colour, the rest of his informant's description resembles the Takhi: 'tan', 'isabella' (a kind of dun palomino) with white winter coats (both Tarpan and Takhi turn paler in winter), 'small and malignant' eyes, a prominent forehead, a thick black mane and tail that was bristly at the top and the appearance and demeanour of 'vicious mules'.

The Cossack said the horses still grazed near the Tom, where Bell had seen them a century before, but also south of the Aral Sea (now Uzbekistan), on the Karakum Desert in Turkmenistan, along the Syr Darya in Kazakhstan, in the south-eastern Ukraine, and the Gobi itself.

The 'vicious mules' had only to see 'the point of a Cossack spear, at a great distance on the horizon ... behind a bush', before a lookout stallion would shriek, and the entire herd – sometimes a few hundred strong – would 'disappear as if by enchantment, because with unerring tact they select the first swell of ground or ravine to conceal them until they reappear at a great distance'.

'Sultan-stallions' fended off bears by striking them with their fore-feet, and if wolves came into view, the horses would make a defensive ring around the foals while the stallions charged and roared outside. Domestic horses were also attacked and killed by the Takhi. They were never seen lying down, and had the ability to recognize and avoid swampy ground. Captured, they 'always die of ennui in a short time, if they do not break their own necks in resisting the will of man'.

Both the Tarpan and the Takhi were in retreat even by this period – the Takhi had not been recorded on the Mongolian steppe since 1200, and at either end of the grasslands man was making inroads into the wild horse's last retreat. *Equus sylvestris* was the first to vanish. Von Brincken had written that there were none left in Bialowiecza, but a small number had been trapped and taken to the estate of a Count Zamoyski in south-eastern Poland. They were kept as game until the early eighteenth century – Zamoyski would sometimes pit them against other animals in fights – when the count distributed them among local farmers near Biłgoraj, and they melted into the domestic horse population.

The South Russian steppe was filling up with human settlers, and the Tarpans there were pursued and killed for stealing domestic mares, or for falling on the stooks of hay painstakingly gathered to last the winter. They competed with cattle for the dwindling lake water, drained for agriculture. They could be attacked at haystacks or at their watering holes, and those which evaded capture often broke legs on uneven ground as they fled.

Few were tamed – generally the rewards of semi-taming one were less than the risks involved. Some were bred into the domestic population, where their offspring were valued for their speed and hardiness. As the herd numbers plummeted, so too tumbled the Tarpan's ability to recover from the periodic severe winters that decimated steppe animal life, domestic and wild.

A new scientific narrative of scarcity and vanishing provoked naturalists to find earlier 'primitive' versions of animals that mankind had brought to perfection, whether as a heavy draught horse or a lean, silky-skinned thoroughbred. In the previous century, Georges Cuvier's Theory of the Earth had proposed that, rather than all animal-kind coexisting in an unchanging divine plan from the beginning of time, creatures like giant lizards and reptiles had come and gone, leading to the suggestion that some species should be preserved artificially by man before they, too, were lost.

By the mid-nineteenth century and the publication of Darwin's *Origin of the Species*, animals had come to be mutable creatures, changing over generations and not just at the intervention of man, but through a natural process that came to be known as evolution. If man's ancestor was some low-browed and hairy ape, so the horse was descended from the large-headed, small-eyed and abundantly furred Tarpan or Takhi. The French doctor Joseph-Emile Cornay even proposed that a 'sacred breed' of French horse could be created by reconstituting an ideal 'primitive wild horse' whose characteristics had been scattered among the domestic horse population. Illustrations of the period show Tarpans with shaggy mutton-chop whiskers, tiny bat ears and petty, grimacing muzzles. The only one drawn from life is of a foal, with a tapering skull so broad at the brow that it is almost triangular, and long ears pressed flat back on its neck.

III

In 1828, 2,886 merino sheep were herded from Anhalt in Germany to a vast new estate called Askania-Nova near the southern stretch of the Dnieper river in the Ukraine. The next year 5,000 more were transferred, along with a hundred or so German workers, some cattle and a handful of horses. The dukes of Anhalt-Köthen had been granted 300 square miles of steppe to begin an imperial Russian wool industry to rival that of their homeland, but despite their rigorous planning and attempt to transplant the German system directly on to the steppe, within a short time the project was a disaster. The climate was too arid in summer, and so harsh in winter that the sheep had to be kept indoors. Askania-Nova was purchased from the dukes by a wealthy Russo-German called Friedrich Fein whose family had left Saxony for Ukraine in 1807 to build textile mills, and who had risen to own a wool-growing empire of 750,000 sheep. Fein had no son, so when his daughter married his German business partner, Johann Pfalz, their surnames were combined to become Falz-Fein.

The grandson of this union – also Friedrich – drilled artesian wells, transforming parts of the flat, dry steppe of Askania-Nova into fertile black earth, and overhauling its agricultural fortunes. He also began to collect animals, birds and plants for a rich and strange nature reserve in a sea of grass: two hundred types of tree – mulberries, spruces, birch, limes, elms, maples – were protected by straw matting against the extremes of the steppe. The cries of nightingales, hoopoes, storks and ostriches filled the air, while camels, yaks, zebras, water buffalo, gnu and no fewer than four kangaroo species were fetched to colonize Askania-Nova, stalking through the grasses and scratching their rumps on stone monuments left by the Scythians. Thousands of visitors came to Askania-Nova each year, including one Ludwig Heck, head of Berlin Zoo and a close friend of Falz-Fein.

The area around Askania-Nova was one of the last holdouts of the Tarpan, and by the late 1870s, when Friedrich was a teenager, the entire district was aware that the one surviving band on the Rachmanov steppe was dwindling year by year.

Finally a lone mare had been spotted by herdsmen hovering near domestic horses – presumably yearning for company. She shadowed the other horses, fleeing if men came too close, but over several years she chose to have two foals by the herd's stallion. One day she followed the band into the stables of a landowner called Durilin, and his men, looking for a reward, rushed to separate her from the others. The fight was so brutal that it cost the mare an eye. She raged, trying to climb the walls of the box they shut her in, striking at them with her heels. Locked into the stable for months, she eventually settled enough to give birth to a third foal, and even to wear a halter, but when spring came and the horses were loosed to graze on the steppe again, she fled them – leaving her foal behind.

That autumn she was seen 24 miles from Askania-Nova, appearing like a challenge. The following winter, the locals decided to test the speed of their horses by pursuing her. They waited across the steppe in a long relay, driving her first to one horseman and mount, and then on to the next, but were unable to exhaust her – she outran their horses, leaping snow banks. But the snow concealed runnels: the mare dropped into one, breaking a leg. Her hunters hauled her on to a sleigh and dragged it into the Agaiman village, where plans were made to create a false leg for her. She died some days later. All that was left of the Tarpan were a handful of half-breds and semi-domestic specimens, one of which lived in the Moscow Zoological Gardens till its death in 1909.

With the forest Tarpan melted away among Count Zamoyski's tenant farmers, and the steppe Tarpan expiring on a sleigh in a Ukrainian village, the only remaining true wild horse was the Takhi, which found itself 'discovered' just as the Tarpan mare on the Rachmanov steppe breathed her last. The reports of John Bell, Samuel Gmelin and Charles Hamilton Smith's Cossack became flesh – or at least a husk of flesh – when a Russian explorer called Colonel Nikolai Michailovich Przewalski returned from an expedition to Central Asia in 1878. At the border post of Zaysan between China and Russia, the local commissioner had given him the skull and yellow-dun folded hide of a wild horse he'd shot on an expedition with Kirghiz hunters in the Dzungarian Gobi. Przewalski took this husk to the Academy of Science in St Petersburg, who christened it *Equus przewalskii* Poliakov 1881 for the colonel and the conservator Ivan Poliakov, formally stamping, sealing and rolling the Takhi into taxonomy. The Russians thought it marked the transition between ass and domestic horse; a French scientist, Fauvelle, cast it once more as ancestor to the domestic horse.

More zoological hunters followed, eager to stake their own pin through specimens. Two Russian brothers, Grigory and Vladimir Grum-Grzhimailo, made numerous sorties, bringing home the bodies of three stallions and one mare from Gashuun Nuur Lake in the eastern Gobi in 1881. They stressed the skill required, for 'Wild horses are very careful at night, during the day they skilfully use the features of the country to camouflage themselves. Chasing a herd 300 or 400 steps behind, we often completely lost sight of it and found it only by tracks.'

Where once exotic new species had been killed and preserved by travelling naturalists, now improvements in transportation led many to try to ship creatures back alive to breed in the menageries, zoos and private parks of the West. To the gentlemen naturalists and explorers a new breed of professional animal collector was added, who undertook at all costs to bring live animals back. For the wild horse, these ruthless new hunters would be both an ark and an accelerant to their extinction.

The Russian Academy of Sciences, which wanted to capture its own Takhi, recruited a Russian merchant called Ivan Asanov who travelled regularly through Mongolia to China. Friedrich von Falz-Fein, mindful of the death of the last Tarpan near Askania-Nova, had already attempted to woo the Mongolian nobility with presents, hoping to have access to their huntsmen and lands, when he was approached by Asanov and the academicians for financial backing.

He agreed happily, investing some 10,000 roubles for each year of the hunt.

In 1897, no foals were found as Asanov's men arrived too late in the year. In 1898, they succeeded in chasing down and holding on to a handful of young Przewalskis near the town of Biisk in south-western Siberia. The foals all died after their captors tried to bottle-feed them with sheep's milk.

The hunters refined their tactics: they bought domestic mares, put them in foal, and took them to Biisk. This time the huntsmen did not pursue the wild horses far but instead shot the Przewalski broodmares. The foals of the domestic mares were slaughtered and skinned, and the Przewalski foals wrapped in their hides. With this deceit, some of the domestic mares fostered the wild foals. Once more, the foals died. The second time, Asanov's men secured six fillies and one colt, which were taken by boat to Biisk, and then by train to Askania-Nova. Four reached Falz-Fein's estate alive.

The new wild horse hunt was afoot, and as more hunters followed in the footsteps of Asanov, it became as ruinous for the horses as the old grand game expeditions of the Prussian forests and Mongolian steppe. Many of the vanishing population of Takhi were slaughtered, and some underhanded dealing emerged.

The German wild animal merchant Carl Hagenbeck was one of the professional collectors who took aim at the Takhi. He was already famous for his ability to deliver tigers or polar bears to order, and for the touring 'human zoos' he organized. His agents travelled to Africa, Asia, Lapland, the South Pacific and the Americas not just to capture wild animals but also to kidnap human 'specimens' from local populations and put these 'savages' on display in mocked-up villages in European cities.

Hagenbeck dispatched one wily agent, Wilhelm Grieger, to Askania-Nova. The Duke of Bedford had commissioned Hagenbeck to bring him wild horses, and he knew that other wealthy collectors would be interested. When Falz-Fein would not reveal the source of his Takhi, Grieger bribed one of his servants for information instead, and secured Asanov's name. Hagenbeck intercepted Asanov and bought horses that he had captured for Falz-Fein.

Hagenbeck also sent Grieger to the Kobdo river valley, north of the Altai mountains, with letters of introduction from Mongolian, Chinese and Russian dignitaries. The agent set off in deep winter, travelling as far as he could by rail, and arriving at Ob – 800 miles from the valley. From Ob he endured 170 miles by sledge to Biisk, where he set about gathering a caravan of local men, camels and horses, whom he paid by breaking off chunks from great, flat silver pieces weighing 12 lb each. Now came a 600-mile trek of snowdrifts, numbing cold that froze the milk brought for the Takhi foals, and eventual arrival in Kobdo, where he set up camp and waited four months until the wild foals would be old enough to separate from their dams and be introduced to tame mares who would act as 'wet-nurses'.

To the horror of the locals, he fished trout and had to be stopped from shooting game birds – Hagenbeck marvelled later that the Mongolians intervened not because they worried about hens being shot in the breeding season and preventing their own sport, but 'from a genuine pity for the hen-bird with her young'. Grieger also recruited huntsmen, and by early May they were ready to begin.

He watched as the Mongolians on their own horses stalked the Takhi patiently as they came down to the river to water, switching their tails against the biting gnats.

The Mongolians give chase, and after a time brown specks are seen at intervals in the dust-cloud. As the chase continues the specks become larger and turn out to be the foals, which are unable to keep up with the older members of the herd. When at last the foals are quite worn out, they stand still, their nostrils swelling and their flanks heaving with exhaustion and terror. All the pursuers have then to do is to slip over their necks a noose attached to the end of a long pole, and conduct them back to camp.

Grieger's success was unprecedented. Six Takhi were asked of him, and he caught thirty, then fifty-two, many of whom, separated from their dams and plunged into a bewildering new existence, died on the long journey home. It took eleven months for Grieger to get his booty back to Hamburg. The foals, unable to walk far, were bundled into packs on the sides of camels as their foster mothers paced alongside. The camels kept bolting and having to be retrieved. The Mongolians mutinied and said they could not go on, at which Grieger, assuming they were trying to wring more money from him, beat them with his whip till they begged him for forgiveness. He eventually delivered twenty-eight Takhi foals to Hagenbeck – a tiny population of startled, woolly youngsters with drum bellies.

For some, Hamburg was only the start of a still longer journey; Hagenbeck sold them to zoos across the world as the century turned. One beleaguered stallion and mare arrived at the Bronx Zoo in New York in 1902 – swapped for a pair of buffalo – and the zookeepers tried to break them to halter with a view to getting them to pull a carriage, but they fought so wildly that the keepers gave up. Some fifty-three Takhi caught in this period ended up in America, Germany, Britain, France, Holland and Russia. Falz-Fein had collected nine Przewalski horses altogether, but only two of them produced foals, and only then after the Tsar gave Askania-Nova a stallion called Kobdo I. There is a photograph of one of the stallions being ridden by a man in a Cossack hat and boots, and the horse, its neck almost as broad as its barrel, has its ears pricked and one hind leg raised to kick.

From their dwindling territory in the Gobi, the Takhi found themselves in an accelerated version of their own extinction: tiny groups, sometimes founded by just one stallion and one mare who might have been half siblings, existed in pens in city zoos, or wandered parks that, though open, were far from the steppe or desert. Measured, weighed, watched, they did not breed well, and though horses were exchanged between breeders, pharaonic levels of incest became commonplace in Askania-Nova and elsewhere as collectors struggled to develop herds. Some were bred to domestic Mongolian horses, or even to zebras.

The offspring of domestic horses and Przewalski were bred into the thoroughbred and Arab cavalry horses raised at Askania-Nova, with the intention of creating a tougher warhorse.

It remained difficult to capture more horses from the wild: not only were the locations remote, but the numbers, as Hagenbeck had known even during Grieger's mission, were low. The Russian Revolution, First World War and the fall of Mongolia to Communism interrupted the hunt, and the Takhi was divided between Soviet east and European west. The Red Army despoiled Askania-Nova in 1918, destroying the family's tombs and shooting Falz-Fein's mother as he escaped. Three quarters of the animals were lost in the fighting. Friedrich died in exile in Germany just two years later.

IV

Humans have a lamentable habit of loving too late – so it was with the wild horse. Part of the process of extracting the wild Tarpan and Takhi from their homes – in Bialowiecza, on the southern Ukranian steppe, on the slopes of the Altai mountains – and bringing them into the pens and zoos of the West was their incorporation into the Western imagination where, even as the real wild horses faded from the real wilderness, they found themselves transcending their own extinction and becoming mutable, flexible symbols.

While the interest of scientists and animal collectors drove some public fascination with wild horses, it was a Romantic poet who sparked mania. Byron's 1819 eponymous narrative poem drew on the true story of a Polish nobleman called Ivan Stepanovich Mazeppa, who was born in 1644. As a young man at the court of King Casimir, Mazeppa had dallied with the wife of another aristocrat, who punished him by strapping him to the back of his own horse, and turning the animal loose. The horse, as horses do, returned speedily to its own stable, where Mazeppa was presumably freed by his grooms. Mazeppa later left Poland for the Ukraine, where he became a Cossack leader and was eventually elevated to Prince of the Ukraine by Peter the Great. When the Tsar turned against the Cossacks, Mazeppa transferred his loyalty to Charles XII of Sweden, fighting alongside him at Poltava in southern Russia in 1709.

Byron built on later accounts to write something altogether stranger. His Mazeppa falls in love with beautiful Theresa of the 'Asiatic eye', the wife of the much older Count Palatine, and together they cuckold the count. Discovered, Mazeppa is seized by Palatine's men:

> 'Bring forth the horse!' The horse was brought;
> In truth, he was a noble Steed,
> A Tartar of the Ukraine breed,
> Who looked as though the Speed of thought
> Were in his limbs – but he was wild,
> Wild as the wild-deer, and untaught,
> With spur and bridle undefiled;
> 'Twas but a day he had been caught,
> And snorting with erected mane
> And struggling fiercely but in vain

Mazeppa is stripped naked and strapped to the horse's back, whereupon the horse, to paraphrase Voltaire's earlier version, goes back to the Ukraine again – no small distance from Poland. The real Mazeppa's saddle horse is conjured by Byron into a 'swift and savage breed' direct from the wilderness of Tartary. As it gallops furiously they outstrip civilization, leaving behind 'human dwellings' for vast plains, dark woods where tree branches whip at Mazeppa's flesh, and past distant, hostile fortresses erected to keep out the barbaric Tartars. On and on, into the semi-savage lands, with Mazeppa bleeding from his restraints, growing desperately thirsty and exhausted to the point where he slips in and out of consciousness. Still the bewildered beast gallops on:

> Untired – untamed – and worse than wild,
> All furious as a favoured child
> Balked of its wish, or, fiercer still,
> A Woman piqued, who has her will.

Mazeppa is lashed to his own sin – his desire for a woman – and now he will not be spared his wild ride. The horse is an engine of his lust: beyond control, and meting out a punishment with echoes of the amphitheatre, where Romans used wild horses to pull victims to pieces. Mazeppa is still helpless as the horse begins to falter itself, but after plunging through a cold, wide river, it reaches its home-lands, and the exhausted man finds himself in a veritable kingdom of wild horses:

> A thousand horse and none to ride! –
> With flowing tail, and flying mane,
> Wide nostrils never stretched by pain,
> Mouths bloodless to the bit or rein,
> And feet that iron never shod,
> And flanks unscarred by spur or rod,
> A thousand horse, the wild, the free,
> Like waves that follow o'er the sea,
> Came thickly thundering on ...

As the wild horses arrive – an elemental host headed by a black stallion who is 'the Patriarch of his breed', swirling, spooking, ready to flee – Mazeppa's horse rallies and collapses, dead. Still bound to the horse, he lies exposed, waiting for vultures to come. Instead he is rescued by a 'Cossack maid' whose black eyes, like his Tartar horse, are 'wild and free'.

Mazeppa inspired other writers, musicians and artists across Europe and America. Victor Hugo and Pushkin both produced poems that continued his story, and in turn they sparked a frenetic 'transcendental étude' by Liszt in which the wild horse gallops in octaves, and an opera by Tchaikovsky. Géricault painted Mazeppa crucified on the horse's back as it struggles out of the river, and in Delacroix's vision the black-eyed horse and man collapse under a murderous sky nicked with the sinister outlines of vultures. In the artist John Frederick Herring's rendering of Horace Vernet's popular paintings, Mazeppa is pink and round-cheeked as an Englishman, surrounded by snarling wolves as his dapple grey steed leaps out of a dark wood and into the river.

Mazeppa was a popular cultural phenomenon too, lending his name to yachts, steamboats, racehorses, pubs and locomotives and his image to Staffordshire porcelain knick-knacks and carved pipes. The story remained a trope well into the twentieth century, and even in the 1930s *Punch* magazine featured a cartoon of a moustachioed horse with a flopping forelock – Hitler – carrying Germany, in monocle and Prussian uniform, to oblivion.

But without doubt, Mazeppa's most sensational incarnation was on the stage, where it became one of the most popular, widely performed pieces in the repertoire of the hippodrama – a play in which the cast rode and performed with real horses. In January 1825 the Cirque Olympique in Paris opened with its version, called *Mazeppa, ou le Cheval tartare*, and soon Astley's Amphitheatre in London followed with Henry Milner's *Mazeppa, or the Wild Horse of Tartary*, which souped up the action as far as theatrical machinery would allow. The horseman Andrew Ducrow was producer, and he added swooping equine dramatics to the more conservative French staging.

In Ducrow's show, Byron's plot was tweaked vigorously into a melodrama complete with a foundling Mazeppa, a phantom horseman, Tartars disguised as gypsies and the love interest, Olinska, nearly killing herself at her hymeneal altar. Tartary itself became an exotic, oriental landscape of mystery – including, for unfathomable reasons, palm trees and occasional zebras – filled with horsemen tricked out in tinsel and Scythian-style spotted trousers. Tartary also developed a mountain landscape garnished with waterfalls and rickety bridges, which was constructed from platforms that could be raised and lowered with stage machinery.

The actor John Cartlich was tied to the back of the 'wild horse' and exited the stage to be replaced by another horse with a dummy strapped on board. This horse and dummy then careered up the helter-skelter mountain course, crisscrossing the stage left to right and back again, all the way to the flies, with offstage thunder, lightning and rattling 'hail' and the orchestra in full swing to drown the hollow drumming of the galloping horse on the scenery. Cartlich and his original mount arrived in Tartary via a painted, rolling panorama of the Dnieper river at the beginning of Act II, and at the climax of the wild ride, the horse crumpled to the boards 'dead' and a mechanical vulture swooped down to menace the hero with clacking beak.

Cartlich would complete over 1,500 performances before retiring to run a saloon in Philadelphia. His 'wild horse' not only had to be sure-footed, but also to act the part of the savage animal convincingly. In 1878 Astley's were even claiming they'd brought the 'wild horse' 'direct from the Ukraine' for added verisimilitude. American actress Clara Morris recalled one mare in a New York production, Queen, which was later killed in an accident in another hippodrama when an inebriated actor steered her off a piece of scenery:

> When she came rearing, plunging, biting, snapping, whirling, kicking her way on to the stage, the scarlet lining of her dilating nostrils and the foam flying from her mouth made our screams very natural ones ... She seemed to bite and tear at [Mazeppa], and when set free she stood straight up for a dreadful moment, in which she really endangered his life, then, with a wild neigh, she tore up the 'runs' as if fiends pursued her, with the man stretched helplessly along her inky back.

Another illustration shows the moment one 'fiery untamed steed' caught stagefright and froze, with Mazeppa upside down on its back and two stagehands hauling the horse back under a rapidly descending curtain by the tail. One horse was so quiet that it had to be harried up the runs by barking dogs. Later actors made the ride themselves, eschewing the dummy; one was killed during a performance in St Louis.

A year after the Astley's opening, Andrew Ducrow had his own amphitheatre on Great Charlotte Street, London, where he presented his own version of *The Wild Ukraine Horse of Tartary, or the Fate of Mazeppa*. Ducrow also had a smash hit with his own equine tableaux vivants or 'pantomimes on horseback' including 'the ne plus ultra of the Equestrian Art, the Wild Horse and the Savage', featuring the grey horse that was his finest actor. The *North Wales Chronicle* gave a breathless account:

> A few palm trees, made by the carpenter and the painter, are disposed of in the circle; a zebra or two pass across to assist in exciting an idea of Indian wildness [sic]. Mr. Ducrow enters as an Indian with bows, arrows, and club; a leopard skin is thrown across his shoulders ... a wild horse rushes towards him ... [The horse's] exertions to seize the Indian with his teeth are frightful. With his ears down, and his head outstretched, pursuing the Indian amid the trees, he is terrific, and snorts and kicks at being foiled. At length the Indian ventures into open ground, and a fierce encounter ensues. The horse lashes his tail, elevates his mane, rears, and endeavours to strike down the man, who dextrously avoids every attack, every now and then dealing such as seem tremendous blows on the animal. He rages, till, from a powerful blow on the head, the horse reels and falls, the Indian leaps on the body, and dispatches his adversary amid the shouts of the audience.

In the early 1860s, a New York entrepreneur called Captain John B. Smith had the happy notion of casting a woman called Adah Isaacs Menken in the leading role: 'a very young girl, thrown on her back, held by one foot only, her loosened hair dragging in the sand, and in this dangerous position leaping with her galloping horse an arrangement of several barriers'. The actress became a *succès de scandale* in her racy, short tunic and flesh-coloured tights – in a burlesque that mocked the production, it was claimed 'the classical style of her dress does not much trouble the sewing machine'. This racy update also ran and ran. The message was still cruder than Byron's hero strapped to his galloping desires: sexy girls on sexy horses in erotic oriental lands.

But beyond the sensation and the burlesque, this idealized wild ride into barbarian lands flourished at a time when the Western world was becoming increasingly urbanized. As the hillsides of Europe began to sprout factories and mills, the Romantics led the Western imagination to the notion of an authentic, Edenic and pure landscape untouched by man, whether it was in unfathomable Tartary or the Lake District.

Industrialization and urbanization also meant that the number of domestic horses was increasing sharply, and most of these animals were humdrum city work-horses. They hauled wagons, cabs and carriages, were boxed up, clipped, shaved and mollycoddled or beaten. The wild horse of the imagination stood in opposition to this, refusing to submit to the yoke – literally – as it shook its long mane and snorted in fury. It was a perfect Romantic emblem: a pure child of nature, returning to a land inhabited by noble savages both human and equine.

By 1893, it had even become the paradoxical fashion to have 'genuine' wild horses to pull your carriage, and W. J. Gordon reported that one blade was dashing about London with a carriage drawn by two 'mustangs' imported specially from America, while another was claiming that his pair were 'Tarpans' from South Russia. But of course, the 'wild horses' that raced up the wooden mountains or pulled landaus around Mayfair were doing nothing for which they had not been trained and then cued, and meanwhile the real wild horse was being leached from the steppe and placed in zoos where curious Westerners could observe it.

In nineteenth-century America, far from exotic Tartary – or the tinselly stage version – a new nation discovered its own wild horses on its own limitless plains, ready for conquering by frontiersmen who doubled as explorers. When the horse had been reintroduced to the Americas by the conquistadors, some escaped, and as more horses were brought to the continent, more joined the growing herds of mustangs. In these animals America found an embodiment of its self-image as free and spirited, independent of the Old World.

Josiah Gregg was crossing the American plains in the 1840s when his party decided to capture the striking stallion leading the herd which had approached their train of packhorses. A Mexican snared the horse with a lasso, and, Gregg said, 'as he curvetted at the end of the rope or stopped and gazed majestically at his subjectors, his symmetrical proportions attracted the attention of all; jockeys at once valued him at five hundred dollars'. Quietened and brought into the *caballada*, the horse accepted the saddle which was thrown on his back and submitted with good grace. It was clear that the $500 stallion had been a tame horse not long since, and thus he was transformed in the men's eyes, for 'in a few days [he] dwindled down to scarce a twenty-dollar hackney'.

The folklorist J. Frank Dobie relays the tale of a surveying party in Texas who saw a great line of one hundred mustangs galloping towards them like Byron's thousand, from patched paint horses to red bays, buckskins, dusty blacks, fox chestnuts, silver greys and frosted roans, which slid as one to a halt, 150 yards from the humans, then span and raced away across the grasses. Wrote Dobie, 'No one who conceives him as only a potential servant to man can apprehend the mustang. The true conceiver must be a true lover of freedom – a person who yearns to extend freedom to all life.' This love of freedom belonged to the horses too, for as Dobie claimed, 'One out of every three mustangs captured in south-west Texas was expected to die before they were tamed. The process of breaking often broke the spirits of the other two' – just like the Takhi and Tarpan.

Folktales concerning a white mustang stallion that evaded capture spread across the American West from the early 1830s onwards, growing ever more elaborate and sentimental. If caught, it was said, the white mustang would pine and die, and yet men yearned to have him for themselves. He was *Moby-Dick*. He is mentioned in the novel: 'Most famous in our Western annals and Indian traditions is ... the White Steed of the Prairies; a magnificent, milk-white charger ... [I]t was his spiritual whiteness chiefly, which so clothed him with divineness; and that this divineness had that in it which, though commanding worship, at the same time enforced a certain nameless terror.'

As the nation clipped away piecemeal with barbed wire fences and homesteads at the wilderness that remained, it fell in love with the wild – and sought to possess it. But like the Przewalskis in harness at the Bronx Zoo, the lacklustre mustang under saddle and the uncatchable White Steed, possession was impossible for such a greedy, consuming embrace.

V

In the First World War, the German and Russian armies in Poland requisitioned small, tough, local farm horses to pull artillery and supply wagons. The Germans called these horses '*panje*' from the Polish word '*pan*' or 'sir', having heard the word repeatedly from the farmers as they handed over their animals. Three years after the war ended, two Polish scientists called Jan Grabowski and Stanislaw Schuch published a study of the horses, commenced in 1914 and interrupted by hostilities; they suggested that they might be descended from the last forest Tarpans, *Equus sylvestris*, dispersed by Zamoyski over a hundred years ago. The notion was galvanizing: perhaps an authentic, elemental part of battered Poland – from a time when nobles were free to hunt among the Bialowiecza trees – was not lost at all.

In the 1920s several Polish studs were breeding these *panje* horses as work animals, deliberately singling out what they believed to be the most 'wild'-looking animals, and one zoologist, Taduesz Vetulani, took a step further, intending to reverse their domestication. He studied animals near Biłgoraj, where the last forest Tarpan had been given away to local farmers, and selected five mares and later a stallion called Liliput. In 1936, he was given permission to place these specimens and further handpicked horses – which he called 'koniks' from the Polish word for 'small horse' – in a large enclosure in Bialowiecza forest, where he intended them to undo their domestication by living with minimal interference from man. In this he got off to a disingenuous start: several of his pairings were, in fact, incestuous, as a result of his desire to reproduce certain 'wild' characteristics like a coat that turned white in winter and a coarse mane. But the programme was successful, and on 1 September 1939 there were forty horses in the Bialowiecza reserves: five stallions, eighteen broodmares and seventeen youngsters.

Vetulani's efforts were watched with interest by two German brothers, Lutz and Heinz Heck. The Heck brothers had grown up among the cages and enclosures of Berlin Zoo, where their father, Ludwig – Friedrich von Falz-Fein's visiting friend* at Askania-Nova – had been appointed director in 1888. Lutz took his father's post in the 1930s. After working at the zoo established by Carl Hagenbeck in Hamburg – and becoming his son-in-law – Heinz became director of the Tierpark Hellabrun in Munich.

Not only were the brothers steeped in zoology, but also the German nineteenth-century *Völkisch* tradition that grafted folk-inspired nationalism on to Romanticism. Germany was a young country – only drawn together in 1871 after a century that involved a long and humiliating conquest by Napoleon, rapid change and at times violent unrest. The *Völkisch* movement tried to buttress this newly built nation with a crafted history reaching back to Roman times, when the German people were the unconquerable forest tribe described by Tacitus as 'free from all taint of inter-marriages with foreign nations' with 'fierce blue eyes,

---

*   One Nabokov family wedding photo shows Lutz Heck with the Falz-Feins in Berlin in 1920, and Ludwig Heck provided the foreword to a book about Askania-Nova written by Falz-Fein's son, Woldemar. Lutz and Heinz were contemporaries of the Falz-Fein children and were often guests at their apartment near the Kaiser Wilhelm Memorial Church, close to the zoo. Anna Falz-Fein later described an older Lutz as condescending and arrogant, and a committed Nazi.

red hair, huge frames'. They had resisted even Rome: led by the hero Arminius or Hermann, they slaughtered their effeminate southern neighbours at the Battle of the Teutoburg Forest in 9 AD while wrapped in animal hides and brandishing spears. The essence of Germany, the Völkisch thinkers said, had existed from before recorded time, and even as Napoleon helped himself to chunks of nineteenth-century Prussia and as the ancestral forests fell to industry, it would endure again. It was of blood, or blut, and land, or boden: a natural cult tied to the soil.

Thinkers and artists like Johann Gottfried Herder conjured up a true German people, their folklore and language, and what Simon Schama calls 'the Teutonic romance of the woods'. Arminius became a cult figure, with a colossal statue erected to his memory in the Teutoburg forest. The imaginary Urwald of these early Germans was replanted with folk stories gathered by the Brothers Grimm in the eighteenth century, and Teutonic myths revisited. In this tradition, wildness meant authenticity and strength, which many came to believe was being lost to the corrupting and cosmopolitan influence of German Jewry.

The imaginary German Ur-forest had also contained the totem animals of Teutonic and Ring cycle myths, and now few remained – just deer where once there had been wolves, bears, wild auroch cattle and horses. The Heck family, in the spirit of the times, believed that these extinct creatures could be reborn by gathering the scattered fragments of wild genes and concentrating them. Ludwig Heck had scoured the Nibelungenlied for clues, totting up Siegfried's hunting bag: a bison, an elk, four aurochsen, 'and a grim schelch'. This, he decided, must be Equus sylvestris – as the word sounded similar to 'Beschaeler' or stallion.* His son Lutz later wrote, affirming his father, 'The adjective "grim" is entirely appropriate, for there is scarcely any fiercer and more dangerous animal than the wild horse.'

The gods of Teutonic mythology had spectral horses to carry them over a bridge made from a fiery rainbow to their home, Asgard. Tacitus' ancient Germans were mounted on ponies. What good was an Ur-man

---

★ It was probably an elk.

without an Ur-steed? Heinz would have seen Takhi in his father-in-law Hagenbeck's collection, and both brothers knew through their father's visits to Askania-Nova that the Tarpan was believed lost.

Heck Senior had given his sons their method: breed a domestic animal with the right characteristics to a wild species, and the vanished species would be reborn. He had bred a domestic goat to an ibex and produced what he thought resembled an extinct paseng or bezoar. 'No animal can be extinct whose heritable constitution still exists,' wrote Lutz, fundamentally misunderstanding inheritance.

In the wake of another humiliating national defeat delivered in 1919 in the Hall of Mirrors at Versailles, the *blut und boden* tradition was reinfused by a fresh, urgent need to reassert German nationhood. When the Nazis came to power in 1933, they threw the resources of the state into the pursuit of the pseudointellectual and physical proof of an authentic German race, taking Tacitus' writings as ethnographic accounts. As a later SS pamphlet put it, 'It must fill us with pride that in our own homeland, in Germany, culture has bloomed in unbroken lines for more than 5,000 years, created by people of our blood, our nature, our ancestry.'

This, of course, was the murderous ideology that whipped German chauvinism and anti-Semitism into a systematic classification of humans by the length of the skull, the breadth of the bridge of the nose, the shade of the iris and hair. The same Enlightenment thinkers who had counted species of animals and theorized about their relationship to one another had also tried to explain how the white race came to be 'superior' or even to spring from a separate Adam and Eve to those of the 'inferior races'. Under the Nazis, the pedigrees of humans were studied like those of farm animals, marriage restricted for reasons of 'racial hygiene', and Jews, Sinti and Slavs isolated in a bureaucratic process that would end in ashes and genocide. 'The history of every great nation shows a clear idea of its uniqueness and a rejection of foreign races,' went on the SS pamphlet. 'This attitude is as innate in people as it is in animals ... Humanity is not equal. Just as plants and animals are of different types, so, too, are people.'

Heinrich Himmler and Hermann Goering were the most prominent Nazis directing this grotesque obsession. Himmler's *Ahnenerbe* or
'race ancestry' institute pursued the human Aryan through biology,
ethnography, ancient texts, archaeology and creative speculation,
patching together a backdrop for the master race that was as artificial
as a museum diorama. Goering, who had appointed himself Master
of the German Forests and Reich Master of the Hunt, pursued the
ancient beasts of Hermann's mythical forest to populate that diorama:
here is the broad-shouldered, blond hero, draped in the skin of a wolf
he has killed, standing before a painted backdrop of stolid, gloomy
oaks. To his left, a stuffed bison paws the ground with a plaster hoof,
while behind a shrub with paper leaves there is a lynx with glass eyes.

In Goering, the Heck brothers found their patron, and their fantastical, Aryan-faunal dreams began to take flesh. For Goering they
would make the dead live: the huge sides of the auroch would rise and
fall again, and the Tarpan would whiten in winter. The extinct would
gallop eternally in Schorfheide, a park to the north of Berlin where
Prussian kings, the last tsar and the Kaiser had all hunted. Goering
intended it to be the equal of the great American national parks, and
in partnership with Lutz and Heck, set about repopulating it with an
authentic bestiary.

There was an enclosure for the seventy-odd bison-cow crosses
that were meant to generate aurochs, and moose calves were shipped
in from East Prussia. There were lions too, whom Lutz Heck travelled
from Berlin to care for. In 1937 a Dülmen pony – one of a breed raised
by the dukes of Croy and considered 'wild' since the 1300s – was
thrown in with the bison. Lutz joined the Nazi Party that year, and in
1938 was given an honorary professorship on the Führer's birthday.
Two years later he would be awarded a silver Leibniz Medal for his
contributions to 'science and public education'.

Goering built a mansion named Carinhall after his late wife in one
section of the 141,200-acre park, part 1930s modernism, part *gemütlich*
grotesque, with heavy beams and wooden chandeliers, and walls bristling with antlers and decked with tapestries. There were friezes of

men spearing boars, and wolfskins covering the flagstones. Statues of deer, wild boar and the goddess Diana on a pony – bareback and brandishing a javelin – looked on as Goering entertained dignitaries or organized hunting parties.

Not everything at Schorfheide worked out according to fantasy: a bison bull, brought with great fanfare from Canada, was so terrified of the females in the enclosure that he hid in his travelling box. A moose broke into the gardens at Carinhall and ate the flowers. The deer stocks, pumped high for Goering's shotguns, overbred and damaged the forest.

Meanwhile, in their respective zoos, Lutz and Heinz laboured to mould aurochs from Highland cows, Spanish fighting bulls and Brahma cattle, and to recreate the wild Schelch for Schorfheide. Under Goering's aegis, they could brush aside the concerns of their critics in the zoological community. To them, the Tarpan – which they had never seen – was not just an extinct wild horse, but the most elevated example of it – a Teuton among horses: 'the progenitor of the breeds of our swift riding horses which attain perfection of physical beauty in the pure-bred Arab', because it was finer-boned and had 'prominent eyes' and 'a more pleasing expression'. Tacitus's early Germans found their equine echo in the Tarpan, a tap root for superior modern horses like the large, athletic but far from nimble German warmbloods* that the Nazis insisted the German polo team rode at the 1936 Olympics instead of nippier ponies, with disastrous results. The Przewalski horse, Heinz added, was merely the ancestor of the carthorse, as it was heavy-skulled and ugly. Nevertheless, they needed the 'the russet horse of the steppe' to cross with their domestic horses with dormant Tarpan genes, following their father's method.

The Hecks sourced their 'primitive' thatch-maned pony mares from Gotland and Iceland and put them to the Przewalski stallions in their zoos. Initial results were disappointing, but when Heinz bred

---

* Today warmbloods are the sport horse par excellence. The term implies a breed partway between a 'hot blood' Arabian, thoroughbred or Iberian horse and a 'cold blood' heavy horse.

one of his crossbreeds to another, they gave him a mouse-grey foal on 22 May 1938. 'We already had our first primeval horse!' exclaimed Lutz. Tarpan production was underway. There were other foals too, and if they didn't match Heinz's notion of a horse type he had never seen and of which few images existed, he excluded them. 'They come into the world the kind of blonde brown of a loaf, without zebra stripes on the legs and without any of the other markings, and they change colour only after a few months,' wrote Heinz. What's more, they had 'hoofs of steel'. Heinz once put one to a cart and had it driven over a thousand miles to Munich and back, unshod.

Some awareness that the Hecks were not, in fact, doing as they claimed, crept in, resulting in muddled justifications from Heinz: 'Through artificial selection, we are able to call forth forms of animal life more quickly than could be the case by means of natural selection ... Nature ... has endless time at her disposal, whereas the life of a man is almost too short for the making of new animals or for the constitution of those that have already disappeared.' They rejected foals with the 'too long head' or 'heavy skull' of the Takhi, arguing that 'finer bones' had some kind of affinity with the mouse-grey colouring. As time went on, not just Shetlands but even Arabians were added to the soup. What they had made, of course, in their pursuit of authentic purity, was a mongrel, a fake – not a reconstructed Tarpan, but a breed as artificial and mixed as any thoroughbred.

The brothers' accounts of Tarpan building were all written in the 1950s, when the Hecks continued in their posts, despite denazification. The war is mentioned seldom by them, and there is certainly no commentary on the racial ideology that underpinned their work – or its other consequences. They are unselfconscious about the parallels between their skull-measuring for the Tarpans and the treatment of the millions of humans whom the Nazis reduced to animals – and worse. In one essay the oblivious Lutz even calls the near wiping-out of the North American buffalo 'one of [the] most shocking chapters in history'. The brothers certainly do not dwell on what the Third Reich's expansion meant for their own project.

VI

Julius von Brincken's early nineteenth-century documentation of Bialowiecza had conjured up an authentic *Urwald* of the kind that true Germans might have inhabited. The real forest people of Bialowiecza were pagans living among the trees on venison they hunted themselves, and the forest itself was free from interfering regulation: its trees seeded, grew and fell in a chaotic, self-regenerating tumult of pine, maple, oak, ash, linden, birch, hornbeam, elm and willow. Later German scientists had written that the ancient forest was the nearest representation possible to the Teutonic *Urwald*: the glowering, fathomless Hercynian woods that once stretched across northern Europe, and at whose easternmost edge the Scythian steppe began. It was another realm, another space and time – beyond the narrow borders set by the Versailles Treaty – with strange deer and fantastic bison. Its wisent, many had speculated, were the aurochsen. Askania-Nova was a curiosity cabinet carved from the steppe; Bialowiecza was a *heimat* close to magical for Nazis. Goering himself hunted there, and stayed in the Tsar's lodge.

When the Nazis invaded Poland, Lutz Heck came hot on the heels of the army to Bialowiecza, and took twenty of Vetulani's thirty-nine grey koniks. One was taken to Schorfheide, where it joined the Dülmen pony and some Polish zebras raided from Warsaw Zoo, which had been heavily shelled in the invasion. Lutz took the zoo's sole home-bred Przewalski to Vienna, promising that the animals were in safe hands – merely 'on loan' to Germany. All these horses joined the Heck brothers' Tarpan production system, where Vetulani's Tarpan fragments were once again absorbed, this time into an inbred race hurried into generations.

Lutz made so many journeys – often by air – to source animals that after the war he was suspected of using the opportunity as cover for aerial reconnaissance for the army. Four Przewalskis were added to Schorfheide in 1940, and can be seen in a blurry black and white photograph with the bison, the striking white 'Ws' on the underside of their bellies emerging under the gloomy pine trees. By the end of the war, there were sixteen 'wild' horses sharing the 1,500-acre enclosure with the bison and bison crosses, and scores of the Hecks' horses at different locations across what was left of the Reich.

Lutz had an agent or representative, it seems, in Askania-Nova – although he did not make it there himself until 1943 – and this person must have made the arrangements for the small dun stallion and mare to be shipped to Germany in the summer of 1942 as the Wehrmacht headed east. One Russian source refers to him as 'Baumgarten'. The stallion was named Charzis, for a Chechen hero who had fought against the Tsar in the nineteenth century. The horses vanish from the record for over a year, but the mare arrived in Schorfheide on 19 October 1943, and Charzis arrived separately on 16 February 1944. Askania-Nova fell to the Red Army and the Soviet troops discovered the remains of Falz-Fein's zoo wandering loose, liberated from their destroyed cages, the arboretum felled, the laboratories smashed, the library stripped and even the collection of stuffed animals missing. Three hundred villagers had been deported to Germany, and over the course of the German occupation, 200 shot. None of the Askania-Nova Przewalskis survived the war, although which side shot them or allowed them to die of neglect is, as the Germans put it, '*kriegsverlust*' – lost in the war.

As for Vetulani's Bialowiecza herd, as the German occupation wore on, the forest – now under Goering's personal protection – was raided for more koniks – seven in February 1942, two the next year, three extracted in 1944 for a Luftwaffe major perhaps eager to emulate his commander-in-chief. The stallion Liliput was slaughtered so that Heck could have his skeleton. A German forest ranger killed and ate a mare called Liiputka, saying she was sickening. More 'sick' horses died, and their hides appeared slung over the shoulders of German foresters. Vetulani's programme was stripped by the end of the war and destroyed.

The human community that had lived in Bialowiecza was also

exterminated: Himmler's Einsatzgruppen ensured the Jews were shot or dispatched to camps, the Polish farmers and foresters executed or deported east. Their homes were razed. Partisans and Jews who had fled the urban ghettos of Poland occupied the forest and harassed the occupying Germans until their hasty withdrawal in July 1944, when the Red Army took over.

As the fighting grew more chaotic, the Heck brothers' programme persisted somehow amid the destruction, although the zoos and parks where the Takhi were kept were damaged badly. Hagenbeck's zoo in Hamburg was laid waste in July 1943, Berlin Zoo's swastika'd enclosures were gravely damaged in the November 1943 Battle of Berlin as firestorms scorched the city, and Heinz's zoo in Munich was also hammered by the end of the war and all but wiped out.

When the Red Army swept in from the east, panic set in. Goering began to remove his art collection from Carinhall, shipping it away to safety. The house was laced with explosives by the Luftwaffe and dynamited as the Soviets appeared in the neighbourhood, some of them riding Mongolian horses. The bison were shot.

The incoming army were hungry and short on supplies. They took timber from Schorfheide to burn, and drove tanks through the underbrush, flushing the deer out before machine guns. The horses were either shot by the Germans along with the bison or killed by the Red Army along with the deer. When it was finally possible to count the number of Takhi still in captivity around the world after the peace, just thirty-one were left.

You do not need to travel as Grieger did to see the Takhi. In Berlin I am spoiled for choice. The sprawling, miserable zoo in the eastern side of the city has a small herd, and west, in the zoo rebuilt after the Second World War in the old Tiergarten park, a small band of five live on less than half an acre. I went to see them one winter, and they bucked in the snow, making squeaky snorts that sounded like high-pressure air pumps. The half-acre enclosure was bare, with thin patches of grass picked to millimetres and surrounded by a large, sloping ditch covered with cobblestones. It was furthermost from the main entrance, and few visitors made it that far. My walk took me past several placards about members of the Heck family and a red-brick house that was still pitted with bullet holes from the final days of the war.

One summer I went out to Schorfheide, whose Russian conquerors had ensured it was part of East Germany, and where a Takhi herd had been restored during the Cold War. There were thirteen in a 23-acre field of lush grass, dozing under the trees, or moving lazily about in single file. Some mouse-grey koniks grazed in a separate, shadier paddock. Their keeper told me she'd acquired the Takhi stallion from the zoo, and when she released him he did nothing but gallop about the field. The next day he didn't move. '*Muskelkater*,' she smiled. Muscle hangover.

The Takhi were safe, but they were no longer wild – at least, not in Europe.

VII

At some point in the dark after Moscow I looked out of the aeroplane's window and saw a net of golden lights – city after city linked by beads of glowing sodium. Glitter poured down the roads that connected one town to another, and another, each spreading like crystals dense at the centre and trailing into darkness at the edges, where a fine golden thread linked them to the next bright cluster. I slept and when I woke we were above the steppe and it was morning.

Mongolia was woven from a single cloth: an unspooling landscape of rounded hills with slopes that fell away in folds, as if the peaks had been gently pinched into shape by hand. There were small patches of bright moss green that blended to khaki, which in turn gave way to a dull dun and that to purple in horizontal slivers, as though knitted into the weft of the hills. Some of the peaks had rocks scattered haphazardly about them, and there were balls of green, bright yellow and pale yellow fuzz in the folds of the hillsides – trees gathered in sheltered crannies. Rivers wound through valleys, sloughing off old oxbows and courses still visible from the air. From time to time there were large rectangular strips of crops, as artificial as a skin graft, but most was unrelieved steppe, with tiny white dots – the traditional 'ger' tents – sitting on valley sides or in the folds of the slopes.

The Great Steppe I saw from the Mongolian Airlines plane with a winged horse on its tail still reached some 5,000 miles from Moldavia across the landmass of Europe and Asia almost to the Pacific, merging into taiga in Siberia and to desert in the Gobi. In the Bronze Age, when the Botai's heirs were on the move, it was a medium for rapid cultural transmission – like a fluid that transmits electricity more efficiently – long before the Silk Road was mapped out in trading stations in the first century before Christ. Later still it was Tartary, home of barbarous tribes and wild horses.

While the fate of the Takhi captured between 1898 and 1903 was dire, the scattered, ragged herds left behind in Mongolia fared worse. It was left to the scientists of the new Mongolian People's Republic to helplessly chart their decline. The Dzungarian Gobi was a hostile habitat – there were few watering holes where the herds huddled, and they often lay up to 30 km from their sources of food, yet this was their last refuge, specifically in the Yellow Mountains of the Wild Horse, or Takhiin Shar Nuruu, where the desert straddled the China–Mongolia border. Although the hunting of Takhi was formally banned in Mongolia in 1926, and the horses came under special legal protections in 1930, many were killed by ethnic Kazakhs who did not share the Mongolian Buddhist taboo on killing more than was needed. In harsh winters immediately after the Second World War, the skins of Przewalskis were seen stretched out near Kazakh rebel encampments – stallions in particular were believed to have fattier meat. Punishing winters where every blade of grass was barbed with frost struck in 1944, 1945 and 1946, then again in 1955.

One last mare was retrieved from the Baitag Bogd mountains in

1947. She was christened Orlitza III and taken to Askania-Nova – the last living ark of Takhi genes to be added to the captive population.

The rare sightings of the 1960s were so important that they often appeared in the Mongolian press, but the last Przewalski in the wild was seen in 1968 by a bus driver from Ulaanbaatar who was hunting black-tailed gazelle at Khonin Usnii Gobi. He saw a stallion with no herd. In 1969 the Przewalski was officially designated 'extinct in the wild'.

In Europe, 1,800 koniks were returned to Poland by the British in 1946, and a domestic breeding programme re-established. Today many koniks are used in 'rewilded' nature reserves across Europe, where their toughness is appreciated. Genetic tests have shown that they are of domestic origin, with no real evidence that they are part-Tarpan or a trace remnant of Equus sylvestris. The Heck brothers' phoney Tarpans became known, more fittingly, as 'Heck Horses', and still exist in various zoological collections and grazing projects. Lutz's son, also called Heinz, emigrated to America in 1959 and became director of a game farm in the Catskills where Przewalskis and Heck Horses were bred. More recently, a couple in Oregon claimed to have reanimated a new Tarpan by selectively breeding mustangs that had a 'wild' appearance. People still dream of resurrection.

For the thirty-one Przewalskis that survived after the war, the problems of the interwar years persisted. Small groups were bred in isolation, and although individuals or pairs were sold to zoos eager to have their own specimens, they were often so consanguineous that even Orlitza III's influence was not enough to straighten out pedigrees that curled back on themselves like twisted willows. As in many an aristocratic family tree, there were abrupt terminations to entire branches, individuals whose behaviour went beyond eccentric and into vicious, and attempts to cut out insufficiently blue-blooded family members. Heinz Heck killed several horses that were partly descended from domestic Mongolian animals, thus squeezing the gene pool further and causing a rift with his chief breeding partner, Prague Zoo.

The Munich and Prague Takhi families grew more inbred as a result. As before, the Heck brothers' notion of what counted as authentic and pure was later proven wrong – in the 1980s tests revealed that several of the Takhis of what Heinz considered pure descent had smuggled domestic horse blood in their veins.

In 1950, a stud book was established to try to keep track of the horses, dispersed as they were across thirty-two zoos and private parks from Havana to Sydney, Moscow, Rotterdam, Talinn and Schorfheide, although not Mongolia. By the mid-1960s, the Takhi was scrambling back, and there were 134 horses globally. Plans were made to reintroduce some to the wild in the Soviet Union. Kazakhstan and Mongolia were proposed, and efforts were made to move some horses to semi-reserves that would take them closer to the experience of being wild.

Testing revealed that the Takhi was distinct from the domestic horses, having sixty-six chromosomes to the latter's sixty-four, and in 2011 geneticists discovered that the Takhi was a distinct species that branched off from the wild horse which became the domestic animal at least 38,000 years ago – long before the Botai or their neighbours first used a crude leather hobble or bridle. As more cave paintings and artefacts with the Takhi's blunt head and dun sides were discovered, it became clear that it was a horseshoe crab of a creature rather than an ancestor – a rarer, more fragile root altogether.

A philanthropist Dutch couple, Jan and Inge Bouman, who had seen a handful of Takhi mouldering in Prague Zoo while on their honeymoon, threw their money and passion into the campaign to return the horses to the wild. Jan devised a computer programme to untangle the family trees, and they persuaded the Dutch government to become involved before it was too late.

Although the number of horses was increasing, the Takhi was dying in captivity. Death rates increased by the year, and two thirds of the horses' genetic diversity was lost. The Takhi began to alter before the scientists' eyes, its torso and skull changing, its black legs fading as recessive genes dominated. The Soviets built enclosures at Bogd Khan, just outside Ulaanbaatar, but in 1989, when there were 961 horses worldwide and several generations had been bred in the semi-reserves ready for release, the Soviet Union began to come apart, and it was a Dutch government scheme to release the horses in Hustai, two hours from the Mongolian capital, that prevailed.

On 5 June 1992, the Takhi arrived in the belly of an Ilyushin transport plane at Chinggis Khan International, where news crews and a crowd of hundreds pressed close to touch the wooden crates they were sealed in. The Mongolian minister of nature and environment sprinkled fermented mare's milk on the crates as a blessing, and tied on blue silk prayer streamers that represented the wide heavens of Mongolia.

The wild horses arrived as Mongolia was both awakening from a long Soviet cultural sleep and fighting for its survival. The newborn independent state stood as a nation of fewer than 3 million people in a territory the size of France, Germany and Spain put together, sandwiched between its enormous, resource-hungry neighbours Russia and China, which had swallowed Inner Mongolia whole a few decades before.

Mongolia was rediscovering its heritage, digging it out, piecing it together, and reinflating Chinggis Khan like a Thanksgiving Day parade balloon, remaking him as a hero from an age when Mongolia had dominated the globe, a unifier of tribes and father of the nation. Shamans began to publicly practise rites that had remained hidden during the atheist years, and the country's Buddhists started to repair the shattering impact of the 1930s purges, when monks were faced with a choice of death or apostasy.

The horses that the crowds were so excited to greet at the airport at Ulaanbaatar were fragments of a past within living memory, charms to store up against fragility, to be held real and solid in the hand as confirmation of a time when Mongolia was great. The crates were loaded on open-bed trucks and driven 100 km to Hustai, an area of semi-mountainous steppe at the foot of the Khentii mountain range, where the Great Khan is thought to sleep in a tomb trampled by 10,000 horsemen.

Two days before the Takhis arrived in 1992, the same minister of nature and environment who greeted them had been in Rio, suggesting to the UN Conference on Environment and Development that his entire country be classified as a UN Biosphere Reserve. He encouraged delegates to think of the nation as the patrimony not just of Mongolia but of the world. Mongolia was offered as a vast, pristine zoological park for a polluted, beleaguered planet trying to understand that its natural defences against the sun were breached and its climate perhaps irredeemably kicked off-kilter.

But in reality the Takhi arrived when a slow-burning disaster was taking hold in Mongolia itself as the collapse of the creaking Soviet sun crippled its tiny satellites. The country's inheritance was being raided and depleted. In the nineteenth century, the Mongolian hunters hired by Grieger would not kill hens and trout, but that had changed.

Under communism, the Soviets gave herders production quotas and regular salaries, but the country's infrastructure remained primitive, and when the Mongolian People's Republic dissolved in the early 1990s, Mongolians were forced into a subsistence economy. The resources of the collective farms were problematic to redistribute, and the social framework weakened. One Mongolian in the western Altai region described the end of communism as 'the breaking of the great bowl'. In just two years, real wages plunged by half, and by a further third in 1993.

Mongolians did what they had always done: took up herding, and the number of sheep, goats and cattle on the land shot up almost twofold, at the expense of the grasslands, which were cropped and desertified. This new, redoubled rural population also fell back on hunting wildlife, a birthright that the new constitution maintained was held in common by them. Although the hunting of certain species was restricted or even banned outright, the country was both too vast and too poor to police effectively. A black market opened up with China: when marmot skins crossed the border they were transformed into sables – now a family would kill fifty rather than five, and in twelve years their population had fallen by three quarters. Maral red deer were slaughtered for their antlers, bones, genitals, tails and tongues, which were ground into powders for Chinese men concerned about their libidos. From a population of 140,000 deer in the 1980s, only 10,000 were left by 2013. Wild argali sheep began to vanish. The wild ass ate too much of the grazing of hard-strapped herders on the edges of the Gobi, and ended up as sausage in Ulaanbaatar. The migrating gazelle that once crossed the steppe in herds of hundreds of thousands was thinned by hunters' guns and the barbed wire unrolled along the country's borders. The global reserve was stripped.

Things weren't much better when I touched down twenty-one years later. Mongolia was experiencing a fresh environmental and political crisis. By then, 90 per cent of the country was at risk of desertification because of climate change and the spread of those post-Soviet herds of cashmere goats, sheep and beef. The land was also now meant to provide a new patrimony for the world and the struggling nation: $1.3 trillion of copper, gold, oil, iron and coal was to be extracted from under the skin of 'Minegolia' – grave goods set by in prehistory for some future nation. There was not enough home-grown technical skill nor capital to manage the great extraction process independently, but when multinational mining conglomerates and Chinese firms acquired stretches of the Gobi and set to work, the idea that foreigners could own swathes of a fenceless country was hard for Mongolians to accept.

The country was still poor, with almost a third of the population below the poverty line, and the percentage that was to be creamed from Rio Tinto et al could be transformational. One copper and gold mine alone – Oyu Tolgoi or 'turquoise hill' in the southern Gobi – was expected to boost the economy by one third in 2020 – if the billions of tugriks printed with the impassive face of Chinggis did not disappear into the pockets of a few. While some efforts were being made to claw back control of proceedings, there was barely any trickledown, and the new, promised great bowl was nowhere to be seen. Politicians had been liberal with cash before the last elections, but the money was borrowed against coal that was still in the ground, and when the cost of extraction and transport bucketed violently up, the bounty was in arrears.

When I arrived in September 2013, there was still a dirt road from Chinggis Khan International Airport to Ulaanbaatar, and it coursed with pale, metallic cars bought secondhand from Japan and Korea, which negotiated the potholes with more care than they negotiated other cars, slipping and jostling like polo ponies, honking like geese.

My first prospect of the city renamed 'Red Hero' by the Socialists wasn't inspiring: as my taxi bounced over a rise on the rutted airport road I saw a dirty sprawl of blocky Soviet-era buildings crouched in the Tuul river valley, and three slim chimneys pouring smoke like cigarettes, feeding the low-hanging smog trapped by the surrounding mountains. Like many historic nations of the steppe, Mongolians did not often sink foundations – the city was previously known as Ikh Khüree or 'Great Camp', and did not take root until the late 1700s, having wandered around the district since the Palaeolithic. Even now it seemed less than permanent.

We drove through industrial areas with grubby gers parked in stockaded yards alongside lorries stacked 15 ft high with fleeces, and past billboards advertizing diggers, excavators and credit cards. In January 2010 Mongolia underwent a dzud or 'white death' when a summer drought was followed by a winter with temperatures as low as -50°C: snow fell continuously up to two feet deep, and went on falling for months, smothering the grass the herds needed to survive, or freezing animals where they stood. When it finally receded the next May, one fifth of the nation's livestock were dead. Nomadic herders who had lost everything had moved to Ulaanbaatar and joined a shanty town of gers, and they had not yet left by 2013. A third of the nation now lived in the city, its homeless in the sewer system. Jerry-rigged electricity lines hung between gers over gullies where rubbish gathered, and where dogs pottered with used nappies in their mouths, tails waving.

The gers themselves seemed more stolid and self-contained than the crumbling or half-built concrete shells of buildings. Even in 2011, 45 per cent of Mongolians had still lived in a ger. Mongolians were asking their government when their share of these newly exploited resources would arrive. There were still barely any paved roads in the country and some districts weren't even connected to mains electricity.

A new 'Law on Budget' meant Rio Tinto could gift local and state officials with Land Cruisers, and give money directly to the president's office. Just a year previously, the former president Nambar Enkhbayar had been jailed for embezzlement. The prime minister who brokered the Oyu Tolgoi deal resigned less than a month later and almost immediately bought an apartment in New York.

Meanwhile, herders were losing their land and animals. Twenty calves near one French uranium extraction plant had died in 2010 and though there was a dzud, hungry wolves wouldn't touch the carrion. The State Veterinary and Animal Breeding Agency found evidence of radioactive poisoning, but their results were hushed up. NGOs had been threatened with lawsuits by multinationals and intimidated by state security forces. Elsewhere, foreign firms mining illegally had their machinery peppered with bullets or dinged with arrows. There were wild rumours circulating that the Americans and Japanese wanted to fill a hollowed out Oyu Tolgoi itself with radioactive waste.

When the steppe is degraded or overgrazed and the grasses and shrubs no longer knit the soil together, there is nothing but dust. If you do not move on with your livestock, they will starve. Long since stripped of grass, Ulaanbaatar seemed to be coming apart that autumn. Overground pipes were wrapped in what looked like thick, cracked clay against the winter. Pedestrians negotiated long, dusty ditches dug for but not filled by underground pipes. Electric cables lay coiled on the roadside where they'd been ejected from the earth, ends bare. The buckled pavements had shivered and undulated as the ground below expanded and contracted with the steep temperature switches. Like navels, manhole covers either protruded from the sidewalks or were sunk into them. The air was gritty with particulate – exhaust fumes, smoke from ger stoves, dust from building sites, smut from the coal-power stations – and thickened in the lungs.

The day after I arrived, there was a protest in Sükhbaatar Square in the centre of the city. Members of Fire Nation, a collection of environmental and human rights NGOs, gathered outside the parliament armed with old hunting rifles. Inside, the Law to Prohibit Mineral Exploration and Mining Operations at the Headwaters of Rivers, Protected Zones of Water Reservoirs and Forested Areas – known, unsurprisingly, as 'the Law with a Long Name' – was about to have its teeth pulled. For Fire Nation, the law was a last thread: mining had already caused pollution, and firms were able to refuse inspections with impunity, and move herders on from their land.

A shot was fired, and police moved in to arrest the leaders: six men in their fifties and sixties who were later sentenced to twenty-two-and-a-half years in jail, even though video showed that they were already being manhandled by the police when the shot rang out. At their trial, it was shown that their guns were unloaded or had chambers full of blanks.

VIII

I took a minibus to Hustai National Park with twenty children from the district, all of whom went to school in the capital and travelled to their parents at weekends or for summers. Gracious hosts, even on the interminable grind out of Ulaanbaatar's rush-hour traffic, they fed me cheese popcorn and lemon sherbets, and were wildly amused when I finally fell asleep with my neck at right angles and snored.

Hustai is the only conservation park in Mongolia run independently of the government, and is now entirely self-sufficient. In the early twentieth century it was the hunting park of the last khan, an unfortunate character sandwiched between the Manchus and the Soviets, whose musty, Russian-styled dacha and snow leopard-hide ger I'd seen in Ulaanbaatar. Hustai was later collectivized by the Socialists, who kept the 121,000 acres as reserve pasture for the one hundred or so local herding families.

To compensate them when Hustai once more became a Takhi reserve in 1992, its European and Mongolian organizers constructed a new cheese factory, workshops, community centres and wells. The children on my minibus had been born into and grown up in its community; their parents and siblings worked as rangers, cooks, housekeepers and drivers at the park. Their lives centred on the wild horses I had come to see: as one teenage boy told me, 'We are very proud of them.' He would, however, be studying mining at university, he self-consciously added in English.

Just inside the northern boundary of the park was a tourist area consisting of three large concrete gers and about thirty traditional gers that served as accommodation, and a central brick building with a restaurant and offices where I was installed in a gelid, north-facing room. The gers, whose yellow-painted doors faced south, away from the north wind, looked back at me in placid surprise when I peered out of the window into the night.

The day after I arrived, the manager told me that perhaps I could join two French ecovolunteers who were studying the Takhi, or else I could walk into the park alone. Most tourists arrived with guides and SUVs and drove in. There was no map available apart from the one on her office wall, but she told me I should photograph it with my phone, follow the road, and find the horses in the Tariat valley. So with no idea of distance or time I set out, passing under a faded blue metal archway decorated with photographs of summer flowers. This was my first sight of the steppe in daylight since the flight.

Up close, its smooth contours were a complex biome: the richness of the steppe that fed the horses and drove the nomadic cultures lies in these multifarious grasses. A few flowers were left among the long, spiky green blades of *Achnaterum splendens* with dirt yellow tips, and grey-blue grasses with minute, petal-like leaves. The unpaved road had the first fissures of erosion running down its centre, and in places a new road had been carved and the old given up. Bead-like pebbles crammed the rivulets left by running summer rainwater. Crickets ground away in the grass, and some large and black insect suddenly swooped over my head making a rattling, papery, urgent noise like a party favour.

The original shamanic faith of the Mongolians was a built-in environmental protection system which assimilated easily into the Mahayana Buddhism that arrived from Tibet in the medieval period. The world's first nature reserve – long before Schorfheide or Bialowiecza – was the mountain Bogd Khan, which appeared, massive and stunning, one day when I was wandering around some building sites in Ulaanbaatar. What I had earlier mistaken for Bogd turned out to be its foothills; the true mountain consumed the sky behind them. In the eighteenth century it was guarded at the Manchu emperor's command by monks with clubs, looking for poachers, and by an eagle spirit called Khangarid. Elsewhere, mountains, rivers and lakes were patrolled by spirits or nagas who must be respected by locals, and who might forbid fishing in their river, cutting trees, destroying grass at the root, disturbing bird nests or the hunting of a certain type of animal across their grounds. Break these rules and the nagas would punish your settlement with food shortages or a flood.

When the Takhi arrived at Hustai they first spent months in several large enclosures to acclimatize themselves to the steppe. Before they could be released, local lamas were consulted and determined the most auspicious date by astrology. As the last gate of a long journey opened, the monks tied blue *khatag* scarves to the fences.

I moved up the road from the camp through a large herd of domestic cows and local horses which grazed steadily. Hustai's own naga-haunted peak lay ahead of me, west of the Tariat valley. Neither herd should have been in the park, and rangers would eventually drive them out, but for now they were feasting – the grass in the park was better than the grazing outside its protective net. The cows were a bobbery pack, the result of several decades of imported specimens aimed at breed improvement under the Soviets. Slab-sided, with backs like ridges and broad foreheads, here was one that looked almost like an Aberdeen Angus, and then a polled Hereford with dark gingerbread fur and a white face. Another creature had long horns and orange, wiry curls that gave away to reveal sleek, deep brown sides, as if it had been part-peeled.

The horses were shaggy and sceptical, moving rapidly away from me. They too were a jumble of colours, more extravagant than the cows: black spots on white, a bay tobiano with white streaks in its mane and tail, smokey duns, fox chestnuts, bays, dark-eyed greys. Mongolian domestic horses are small, tough and largely disinterested in humans. Herdsmen can own hundreds of horses, many of whom they will never ride or even milk – the more horses, the greater the status – and riding begins at toddler age when children help to social-ize foals. They roam freely, often not ridden or handled for months at a time, and are sometimes raced, with child jockeys, for up to 30 km across country.

These strays at Hustai trespassed insouciantly, pregnant mares balancing their bellies like milkmaids their yokes. They raised their heads and watched me through thick, trailing forelocks, jaws grind-ing. I snapped a few photos and left them behind, following the road over the low rises in the landscape as the hills rose up alongside me, building towards something. After forty-five minutes I turned in a slow circle to try and cram the scale of the landscape into my vision and saw that although the sky over the valley ahead was the bright blue of summer, the sky over the distant, dark rocky curtain behind the camp was deepening to indigo, and the grass the horses and cattle grazed was darkening. I abandoned the march and walked back.

IX

I woke in the middle of the night and heard the wind howling. In the morning the sky was occluded and snow poured off the flatly conical ger roofs like sand, driving thick flakes wetly against my window. A fringe of stunted icicles protruded from the restaurant eaves, sloping where the gale had drawn them drop by frozen drop. The main door of the building was unlocked, and light and cold air sliced around it. When I stepped outside, the camp was silent. The wind burned my cheeks and ran slick and frigid as a cold silk scarf round my bare neck. I thought it would peel the skin off me, one layer at a time. It was moving the settled snow, pushing it on, not letting it rest till it caught at the edges of the concrete paths in the camp, snagged in the grasses, and melted into thin ice sheets like thick cobwebs that gathered more rolling specks of snow.

The gers sat fatly with their backs to the wind, each girded with three black bands like belts that contained their thick felt coat. At the base they had been swagged with baggy plastic against rain. Smoke chugged from a handful of the chimneys before being blasted sideways by the wind; the antennae collection on the roof of the visitor centre swayed and rattled. The only sign of life was the sparrows which were noisily rejoicing in the hot-air outflow pipe from the kitchen.

In the car park I found the park's horseman and a ranger, whose dogs demanded to have their ears rubbed. One was like a smaller, fluffier German Shepherd, the younger one black, with a white bib, a short coat of incredible density, tawny eyes and gingery eyebrows – a 'four-eyed' Bankhar mastiff, bred to live with herdsmen's flocks and protect them from predators.

The park had instituted a breeding programme for the mastiffs to assuage local uneasiness about a ban on hunting wolves, which had flourished as the rest of the wildlife collapsed in the 1990s. The dogs were often seen trotting about the park in pairs, or at the doors of gers, raising their huge skulls to be stroked. The wolves also guarded the park, after their own fashion: when a timber thief brought a horse and cart to carry away some of the straggling trees in Hustai, a wolfpack ate his horse.

There was no sign of the ecovolunteers – I later learned they had gone into Ulaanbaataar for the day – so I set off along the road to find the horses again, framed by the hills which were now mottled like greying animals. Where there was shelter from the wind, the snow lay in white blankets, but on the bare slopes and hilltops it had been driven sparsely through the grasses, whose colours were muted – the spiky yellow *Achnaterum splendens* was dirty ochre, the brighter greens were now dark conifer, the blues drained to grey. Snow in the road hollows had melted and refrozen in thick crystals that lay across one another like pick-up sticks.

After I'd been walking for half an hour I looked up to keep my bearings and noticed that the rocky tors I'd seen at the summit of the hills the day before were now smudged with falling snow and low cloud. The long black ridge north of the camp was narrow and chipped as a piece of slate, blurred in charcoal and deepest green with a powdering of white far down where the wind drove it. Three gyres of pale mist spiralled down it as though the clouds were reaching down to the earth. As I turned to look back at the camp, I noticed a distant, snowy mountain range had appeared to the far right of the black ridge – like Bogd Khan materializing in Ulaanbaatar. There seemed something restless and mobile even in the geology of the country.

I walked accompanied only by the power lines, the moon and the noise of my clothes as I moved. A shadow stood behind each pebble, even down to the granule-like stones that gathered in the rivulets. Every piece of foliage that emerged from the ground was silvered with rime; the tiny leaves that struggled through the road looked like glinting crumbs of broken glass. No tourist jeeps came along the road. Two giggling black choughs fooled ahead of me, landing then flying further as I approached, leapfrogging me along the road.

I found deer prints first, then the unmistakable full crescent of horse hoofs – probably left by the herd of domestic horses, but perhaps not. They were shapely, the foals' teacup impressions tiny. Cow or deer prints tracked across them. Next I found a heap of fresh, dark horse dung, not yet caught by frost. I was getting closer to the valley. I could see nothing on the hills, though I found a track of hoof-prints through the snow, dragging powder after them towards the road. There was a smooth, hemispherical meteor of compacted snow with the pinched triangular impression of a horse's frog in it, clearly kicked out of one of the hoofs.

I was tiring now, and unnerved by the empty distance behind me. The road rose as the hills came down to embrace it, and at the lip of the Tariat valley was an *ovoo* – a cairn of stones dedicated to a naga – from which rose a long branch trailing blue *khatags*. The breeze that shivered the silk streamers brought a sudden whiff of fresh horse dung, but when I turned full circle, squinting up to the slopes, looking for the horses, I saw nothing.

A raven was sitting on top of the *ovoo*'s ragged standard. Beyond it the road plunged away into what seemed almost like a canyon after the gentle slopes I'd passed through – the slopes steeper, the sides more crimped and complex. I couldn't see the Tariat valley research centre or any life as the road soon disappeared into the folds rather than revealing a magical valley of the lost horses. The raven kroaked-kroaked and took wing, trident-shaped against the blue. As I was standing still and my clothes were for once silent, I heard a low, whispering rasp, slowly realizing it was the noise of the wind caught in the raven's pinions.

Tired, I turned back. At the camp I looked again at the map and realized I would have had to walk twice as far again to reach the valley.

X

On the fourth day, I travelled with the ecovolunteers, and it seemed as though the landscape had come to life with the thaw. Our Russian-made minibus had two galloping chrome horses on its bumper and tackled the dirt road with enthusiasm – in minutes we covered what had taken me half an hour to walk. We bounced on the old plastic seats like air passengers in turbulence, clutching at the straps, while a golden tassel hanging from the driver's mirror swayed and kicked up its skirts like a cancan girl. Marco and Laétitia, old hands, pointed at the hillsides – 'Marmot!' – and there would be a marmot, endomorphic and alert, anxiously sniffing the wind on its hind legs, ready to vanish into its burrow. Now you see them, now you don't – they blend into the landscape and are gone. In Hustai they are food for wolves and eagles, rather than coats for the wealthy.

We rumbled up past the *ovoo* and swaggered down the steep slope into the valley like a lucky drunk who always catches himself before falling. At the bottom of the slope we hit a dry riverbed where the road split into two choices – one at a 45° angle and the other at 55° – and ground up the other side on the steepest. A little way down the valley, over one rise and down the other side, the driver pulled over and pointed. Marco and Laétitia peered out, and I, with no idea what was happening, looked where they looked.

'Takhi!'

At the top of one of the hills, on the very ridge, stood four horses, side on. These weren't domestic horses: the silhouette was unmistakable – the heavy, full-cheeked head with the powerful jaw, thickly hinged to a short, crested neck. The full summer belly with a dirty white underside that reached up at the points where the surprisingly fine legs joined the torso. Lascaux horses. The fuzzy photographs of Charzis at Askania-Nova. The horses in the shadows at Schorfheide. Even at this distance you could see the tails: thick, short and bristling at the dock, flowing into a long, black trail to their fetlocks. Their dun coats rose upright on their necks, guarding a thick, black cockscomb of a mane. There were four. No, five. Six. Two stood neck to neck in an embrace, scritching at each others' withers with their teeth.

We got out of the bus with rucksacks, binoculars, GPS, a clipboard and Marco's camera, and began to climb up the slope at an oblique angle to the wild horses. One very pregnant mare stood facing us as sentry, watching. There were eight horses – two foals and six adults. As we pegged away from them up the hill, they began to walk, not too hurried, and then they disappeared over the other side of the ridge.

After a few minutes of toiling up the slope, we stood on the ridge where the wind was stronger and looked down. The landscape was empty. The horses had vanished, although there was no cover – no tree or scrub in sight, just bare green hills leading to a small, steep, riverless valley.

Then there was movement, and yellow dun shapes against the green as they emerged at the base of the hill we stood on, having slipped down a fold, hidden from sight. Just as Hamilton Smith had written of their ancestors, they had used their 'unerring tact' to 'disappear as if by enchantment'.

One bore off away from the group, heading up the other side of the valley towards a rich green patch. The others broke into a trot across the gully, their rumps framed in white, like deer, black tails switching below their hocks, their fox ears flattened on their necks. When they had judged that they had put enough distance between us and themselves, they turned left as a flock and walked after their leader. Laétitia entered their numbers, sex and coordinates on the clipboard as Marco confirmed the details through his binoculars.

We sat on the hillside to observe and I began to distinguish between them. One mare was dark, almost chestnut, her underbelly a bleached-wheat shade. A couple were paler, dirty cream. Most were classic Takhi, the deep turmeric dun and soft white that's not quite cream, and black-dipped legs. This was a harem band of one stallion, a clutch of mares and just two foals still alive. The youngsters had stiff pale coats like those of Steiff teddy bears, and kept close to their dams. They blended perfectly into the drying vegetation.

As the band reached the emerald green patch, the horse lowest on the hill took up the watch, standing side on to us so that those eyes – the largest of any land mammal – could survey us. We sat in silence and watched. It was the end of a good summer, with enough rainfall to push up the spiny festuca, sparse stipe, wild iris and geranium that had fattened the Takhi. They would save the dried, long, hay-like blades of *Achnaterum* for winter, when they would dig down through the snow to feed.

One of the foals nursed a little. Now and then a horse would snort – letting off the tension caused by our slow pursuit. As we watched them through the binoculars and Marco's telephoto lens, they swept slowly, gradually up the hill, spreading out from one another and away from the green sward. The crickets had recovered from the snow and were rasping somewhere nearby. Further up the gully, some puny birches with yellow leaves and white verticals for trunks clung to the slight shelter of another, smaller gully. We moved down out of the wind.

By common accord when both the foals lay down, the band stopped grazing entirely and stood guard over the youngsters. Even though the base of the gully was a richer green, the Takhi kept to the safer slopes, one watching the distant tourist minibuses that would appear stage left and tootle along the road like Tonka toys. There was another herd two slopes over, but our Takhis ignored them, and a third group picking their way over a western ridge, slipping out of view. They shared the landscape with the invisible but very audible maral stags who were in rut, bellowing like dinosaurs. The horses ignored them too, although they share a kinship with them. Some young horses, driven out by the harem stallion, have been known to take up with red deer groups and live alongside them.

A bright yellow tourist bus stopped in view and one of the Takhis turned to watch it – you could tell the degree of the emergency by the number of horses who froze, head turned to the source of disturbance. In the far distance by the *ovoo* a string of rapidly moving black rocks were tumbling down the hill. We tuned the binoculars: domestic horse. No. Sheep. No. Cows? Dark, livery chestnut blobs. What where they? We made notes and crossed each possibility out.

An hour or more of the afternoon passed. Shadows began to creep down the hill to where we were sitting, draining the warmth from the wind and filling in the valley in a slow creep. We moved along a track the horses had made, and saw the cow-sheep shapes coming down the road towards our hill. Five of the Takhis broke off grazing to watch them, then all six of the adult horses, turning their white rumps to us so we could see the black zebra stripes on their hocks and the hair fringing their oval ears as they strained to hear.

As we reached the road where the driver would collect us, the shadows were moving faster than we could walk. Suddenly there was a squeal and a snort from the stallion; as I turned I saw him, ears pinned back, rounding up his band with nips and shoves and pushing them away up the hill like a school of fish startled into a tight ball.

There was a rumbling which I thought was the Takhis galloping till I saw the extraordinary creatures that came charging up the hill towards us. They stopped, drawing up in a long, ragged line and goggling back. It was the cow-sheep: huge quadrupeds with long horns and shaggy coats, dipped necks and thick-boned brows. At one end was a colossal beast that looked like a clipped bear. Could they be? Were they? Yaks. We stood and stared. They stood and stared. Finally they broke away and went capering up the hill towards a small gully, snorting and hissing.

Up one side of the gully the yaks galloped, then back down into the dip, waving their stumpy tails high, then, bucking and rollocking, they raced up the other side, and turned and ran down again. Up they went, grunting in effort, down they careened, again and again.

When the bus collected us I looked back up the hill where the yaks were rolling like drunken marbles in a half pipe, and further above saw the Takhis watching warily from the top of the hill.

XI

'The yak were domesticated last,' said Dr Usukhjargal Dorj, the chief Takhi biologist at Hustai, 'so they really are a problem. The cows always go home at the end of the day if they've grazed in the park. The yak don't. The wolves at Hustai attack the deer, the horses ... but never the yak. They're too scared.' Dorj was a serious man in his thirties, struggling with a bad cold as he politely talked to me about his work. He was gently spoken, and his words ended with a soft guttural flutter. A faded red Harvard baseball cap lay on his desk along with plastic files holding pictures of petroglyphs he'd brought out to show me as evidence of the early Mongolians. Behind me was a glass cabinet fully stocked with studies on the Takhi and a copy of *The Secret History of the Mongols*.

The bright, modern office building shared by the three biologists in Hustai was just a short distance from the ger camp. Usukhjargal Dorj had worked at Hustai for ten years. His wife managed the tourist camp and his son was the much-coddled toddler who could be seen being passed from lap to lap every day in her office. They spent all year at the camp, monitoring the horses even in the dzud winters. 'If you want to study the Takhi,' he explained, 'you have to be here for a long time, and that's hard if you're single.'

He knew every one of the 300 horses in the park by sight. 'This one,' he pointed to a stud book card on his screen, 'birthmark on shoulder, shape of an arrow. This one, broken ear. This one, broken ear, no tail … ' Sixty of the horses were bachelors with no harem and the result was a fractious, sometimes violent society. In spring, bachelors would chase a band of mares, sometimes in tandem, for days until the stallion was exhausted by driving them away and turned to fight. They skirmished with teeth and heels. 'These are the soldiers,' said the biologist, bringing up a series of pictures of lone males, whose ears sagged against their skulls, or whose tails had been stripped. One had a huge open wound between its eyes. 'We think the park can take 500 horses, but we only calculate that by the grazing. It could be that socially it's not possible.'

There was no intervention by park staff, although some mares, once exhausted by blocked labours, were helped by rangers or scientists. One theory held that when wolves or bachelors harassed the harems, the mares were chased uphill over rocky ground by their stallions, and this sometimes led to the unborn foals laying awkwardly.

Thirteen generations in captivity did not erase the instinct to drive off a wolf. At the sign of those low, loping forms, the mares formed a stockade with their hindquarters to the foals, teeth and forehoofs front. The stallion and sometimes a mare ran around its circumference, attacking the wolves. Foals are also swifter than wolves, although not over much distance. Still, more than 39 per cent of the foals could be picked off by wolves in a year – many of those in their first week of life.

Dorj was unconcerned: 'We have a good balance between horses and wolves. This is a reintroduction programme and what survives is enough. Some years the wolves eat more and then we scare the wolves off. From mid-April to mid-June the rangers camp near the horses every night – just in their habitat, not right next to a harem – so then there's light, and talking, and the wolves stay away.'

Those 300 embattled Takhi in Hustai were the largest population in the world and the only group to live in what scientists believe to be its optimum habit. There are fewer than 2,000 Takhi in existence today, and most of them live in urban zoos or semi-reserves. One herd roams the post-nuclear wilderness of Chernobyl; another is being nurtured near the Yellow Horse Mountains in China. Only in Mongolia are there three introduction programmes that are succeeding in establishing independent, self-sufficient populations. Two are in the last home range of the Takhi, at the fringe of the Gobi, but the conditions are as harsh to the horses as they were in the last century. In the old khan's hunting grounds the Takhi are truly thriving.

'Only seventeen of the horses are among the eighty-odd that came from Europe. All the others were born here. Over 600 were born here over the years but in the long-term average 40 per cent die in the first year before the end of December.' Dorj clicked through to a graph. 'It's natural causes – mostly we ignore that. It's OK.' The population at Hustai was already sustainable, and its inbreeding quotient was falling – the horses that were brought from Europe in 1992 had been specially selected to be not just tough but relatively unrelated to one another.

'More than 70 per cent of Hustai mares get pregnant – with feral horses it's 50 per cent – so they are working hard to survive. They produce lots of babies but very few per cent increase – but some of the years the foal survival is more the 65 per cent and when it's that high the population increases 12 per cent or 14 per cent.'

I asked Dorj why the Takhi were so important to the Mongols. I'd been told they were not just symbolic but sacred, though I could find nothing in writings on traditional Mongolian religion to support that – horses tend to be transport for gods rather than gods themselves. 'We thought that the Takhi is the ancestor of the domestic horse, and the word Takhi means holy horse or spirit of horse – so you don't touch it. But,' he added, 'I think it's not religious, it's ethnic, because we are a completely nomadic country and domestic horses are important.'

He showed me images of two local villages that had proudly adopted the Takhi as their symbol. There was a folk song about the Takhi, and models and stuffed toys on sale in Ulaanbaatar alongside the horsehead fiddles and Chinggis hats. They had their own room at the natural history museum.

They are still coveted, too: Hustai had to deal with a ranger who bred his mare to a Takhi stallion, hoping for a winning racehorse. The Mongolian government even accepted a gift of seven Takhi from Australia to swing the Olympic voting in Sydney's favour for the 2000 Games. The government likes to bring visiting panjandrums to Hustai to see the success of the programme, and give the occasional Takhi – from other projects – to foreign governments. Mongolian wildlife was once more a diplomatic resource, though shared more selectively than the nation-biome that was meant to be open to the world. Just before my visit, despite objections from conservationists, the government had granted Qatari and Kuwaiti hunters a licence to capture 200 rare saker falcons that would 'act as leverage for prospective cordial relations'.

The next day, down by the river in the Bayan valley, we found a corpse. Marco, Laétitia and I had been sitting on a chilly, north-facing slope riddled with marmot holes, where handkerchiefs of snow still sat on bare patches after a thaw. Below us a dark brute of a stallion stood in the riverbed before a huge boulder, unmoving. His harem grazed on the bank, the foals dozing, one mare straying close to him. In this valley many harems overlapped, generally peacefully. Hoofprints had mashed the softer ground at the water's edge and dragged the longer grass into clumps. By the road were several 'stud piles' where one horse would drop dung and then another come, assess it, and drop more on top, until a large, tumbled pyramid grew – a kind of *poste restante* for horses.

We found the body after the herd had moved up the southern slope, pursued by an overzealous guide and a band of tourists who came too close. It was a yearling, killed by wolves in August, when the cubs were learning to hunt. It had been broken into pieces – skull, quarters, forelegs, torso – and scattered. Its mangled, twisted hindquarters still had some flesh and fur clinging to its tricky lower leg joints, which ended in immaculate hoofs only a few centimetres across. Its tail straggled on long grass. Matted tufts of its pale, collapsed coat were blurred with the dying, pale foliage of the steppe. An intact spine writhed beside a wide blade of scapula, and a few flimsy white ribs lay here and there. The skull had separated into two halves: the lower jaw a long set of tongs, curved up at the molar end to meet the hinge of the upper jaw. The top jaw lay upside down, the ridged teeth staring up. Flipping it over with my toe, I saw the long, smooth skull ending in a jagged point where the horse's soft nostrils and nasal passages – higher and deeper than those of a domestic horse – had once quivered.

Having tracked up the southern slope, careful to keep further from the horses than the tourists had, we stayed with the harem for the rest of the morning. We watched the mares take turns to rub their round butts on the concrete post that supported an electricity pole. The foals, looking too narrow and pale for the coming winter, dozed and grazed. The slope was covered with dried horse and marmot droppings, and Laétitia found a bone that had been nibbled by a creature with sharp, persistent teeth. We dozed too, a little stunned by the wide blue that the moon never seemed to leave, and by the silence.

The stallion alone did not eat. He continued to stand some distance away, closest to the road, his long black tail so heavy and coarse that the wind could catch and move only the last foot of it. While most of the band grazed in a shallow crease, hidden from the road, the mare who had stood by the stallion at the riverside remained near him. As the morning wore on, her motives were revealed. The stallion suddenly gave out a raspy whinny and walked rapidly over to her before climbing on. She moved gently away, and he dismounted and came to stand head to head with her, blowing at her nostrils then peeling back his top lip to better savour her scent. Had the mare lost the yearling that was bleaching in the valley? Or another youngster that summer? Either way, it was a terrible time to conceive a foal which would be born eleven months later and have a few scant months to fatten before the winter arrived.

As the herd worked their way up towards a ridge piled with rocks like an *ovoo*, I noticed the stallion was lame on the right fore. The snow had already warned us it was the tail end of the summer. I wondered if the stallion's little band would make it intact through the winter.

XII

On my last day at Hustai we met at 5.30 am. We drove out to a huge, flat plain, and there, in the far distance, among black, slow-moving, boulder-like cows, raced 300 gazelle, so well camouflaged that it was as though they were transparent – a flickering disturbance in the landscape, ghosts back from the borderlands of extinction. The day before I had seen a golden eagle waddling by the roadside, a sight so unexpected that I took it for some species of overweight grouse.

'Hustai was created for the Takhi,' Dorj had told me, 'but they have become a flagship and an umbrella species. In 1993 they made the first count of animals in the park, and there were only forty red maral deer. Now there are more than 1,000. There were no Mongolian gazelles and now there are between 200 and 400. There were no argali sheep and now there are forty in residence – not just migrating through. We have the highest number of Takhi in the world, the highest number of red deer in Mongolia, the highest number of marmots.' He counted them all off with pride as he showed me how they had they crawled back from the depredations of the 1990s, when the great bowl broke and Mongolians were scrabbling to survive. 'This is the most important wild horse study in the world. I send my data to Cologne Zoo in Germany and they spread it to the world. This is why Hustai is important and famous in Mongolia.'

A few months after I left Hustai, its independence from the government was officially confirmed once more. I thought of the dusty valley beyond the park's limits where we had visited a local family in their gers, and the contrast between it and the sere but fertile Tariat valley shared by the red deer and the Takhi. What would happen when there were too many Takhis for the park's wolves to cull? Where could they be rehomed?

But for now, the success of the Takhis, where the dodos, the quaggas, the Tarpan, the aurochsen, had failed, was to be celebrated. I raise a glass of *airag*, the Mongolian fermented mare's milk spirit, to them. For the first time in their recent history, they had served their own survival, rather than the changing notions of mankind. The horses had made their long, circular journey from steppe to zoos and back to steppe again to become nagas of a kind, defending a tiny refuge at the heart of the country, a small Eden protected from copper mines and deserts and uranium extraction and hollowed-out coal deposits. Like the roots of the steppe grasses, they knitted together the soil, and stopped it turning to dust and blowing away.

CULTURE

Horses Strangely Wise

I shall next set down the method of riding which the horseman may find best for himself and his horse.

On Horsemanship by Xenophon (c. 350 BC)

I

The stage is dark. The boards are covered with black sand, the wings and backdrop are shadowed. In the audience we can sense, if we cannot see, a presence in the darkness, and a dimension to that negative space, reaching back deep behind the proscenium of Sadler's Wells theatre. A single overhead spotlight comes on and a flood of white light carves a strange creature out of the blackness. It has the long, satiny head of a black horse, its ears moving as it sees the people revealed by the chilly light. Below its jaw is the body of a man, barechested, muscular and a little grizzled in the glare, and ending not in legs and feet, but in wide furred legs, hoofs obscured. This centaur is reversed, with the head of a horse and the body of a satyr or a wild man. The centaur raises expressive white hands, palms flat and open, fingers outstretched, the gesture of speech or supplication, but before its lips can move and the horse for once speak to us directly, the centaur folds its hands over its eyes and stands, blinded and silent.

II

The baroque barracks of the Palace of Versailles stretched across the horizon, silly with gilt gingerbread work that glowed through a wash of gloomy, spitting late November mist. Standing with my back to the palace, I looked forward over the empty expanse of the Place d'Armes, to the Grande and the Petite Écuries, or great and small stables, which looked back at me, wide-set and faithful as the eyes of a horse, the Avenue de Paris a broad blaze between them.

I walked down the square past a pedestal on which the Sun King, Louis XIV, sat on a copper prancer with an arched neck and a ribbon on its tail. He had a hunched pigeon perched on his head, and gestured with one arm to the Petite Écurie on the right, but I turned left across the cobbles where the tourist buses had drawn up like waiting carriages earlier that day, crossing the Avenue Rockefeller and passing through the peeling wrought-iron gateway towards the welcoming colonnaded horseshoe of the Grande Écurie. At the centre of its Cour d'Honneur, surrounded by parked cars, was a shallow, rectangular sandpit where, earlier that day, two Portuguese horses had cantered slow circles, their riders growing ramrod-straight from their curved backs like flagpoles on a summit.

Behind the sand school, the doors of the écurie were open and above them poured three surging stone horses with open mouths and boggling eyes, bare of riders and bridles – though one had a bird's nest on its back, perhaps that of the King's pigeon passenger. In the time of the Sun King there was a theatre in the Cour d'Honneur; now it has moved behind the warm orange ticket booth guarded by the stone horses. I spent three November days in its rehearsal rooms and its dressing rooms, dashing down empty staircases backstage and sometimes wandering on to the stage itself – though more often standing aside to watch the performers.

The theatre consists of three spaces. The first of these is the Cour d'Honneur, a sandpit in which horses are schooled, luring the tourists on the Avenue Rockefeller away from the half-empty palace up the hill, drawing them in with their mysterious figures in the sand and configurations of rein and bridle. Here anyone can eavesdrop on the closed-off world of the riders, who talk to one another in body parts, 'Leg straight, seat deep, move your hands.'

The second space is the Carrière, which is reached through a stone archway to the left of the orange booth. It is an open-lidded box of softened red brick and washed-out stone, bordered with flagstones and dominated by another sandpit, a shade more taupe than the brick, soupy in parts with rainwater. Flood lights on green poles wash the dirty white sand with light. A magpie lazily makes its way up to the mansard. There is one two-storey wooden double door, covered with pale, faded and blistered paint, like the trunks of silver birches, in the centre of the south side, which leads to the third space – we will come to this later. To the north, all the doorways are bricked up, but their archways remain visible so that architectural harmony is maintained. Another archway leads east through a high cobbled passage scattered with curls of wood shavings to a covered round pen, tucked away from sight. Behind the sand arena on the western, Versailles side, are three storeys of backstage offices and rehearsal rooms: on the top floor, reached by dark green stairwells with slippery, deep red hexagonal tiles underfoot and elderly, shallow wooden stairs, is a long room that runs almost the length of the Carrière, with one bare brick wall covered with a mirror, bisected by a ballet barre. There's a low, orange divan the colour of the ticket booth, and another in maroon, a rack of épées and some concentric archery targets. At one end is a door, from behind which come the voices of women, raised in song. On the floor below are the white, bright offices overlooking the Carrière, and a warm orange gallery that curves around the inside of the horseshoe and looks across the Cour d'Honneur to Versailles. Below the offices is the common room, divided by a flight of stairs leading from a mezzanine where racks of costumes are stored.

On the day I visited, the common room was a curious and contrary clutter of objects. There was a large tropical fish tank, which bubbled in the silence. A rack of undulating bows as tall as a woman and boxes of arrows. Corkboards with pictures of horses and archers on them. A desk scattered with timetables. A workplace heaped with leather – reins, straps and a fleece-lined human boot, gralloched next to a formidable green metal sewing machine. Scarlet *umanori hakama* – Edo-era Samurai trousers or culottes – hung from the underside of the stairs, and there was a carved red horse head emerging from a pot plant by the door. One long white table, surrounded by chairs like white nutshells. A full bottle of red wine, opened and recorked. A box of instant noodles. A thick, quilted saddle pad on a radiator, and a washing machine chuntering in a side room. At one end was an easel with a flip book on which curlicues representing horses were spaced like some obscure script or punctuation, edging across a stage.

The third space is the Manège. This is the grandest chamber. The end nearest the palace is a sloping ziggurat side of raw pine, the scent of which is sharp and soft in the November cold. Flat orange pads cover the steps. Two rows of five boxes are suspended over the ziggurat. The walls on the two longest sides are covered by large mirrors in rough-timbered wood frames, reflecting one another into infinity. At the far end is another door and two wooden sculptures – the front and back end of a horse – projecting from the wall. The stone is roughened, patched with concrete and, where it is exposed, covered with sketches of muscular Portuguese horses; on the rear wall, seven rotate in sequence, the first with its rump to us, and then each horse turning until the last faces the front in a revolution of musculature that recalls Da Vinci's anatomical drawings. Over the pockmarked sand floor hang fifteen balls of light made from thick, ribbed glass leaves that curl upwards to the wooden vaulted beams.

Voices soared across the Carrière, and the day was on the turn, slipping towards an early teatime darkness. There was a sense of offstage business here, as though I was a perpetual Guildenstern. In the peace of the Grande Écurie, the activity had an esoteric, private sense of purpose, always sealed off even when on display, like the riders in the Cour d'Honneur.

I made my way to the stable.

III

When the horses of the steppe were first domesticated and distinguished from the wild Tarpan and Takhi, their riders had to acknowledge that the animals they hauled into the corrals and the settlements had their own willpower and an intelligence whose features were not so distinct from our own: they could remember, follow a line of reasoning, make decisions according to their own knowledge. To complicate this, horse and human had no shared direct language but that of the body.

The horse responded to physical cues for which it had been prepped – deliberately or inadvertently – by humans, or it followed its natural instinct – sometimes in ways useful to humans, sometimes not. Force and intimidation worked but could be counterproductive. How could man get the horse to do more than simply tolerate his presence on its back, or be spurred or dragged about by bits? How could he negotiate with this other intelligence? Over basic domestication humans began to add a layer of culture that set the horse apart from other mammals – it would be asked to do more than any other livestock or pet, and in ever more complex forms. In time, the horse would even dance.

Two millennia or more of our earliest oral knowledge about the training and understanding of horses have been either lost or dissolved into a wider pool of knowledge. When it does resurface in a way we can pinpoint to a period and location, it's in a set of instructions in Hittite written by the Mitanni horse trainer, Kikkuli, for preparing chariot horses in Anatolia around 1360 BC. These instructions are so detailed that twenty-first century horsemen have used them to prepare their own animals,⋆ but Mitanni's horses are only expected to run fast in front of a chariot – there are no scribbles about persuading a horse to pull a chariot in the first place.

The Sybarites of sixth-century BC southern Italy are often credited with the first sophisticated feat of horse training. Foolish and pleasure-loving, according to Athenaeus they taught their warhorses to caper to the sound of flutes. When they went into battle, their enemies the Krotoniates struck up a tune on their own flutes, the Sybarite horses pricked up their ears and pranced, and their riders were picked off one by one. How they trained the horses is lost; the classical texts relay only the military disaster, perhaps with some exaggeration around a kernel of fact.

Horses, of course, first danced for their own pleasure and advantage long before any human waving a spear came into view. Not to music, or to any rhythmic measure other than the exigencies of their own four hoofs, but with spirit and passion. A stallion going out from his herd to meet a rival, or to seduce a mare, will arch his neck, gather his body together to make it appear more powerful – neck high, quarters rounded – raise his tail like a flag of intent and throw out his legs as he trots. At contact, two males squeal, chins tucked almost to their chests, strike out with a foreleg and half-rear in display, and then repeat their gestures until either one flees, impressed by the other, or a full fight develops. A band will gallop and play exuberantly, kicking their heels and squealing in imitation of the stallions, manes flying.

Like the admirers of the $500 wild mustang who turned out, once caught, to be a $50 hackney, the Sybarites must have wondered – and discovered, unlike Josiah Gregg and his men – how to recreate this extravagance in their own service. A horse's rapid flight and nimbleness were used for warfare, its strength for hauling burdens and its stamina for covering long distances, but the surviving horse training texts that followed the Sybarites sought instead to replicate the horse's instinctive dance to enhance the prestige of its rider.

⋆ Dr Ann Nyland of Australia revived Kikkuli's programme for testing in a 1993 book. As chariots were not available, she suggested leading horses from car windows.

Less than a century after the defeat at Sybaris, Simon of Athens wrote a guide to keeping and training horses that was well known enough to be quoted by the third surviving text, *Peri Hippikes* (*On Horsemanship*), by Xenophon, the Athenian general, historian and follower of Socrates. Simon's text endures only in fragments today, and in a philosophy that was carried over into Xenophon's text, which in turn became the foundation for the following two millennia of horsemanship in Europe.

Xenophon's concerns in *Peri Hippikes* are for the most part practical, as befits someone who had spent years at the rough end of military service. He instructs his readers on the selection, management, training and preparation of horses for war. Such a horse should be capable of traversing all kinds of territory – slopes, banks, ditches, even leaping over walls – but Xenophon also recommends a suppling exercise using what is often translated as a 'volte' or half-circle and a 'career' or dash in a straight line, both combined into a sequence that traced an oval. This was a rehearsal of manoeuvres required in the thick of battle, but it also began the process of bringing the horse from open space and free, forward movement into geometry and discipline.

Peri Hippikes also shows that horsemanship had, two millennia after domestication, become an intellectual endeavour – not just a matter for grooms – and a test of a soldier that went far beyond his physical ability to stay on a horse as it swerved through a battlefield. Throughout *Peri Hippikes*, Xenophon is preoccupied with the temperament of his horse and the way in which a rider must consider the thoughts and reactions of his mount at all times. Xenophon's horseman is a thinking rider, who negotiates with the intelligence of the horse, which should not be irked unnecessarily, nor brutalized.

Xenophon stresses that the rider's reaction can startle a horse when a trumpet sounds or a shout goes up. Forcibly hauling on the mouth of a bolting horse will not work – the horse will just haul back. Treatment must also vary according to individual horses: bit, spur and whip should be used vigorously but not unthinkingly on the 'sluggish' and with careful consideration on horses that are sensitive. His horses may not be anthropomorphized, but they are sentient, responsive beings despite this. Xenophon invokes Simon of Athens to say that 'what the horse does under compulsion ... is done without understanding; and there is no beauty in it either, any more than if one should whip and spur a dancer'.

For warfare, Xenophon advises choosing a horse with a steady character that is not 'high-mettled', but it's the high-mettled, sensitive horse – the dancer – that catches his imagination the most, and its spirit that he seeks to harness as lightly as possible in Chapters X and XI. These passages are dedicated not to creating an ideal cavalry horse, but to the victory parade.

Here Xenophon gives advice on reproducing in the ridden horse 'exactly the airs that he puts on before other horses' – the natural dance of horses at liberty. Handled correctly, 'he will then throw out his chest and raise his legs rather high ... when his fire is thus kindled, you let him have the bit, the slackness of it makes him think that he is given his head, and in his joy thereat he will bound along with proud gait and prancing legs'. This contained impulsion later became known as 'collection', and, rather like some ecstatic yogic state, is meant to be attainable by following a set of straightforward instructions like those in *Peri Hippikes*. In practice it is every bit as unfathomable to the beginner rider as the eighth limb of the Patanjali sutras is to the uninitiated yogi.

The second parade movement for which Xenophon provides instructions is what the old general called 'the prettiest feat that a horse can do' in which the horse brought its hind legs underneath its quarters, shifting its weight back, then lifted its forelegs off the ground, and balanced for a few seconds in suspension: it is a choreographed, controlled, collected movement, and not the vertical, wild reach of a rear.

This, Xenophon says, is the pose 'in which the horses of gods and heroes are always depicted' – man, with the aid of a horse, could remind the mortals viewing him of another order of being altogether. On the Parthenon's frieze, on red and black vases, the horses hung, half on the earth, half in the air, for 'everybody that sees such a horse cries out that he is free, willing, fit to ride, high-mettled, brilliant, and at once beautiful and fiery in appearance'.

So the purpose of all this training was to draw the eye of others, to impress them with the horse but also the man aboard it: like a jewelled sword or golden breastplate, the horse reflected lustre back on to the man. But while any man can be bolted into a golden breastplate, not everyone can ride a spirited horse as it plunges and leaps. It was human skill, not wealth or strength, that completed the fine parade horse's impressive and intimidating display. Through his high-mettled mount, the individual could stake his own claim to public recognition before his defeated enemies and his fellow citizens.

The Romans who followed Xenophon and the Greeks took the parade, combined it with military drill and added spectacle. After major victories and at certain imperial events, the Lusus Troiae or 'Troy Games' – a complex dance of what Virgil called 'sham cavalry skirmishes' performed by teams of youths – was presented. Roman cavalry performed the hippika gymnasia or 'horse exercises', in which two teams attacked one another with blunt weaponry and tried their skill at targets. It was a demonstration of martial might, and a chance to rehearse for the battlefield, but also had a theatrical side: the teams took on the role of historical or mythological peoples such as the Ancient Greeks or Amazons, and wore face masks finished in gilt or highly polished silvery metals, with plumes and hammered metal caps or helms that identified their 'tribe'. One was found on the site of the Battle of the Teutoburg Forest, where Arminius, unphased by the gilt, wiped out the Romans.

The horses, too, wore fine armour, and Arrian, who gives the most complete account of the hippika gymnasia, says that the Romans borrowed from the steppe nomads of Scythia the standards which flowed behind galloping horses: 'When the horses are standing still, these contrivances look no more than a variegated patchwork hanging down, but when the horses are urged forward the wind fills them and they swell out, so that they look remarkably like live creatures, and even hiss in the breeze which the brisk movement sends through them.' The *hippika gymnasia*, though taxing to horsemen, were also undoubtedly good propaganda, and were performed around the empire from Germany to North Africa.

While the parade horse was sculpted in stone under the emperors, generals, satraps and kings, on victory columns and triumphal arches, another troupe of dancing horses occasionally appears in classical art and literature – one trained by the travelling street performers who carried out daredevil feats to earn a living. A stable boy stood on a horse's back, brandishing a sword, as it raced around the hippodrome at Constantinople. One Panathenaic vase shows acrobats dancing on the backs of horses to a flute accompaniment, and *The Iliad* mentions 'a man skilled in feats of horsemanship [who] couples four horses together and comes tearing full speed along the public way ... all the time changing his horse, springing from one to another without ever missing his feet while the horses are at a gallop'.

This horsemanship of vaulting, balancing and leaning from the saddle to snatch things from the ground – war tricks of the steppe – was no less skilled than that of the learned elite like Xenophon and his readers, but was transmitted orally and visually, in person. It was theatre for the people, done not to intimidate like the *hippika gymnasia*, but to entertain. This pattern of military/nobility vs entertainment/ populism plays out throughout the history of horsemanship despite revolutions, wars and industrialization, through different media whose influence would wax and wane, but the essence remained unchanged from Xenophon's desire to show off both horse and man's mastery of the horse.

Peri Hippike was read enough to have survived not just in fragments, but as a complete text which slipped from papyrus to parchment, copied by hands in Greek communities from Constantinople to Alexandria, perhaps translated into Latin and Arabic on its travels. Its exact path is unclear, but when it made its final slip from parchment to printed paper in Florence in 1516, it arrived in a climate that was amenable to its familiar message.

In the standard story told of the history of dressage, Xenophon's book was 'rediscovered' in the warm, sophisticated brew of the Renaissance, and his message of humane horsemanship revived, repackaged to suit the times, and turned as a corrective beam on medieval horsekeeping and riding, which was perceived to be crude, clumsy and brutal. But knowledge is too fluid, and custom is seldom revolutionary: we are always discovering that the tricks we thought we were teaching our grandmothers are ones they knew well for themselves.

Classical texts had many lives and travelled far, but they were read by very few. The hands-on knowledge of Xenophon and other riders of his age, however, had an existence beyond the parchment on which it was written, and was transmitted to a far wider audience of horsemen from a far broader base – Xenophon was unlikely to be its sole proponent – and it then either met with similar philosophies or mixed and mingled with new ones. Horses, after all, were relatively expensive, and poor training could have grave financial and physical consequences for their riders, who hired trainers if they did not personally pore over texts.

This transmission of knowledge was facilitated by the empires, cavalries, traders and migrations that succeeded one another, or overlapped and coexisted in the parts of the world that rode horses. Through this series of canals and locks ran ideas and horses and horsemen, practices were defined and developed, and cultural phenomena appeared in different guises. A Greek and Macedonian diaspora of cavalries reached from Spain to the Black Sea, deep into India, and along the northern coast of Egypt and Libya. The Romans and their *hippika gymnasia* marched from Scotland to Babylon, Sudan and the Pillars of Hercules in Iberia. When the Byzantine empire rose, it covered the Mediterranean and Anatolian territory of the Romans, and maintained libraries of classical and later texts on horsekeeping. When it collapsed, it sent west the scholars who brought texts like *Peri Hippikes* to Italy.

In Europe, the *hippika gymnasia* and its splendours resurface in the Carolingian cavalry games and mêlées, before reaching its familiar form in the twelfth century as the courtly tournament complete with jousting. In the Levant, where the Mamluk warrior caste rose through Muslim caliphates and sultanates and withstood both the Crusaders and the Mongolians to dominate the Middle East, India and Egypt, the *hippika gymnasia* shed its silver masks and was reincarnated as the more workmanlike *furusiyya*, with Turkic, Arab and Persian influences, and mounted martial games grounded in local raiding tactics – charging with lances, swordsmanship, archery and polo-like games. *Furusiyya* training was done at a meydan built over the earlier hippodrome at Constantinople, where the stable boy had stood on the galloping horse's back.

The Levantine horsemen who clashed with the European Crusaders shared a similar ethos – *faras* and *cheval* meaning horse, and *furusiyya* and chivalry horsemanship – which became an idealized, if seldom achieved, code for good Muslim warriors in the Golden Age of Islam or Christian knights, who were known, like Greek and Roman nobles, by the words for horseman: *chevalier, cavaliere, caballero, Ritter, cavaleiro.*

The Berbers and Arabs of North Africa took *furusiyya* and earlier Arab horsemanship with them on their repeated invasions of the Iberian peninsula, along with their light Barb horses and their own Berber philosophies of horsemanship. Their military games were enthusiastically adopted in Spain too: in 1390, Juan I of Castile would be killed while riding in a *furusiyya*-like game, surrounded by horsemen dressed in Arab costume. *Furusiyya* in turn met the Iberian mounted bullfight performed by local aristocrats, which had itself arrived via the Roman circus and its gladiators.

The texts that accompanied this horsemanship travelled along the same routes. The earliest equestrian book to surface in Spain, in 1185, was devoted to the *furusiyya* and had been written in the ninth century by Ibn Akhi Hizam, chief equerry of the stables of the Abbasid Caliphate in what is now Iraq. In a Sicily influenced in turn by Byzantium, North African Muslims and Normans, Giordano Ruffo di Calabria counselled the calm and gentle handling of horses in his thirteenth-century manuscript, *Mariscalcia equorum* (meaning roughly *The Keeping of Horses*). Dip the bit in honey to make it more palatable, he advised, and ride a horse in heavy ground to give it an elevated trot – an ill-advised shortcut that would have damaged the horse's tendons. The Portuguese King Duarte I wrote the *Livro da Ensinança de Bem Cavalgar Toda Sela* (*The Book on the Instruction of Riding Well in Every Saddle*) in 1434, considering the ways in which a rider could overcome a natural fear of riding with reason, knowledge, willpower and training, but above all a sensitivity to the horse.

There is plenty of evidence, too, that quite advanced horse training existed long before the Renaissance, at all levels from the battlefield to the public square. Totila, the king of the Ostrogoths, inspired his troops before a battle by making his great horse canter rapid circles on one rein and then the other before them as he tossed his javelin from hand to hand, 'like one who has been instructed with precision in the art of dancing from childhood', according to Procopius.

Medieval warhorses were also taught to perform defensive movements for battle – the *Luttrell Psalter* of fourteenth-century England shows an eager, riderless grey horse performing a leaping kick, legs flying, with a trainer behind deflecting his hoofs with a small round shield and a raised sword. Mamluk horses could pick up their riders' dropped weapons. In one French poem a knight expresses his joy at the notice a fair lady has given him by making his horse 'leap, spin, and turn'.

In Iberia, local riders had been influenced by their Berber conquerors, whose light *gineta* style encouraged nimbleness and rapid manoeuvrability in battle, as opposed to the *la brida* form of horsemanship, largely employed for heavy, knights-in-armour equestrianism. Some Berber riders, it was said, barely needed their reins to steer their horses in battle, leaving their hands free for their weapons. *Gineta* was suitable for fighting bulls on horseback, a practice which King Duarte also sought to codify: as historian Carlos Pereira points out, the bullfight became a matter of skill and horsemanship rather than brute chance when Duarte decreed that the bull should be struck at a particular point between its shoulders – to be close enough, a rider needed a horse that was obedient but also sufficiently agile to clear the area rapidly when the wounded bull tried to fight back. Building on texts that had arrived with the Moors, the first printed work on horses was produced not in Italy but in Spain in 1495: Manuel Diaz's *Libro de Albeyteria* (*Book of Horsekeeping*), commissioned by Alfonso V of Aragon.

Along with this cosmopolitan flow of games and theories and texts and horsemen went horses themselves, whether as cavalry or in victory parades like King Alfonso's arrival in Naples via Sicily in 1443 in a chariot drawn by four of his finest animals. The Iberian horse in particular was considered the foundation of a royal stud – they were diplomatic gifts of the first water long before the Renaissance. Descended from cavalry horses admired by Xenophon himself, they were powerful yet swift, with crested necks, rounded quarters and fine limbs. Some argue that they are a pure tap root, descended from a now extinct 'ram's head' wild horse that roamed Iberia in the Holocene, while Tarpans and Takhi ended up on the Eurasian steppe. The Moors who arrived in the first millennium AD so admired the local horses that they shipped a quantity back to North Africa, where they perhaps influenced the development of the Barb – or vice versa, as their DNA is so interlaced.

These horses are not very tall but they have been bred to dance, with compact, strong frames that find it easy to 'bound along with proud gait and prancing legs' in collection or support their weight on their hind legs in the pose 'in which the horses of gods and heroes are always depicted'. One Renaissance riding master said they could lower their quarters 'almost to the ground', and the English Civil War general the Duke of Newcastle said that when he rode one of his Spanish horses in a pirouette, 'I was so dizzy, that I could hardly sit in the saddle.' They were, he went on, 'far the wisest, and strangely wise, beyond any Man's imagination'.

Like equine Habsburgs – with better conformations – Iberian and Barb horses were imported by royal and noble European houses

across the continent to add blue blood to their own equine dynas-
ties, from the celebrated Mantuan horses to the Neapolitans bred
by the Gonzagas, Vienna's Lipizzaners, the black Friesians of the
Low Countries, Denmark's royal Frederiksborgers and the Czech
Kladrubers, and perhaps even, later, the English thoroughbred via the
'Royal Mares' bred to the founding oriental stallions.*

IV

Versailles had its own Iberian and Barb horses, housed in the Grande
Écurie whose empty Carrière and Manège I was exploring, along with
the other hot-blooded horses that were literally closest to the king
– the thoroughbreds brought from England for hunting deer in the
park and the broad-chested warhorses who stiffened at the sound of
a trumpet. The Petite Écurie, now converted to offices, was small in
name and status only – it had housed the carriage horses and cold-
blood, thick-set working draught horses from northern Europe.

In the palace itself, I'd seen Iberian horses capering across vast
canvases in heavy gilt frames; in the squares and parks of the town
they shared pedestals with kings. In engravings in 500-year-old books
they hung suspended in mid-air, forelegs neatly tucked under, hind
legs horizontal behind them, floating like moored balloons. Their
manes fell in braids woven with ribbons, or tumbled with colourful silk
adornments that echoed the extravagant periwigs of their riders. Their
saddles were cushioned with brocade cloths, and the most precious
horses at Versailles wore gold bridles, the ranks below them silver.
They were grey, piebald, spotted, cream, bay, black, with names like
Cerbero, Grandissimo, L'Amour and Curioso. They had fine, small,
Roman-nosed heads and large eyes, often turned to the artist in an

---

★   The Iberian horse is now divided into the Pura Raza Española (PRE), or Pure-bred
    Spanish, and Portuguese Lusitano. They were considered the same breed until the
    twentieth century, although they have since become more distinct.

echo of their royal or noble riders, who directed them insouciantly with a raised, upright switch, as if conducting music or wielding a sceptre.

Under Louis XV, great-grandson of the Sun King, the two stables held 1,700 horses – 785 of whom lived in the Grande Écurie – tended to by over 2,000 men. The animals stood side by side, tethered to the long stone mangers that jutted from the walls. Posts guarded their hindquarters. Just as the rococo hallways of Versailles were squalid and over-packed with ambitious humans squeezing themselves into every last cranny and cupboard to be close to the king, so the stables, grand as they were – the King of Siam's ambassadors were impressed that such a building was constructed just for horses – must have been rank and teeming. The politics between the horses can't have been much better than that between the courtiers, negotiated in kicks and nips from animals tethered too snugly side by side.

When I wandered from the Cour d'Honneur into the old Grand Écurie that November there were forty-three horses in the curving mahogany loose boxes that ranged down two high-ceilinged, sunlit corridors, set at right angles to one another around the east and north of the Carrière, half of them stallions, half geldings. Each box encompassed the mangers – still set on the wall – of two eighteenth-century Versailles horses. Old wrought-iron brackets decorated with coats of arms dangled lamps over the dull maroon brick floor, and modern light fittings like unicorns' spiralling horns rose on poles over the stall doors. The muted, weathered patina of the Carrière was carried through the great doors: peeling wood panels, pale blond wood shavings, white stone walls gone to cobweb, and the mahogany, now settled into its role for over a decade, scraped by the occasional hoof or set of equine teeth. There was none of the gingerbread dazzle of the palace, but instead a lived-in, harmonious ambience. Nobody was about but me and a groom, who was a distant figure at the end of the row of stables, wheeling a barrow of dung, so I introduced myself to the horses.

Wooden lattices at the front of the boxes protected them from the more pressing attentions of visiting tourists or their neighbours, but these had a square opening through which food miraculously arrived three times a day, according to a chalkboard hung below it. Not surprisingly, the horses were favourably inclined towards anything that reached them through it. Through one of these confession booth hatches I communed with a cream-coloured, sapphire-eyed Lusitano called Uccello.

Uccello the artist painted a chunky double of this Uccello in The Battle of San Romano – sturdy and compact as he leaps over discarded helmets and broken lances against a thicket of white roses and dark knights – although as he stood in his mahogany box before me, Versailles' Uccello looked quite small and ponyish. Angling his ram-nosed head carefully, he pushed his muzzle – peach and bare – through the opening, and I held my hand to one of his nostrils, which expanded to the size of my palm, investigating the skin. Then his lips checked my fingers for food. I had nothing, so we settled for letting me rub the sides of his nose and his cheek, and he rested his chin on my hand on the metal edge of his door.

Normally I watch horses' ears for mood, but at Versailles the horses' nostrils and muzzles had an exquisite expressiveness. Some painters superimpose anthropomorphic eloquence on to less demonstrative, more subtly articulate equine faces, but the Spanish-rooted breeds of the Baroque courts – the Lusitanos, the Lipizzaners, the Kladrubers, the Frederiksborgers – have faces, like old actors whose features are mobile beyond beauty. Muzzles crinkled like foreheads, lips grasped like hands. It is as though they have made an effort to cross further along the bridge between human and animal understanding than other breeds.

Most horses have dark brown irises and black pupils that are hard to distinguish unless you are close, but as a pink-skinned cremello, Uccello had pale blue irises around dark blue pupils. This added to the human look of his face. Cremello colouring was one of the genes that emerged after domestication – as well as the ghostly creams, it made

the buckskins, smokey blacks and palominos found in the early corrals; thousands of years later in Europe, cream horses became linked
to royalty. The Hanoverian Elector at Herrenhausen bred his own
red-eyed creams from imported Spanish horses, using them to draw
carriages on ceremonial occasions. Napoleon stole eight for his own
coronation, and these deep-necked creams were given great plumes
like Roman crested helmets when they led a parade of thirty-two
golden carriages through Paris. According to Tolstoy, Bonaparte was
riding a cremello as he approached Moscow, dreaming of conquest.
One of his creams was still in the stables at Versailles in 1830 after the
July Revolution, and spoken of as 'a favourite of Napoleon's', before it
was sold to a humbler home.*

Uccello was one of twenty cream Versailles Lusitanos inheriting
the mantle of that old pink-eyed stallion of Napoleon, and they make
up one of three corps de ballet. Some have been performing for over a
decade, others are newer recruits, like the juvenile apprentices whose
manes were still cropped like cadets' heads – traditional practice for
horses raised on the shrubby grazing land of the Iberian peninsula,
where manes and tails can rapidly get snarled with twigs and burrs.
Uccello was older, but had only been at Versailles for a few months,
joining the horses named for artists, like cream Géricault and Matisse,
and Pollock the grey Quarter Horse.

A horse – perhaps Géricault, Matisse or Pollock – was drinking
from the little bowl of his automatic waterer, which swooshed as it
refilled, and set off an answering rumble of pipes as more water rattled up to trickle in. There was a long snort of relaxation. The sound
of molars grinding against one another. Soulages,** the big black
Lusitano who danced the solo twice a week, had his eyes half closed
as he dozed under the dusty stucco high over his box. Le Caravage,

---

* The British Royal family had its own 'Hanoverian creams' including George III's
Adonis, immortalized in a painting by James Ward. They were last used in ceremonies in 1921, when the stud was found too expensive to maintain. The Royal
Creams ended up, variously, in a circus, working at golf clubs, doing farmwork,
carrying regimental drums and with private owners, including the King of Spain.

** Pierre Soulages is a French artist famously known as 'the painter of black'.

another soloist, was curled up on the floor of his loosebox, his dappled buckskin coat a chiaroscuro of light and blurred dark, like his namesake Caravaggio's paintings. His chin stretched out on to the soft, deep shaving bed.

Every consideration was made for the horses: they did not have golden bridles, but better, as far as horses are concerned, they worked with and lived next to the same companions at all times, whether in their boxes, in the manège or in a lorry. The head groom, Philippe, a bullet-headed old racing man with a necklace of thick silver links, had talked me through their requirements before I met them. 'No one is like another. They all have their little *manies*, *habitudes*, their little *bobos*, *fatigues*.' Arlequ'un would not travel – his trip from Portugal had upset him so much that the mere sound of a lorry brought him out in a sweat. Adagio wouldn't wear a rug. 'We try to understand, to anticipate, think like a horse,' Philippe said. 'Le cheval, ç'est Roi?' I asked, thinking of the stables' new logo, in which a horse's mane spread like the rays of a stylized sun, as though the Sun King were being supplanted by his own horses' heirs. 'Tout à fait!' Philippe had cried.

Philippe had told me that the horses only truly relaxed when they were no longer listening for the voices of their riders, which they always recognized. This was the contemplative silence I found myself sharing with them in the empty stable. One horse began a rumbling in its chest, and the noise echoed from the rounded brick ceilings, the curves of mahogany, the brick floor, and was picked up by another horse, that rumbled in response, ending in a little whinny like an echo. I walked slowly on, watched by the horses as I passed.

In the other corridor were the second-ranked corps de ballet, the seven Criollos, a small, handy Argentinian type descended from the conquistadors' horses whose black, dun and white hairs blended into Turners of greys from dove through donkey to charcoal, whitening where the parallel bones of their noses rose under the skin. Some had horizontal black bars on their fuzzy ears – little escaping traces of wild ancestors – and their tails were bobbed short. Their names were studies in personality: Nerveux, Curieux, Intrepide, Dormeux.

In blue tarpaulin stables rigged up in another courtyard, I found the third corps dozing: seven small, Tarpan-coloured Sorraia horses with black and cream manes, dappled sides and zebra stripes on their legs, even more primitive than the Criollos. They are tough little horses from along the Sorraia river in Portugal, distant ancestors or relatives of the Lusitanos. At Versailles they were named after the planets – Mercure, Pluton, Saturne – and, unlike the creams, were not fascinated to have me in their stable.

Slipping back through another great door to the main corridor, I went to talk to Uccello again, and he trustingly presented his muzzle once more, so I could gently scratch the side of his flat, round cheeks and talk nonsense to him in a low voice. The apprentice creams watched him and me, their ears pricked, and pushed their long noses between the bars of their stables. Together we waited for the riders.

V

At its simplest, a well-trained horse is an investment against losing face. It is safer, as Xenophon put it, because 'a disobedient horse is not only useless, but he often plays the part of a very traitor'. It was also, as we saw in the hippika gymnasia and the furusiyyas, a display of military might. In the Renaissance, Xenophon's ethos, the gineta, the furusiyyas, the Barbs and the Spanish horses would be put in service of ever more complex, elevated horsemanship with a new significance.

The Renaissance was a time when kings sought legitimacy by linking themselves to the distant golden ages of past empires; they were patrons of scholars, writers and artists who evoked on their behalf the gods, myths and heroes of Greece and Rome. Horsemanship was gathered into this cultural embrace as an art in itself, and much embellished. The great men of the Renaissance also sponsored riding masters, who, even if they were not always of noble origin themselves, were cosmopolitan individuals – wandering equestrian scholars – who circulated between countries, establishing new centres in the courts of Europe. They were often close to their pupils, the sons and relatives of the great men themselves, and it was their job to put a final equestrian polish on callow boys.

Castiglione, whose Il Cortegiano (The Book of the Courtier) established him as the authority in the building of princes, stressed that the young men must 'ride and manege well his horse. To be a good horsman for every saddle ... showe himselfe nymeble on horsbacke.' Moreover, the most desired quality was sprezzatura, 'a certain nonchalance, so as to conceal all art and make whatever one does or says appear to be without effort and almost without any thought about it'. Applied to the riding ring, this meant seamless horsemanship, the rider all but motionless as the horse exerted itself in ever more sophisticated and spectacular fashion.

The first text to set down the new parade movements of the Renaissance was published in Italian in 1550 by the Neapolitan Federico Grisone: Gli Ordini di Cavalcare (The Rules of Riding) was subsequently translated – and plagiarized – in French, German, Spanish and English. The career and volte of Xenophon resurface as Grisone's horses circle and make figures of eight before accelerating straight along the career and, at its furthest point, bringing their quarters beneath them and turning rapidly, almost on the spot, like the Spanish horse that would make Newcastle dizzy. The simple oval dash became trefoils and clover leaves and serpentines as formal and intricate as knot gardens.

The high-stepping, prancing parade movement that Xenophon described had become the passage, a slower, suspended trot. The controlled rear that was 'the prettiest feat' became the levade,* which could be used at the end of a career to turn the horse. Movements were combined and created, padding out a repertoire that now included the 'airs above ground', in which horses hopped, bounded, kicked and leaped. Many of the 'airs' were named for human dances, because

---

★   To help the reader I have simplified my use of the names of the different movements as many changed over time – sometimes becoming another step altogether.

dancing is what the courtly Renaissance horses were increasingly expected to do.

An Italian man dancing a galliard finished each sequence with a leap; a horse performing a *galoppo gagliardo* or lusty gallop completed its bounding canter sequence by kicking out its hind legs. A capriole or goat step involved a human jumping with legs extended, and beating them together then apart as many times as possible. For a horse, it was a rear followed by the hind legs kicking out vigorously behind – a full leap of all four hoofs off the ground, making it into a Pegasus for a second, perhaps related to the battle move performed by the horse on the *Luttrell Psalter*. Both humans and horses performed the *salto* or jump, the *volta* or turn, and *ciambetta*, which historian Giovanni Battista Tomassini thinks was what is now called the Spanish Walk: slow steps with each foreleg in turn raised high under the horse's chin – a faint echo of the showy strike wild stallions make when they clash.

One of Grisone's contemporaries, Cesare Fiaschi, included in his own manual music that the rider was expected to sing as he worked with the horse – each piece matching the tempo required for a particular movement. On his pages, a tiny horse and his rider advance along a guideline next to the musical score, the rider calling out 'Ah! Ah!' at each stride, and then 'Hai!' as his horse leaps clean off the ground.

Scores of manuals on horsemanship were produced across Europe in the early centuries of printing. Their elaborate titles are full of mastery and aristocratic polish: Grisone's *Ordini* or rules, la *Gloria* of the horse, the *kunst* or art of riding, done *con gracia y hermosura* or with grace and beauty, to perfection by le *Parfait Mareschal* or perfect marshal, whose title, like chevalier or Ritter or equerry, links him to the horse (it derives from the Old French for 'mare').

The word *managgio*, from the notion of household, was put into service to describe this new horsemanship, and the first riding schools or *maneggi* or *scuole* – literally, 'schools' – to be built in the wealthiest houses were colonnaded yards or even indoor chambers.

The medieval tilt yard and meydan were roofed over; these expensive horses, who could eat in one day what it took one labourer the same time to earn, were increasingly kept in stables rather than on common land. The horse and horsemanship came indoors into a bounded, cultivated space.

These schools, centred around the riding masters and their patrons, were not just places for riding lessons, but 'academies' inspired by the Athenian olive grove where Plato taught Aristotle. The students lived at the academy as they might later live at the main court, mingling with their peers and creating alliances. Horsemanship was a matter of education and refinement rather than basic training.

In 1594, a Frenchman called Antoine de Pluvinel de la Baume founded his own Académie d'Equitation close to the new royal palace of the Tuileries in Paris. The palace had been built by Catherine de' Medici after the death of her husband, King Henry II, who was killed by a flying lance fragment in a tournament. This catastrophic accident had spelled the beginning of the end of the true medieval tourney: princes and kings became too precious to risk in the crunching of lances striking shields at speed, or in furious mêlées where swords were drawn. Bullfights – still a staple of the Spanish court – were outlawed by the Pope in Italy after one too many young nobles met their death tilting at bulls. Instead, the focus turned to the practice Pluvinel was teaching in the manège, which sat next to the palace walls, under the wing of the court.

Pluvinel was not himself an aristocrat but had travelled to Italy to study under a master called Giovvanni Battista Pignatelli in Naples for six years before earning the patronage of Henry II's son, the Duc de Anjou, who became King Henry III in turn on the death of his older brothers. Pluvinel survived a rocky period of high royal turnover, assassinations and civil war at the side of the royal family, even as the line of descent zigzagged between successors cut short or dying childless. Henry IV, Henry III's brother-in-law, entrusted the care of his son, the future Louis XIII – and father of the Sun King, Louis XIV – to Pluvinel, who became his *sous gouverneur*.

With the creation of his academy, Antoine de Pluvinel founded what would be called the School of Versailles, although at this time Versailles itself was just a village with no royal associations. The year 1594 fell in a period of civil war, but the purpose of Pluvinel's academy was not to turn out battle-ready nobility, but to draw them together in camaraderie after decades of unrest and factions. Under Pluvinel, young sprigs of the aristocracy from France and the rest of Europe learned fencing, military history, dance, history, politics and above all horsemanship alongside one another as true heirs to Castiglione's ideal. In under a century France, and Paris in particular, usurped Italy as the ideal finishing school for European nobility.

Pluvinel wrote a Socratically styled account of his education of the dauphin Louis, called *L'Instruction de Roy en l'exercice de monter à cheval* (*The Instruction of the King in the Exercise of Riding Horses*). It was published four years after his death in 1620, with a series of engravings by Crispian de Pas. Like Xenophon, and undoubtedly under his influence, Pluvinel believed that a horse would never achieve grace in the manège if it was unhappy – its goodwill was as fragile as the fragrance of a fruit blossom, he wrote, and so the horse must learn instead that its obedience leads to caresses.

Pluvinel has a slightly wild look to him in Pas's engravings, his hair springing from his scalp in tufts, and he has the marked chin, nose and eyes of a Baroque horse. His acolyte, Monsieur Le Grand, joins the dialogue to say that one must look a horse directly in the eye to judge its character, stressing the individuality of each horse. Le Grand mentions a sensitive bay Barb called La Bonité, which other riding masters had called untrainable and was later presented to the king as a 'chef d'oeuvre'. Pluvinel replaced his bit with a silk braid and coaxed him into performing an air above ground called the courbette or little crow, in which the horse rose in a levade and then hopped on its hind legs.

La Bonité appears in Pas's illustrations against the backdrop of Pluvinel's Paris academy: part colonnaded Athenian grove, part riding school, where men gather to watch and discuss as a handler draws a levade or courbette from a horse. La Bonité is hairy of heel, his tail curling and loose, his quarters muscle-bound and head tipped with sharp ears. He passages as his groom looks on admiringly, or is lunged around a pillar as the dauphin watches from a throne. Bridles hang on the Corinthian columns, and horses with blankets, hoods or even blinders wait alertly nearby. Unblinded, they are observers too – all learn by watching. Boys stand by with spare switches, which the riders hold upright in their fists as in Fiaschi's illustrations or, dismounted, use to tap the heels of their horses to cue them. These boys would later take Pluvinel's work to courts across Europe. His manège is a space of civilization, between stable and palace, where the horse is, as the dauphin says, 'brought to reason'.

Though Pluvinel's work also deals with the familiar tournament games and jousting, it is the manège horse that is foregrounded: Pluvinel aims, he says, 'to reduce the horse in little time to obedience to the just proportions of all the fine airs and maneiges'. Riding masters admitted that only the basse école of dashes and turns was needed for destriers – the leapers and the piaffers were too temperamental for battle, just as Xenophon had noted. Not only was the tournament de-toothed, but the heavy cavalry of the age of chivalry was in decline, their numbers subsumed by more cost-effective infantry and their lances replaced by guns.

But far from being rendered irrelevant by this change in martial tactics, manège riding had instead grown a significance beyond Xenophon's aesthetic and military pomp, beyond even Castiglione's requirement that a courtier acquit himself with grace in the saddle. In the mastery of the horse, Pluvinel saw a much larger schema: to ride well meant nothing less than to 'attain perfection' but also to exercise a combination of 'patience, resolution, gentleness and force required'.

The same methods used on the horse should also be employed with men, who, though more rational than horses, also wanted nothing more than to learn virtue under a benevolent but strict master. Here in the colonnaded riding school the aristocrats would learn not only camaraderie among one another, but also how to rule their own people in turn. Pluvinel advises his king to found similar academies in Tours or Poitiers, Bordeaux and Lyon to educate an aristocracy that is 'la plus grande force de la Monarchie Françoise'.

Every morning they will practise in the manège, and after dinner play at bague – a tournament game involving catching a hanging ring on a lance tip. On Mondays, Wednesdays, Friday and Saturdays they will have weapons training, dancing, mathematics and vaulting. There will be scholarships for the impoverished. On Thursday after dinner a man of letters would come to teach them history and later they would move on to politics, and the governing of the territories that the king entrusted to them. There would be tournaments once a month, which would include not just the formal games of the tourney, but also retreats, attacks and the defence of a fort.

'It is well known,' said Pluvinel, 'that nurture has more influence on the spirit of man than his birth or natural inclination.' The sand of the manège would, after countless trackings and figures and circlings, produce an aristocracy whose mettle had been tempered by patient consideration of the horse, and the pursuit of collection.

The timetable, which was echoed in many academies and courtly riding schools in Europe, was demanding, and required, perhaps, a degree of horsiness that even the most ambitious young man might find hard to stomach. Two young Dutch nobles at one academy in 1650 complained that after riding four horses a day, they could barely walk. In 1574 Sir Philip Sidney went to the riding school in Vienna to see the Italian horseman Ion Pietro Pugliano, who told him 'what a peerless beast the horse was, the only serviceable Courtier without flattery, the beast of most bewtie, faithfulnesse, courage'. Sidney received this with one eyebrow cocked: 'If I had not beene a peece of a Logician before I came to him,' he commented, 'I think he would have perswaded mee to have wished myself a horse.'

Pluvinel's further refinement was to the modification of the parade that had always preceded the tournament and the bullfight:

what became known as the carrousel, and its most elegant feature, the *ballet à cheval*. The word 'carrousel' came from the Neapolitan term for a shaved head, or *caruso*, which was the name given to the round clay balls that mounted teams threw at one another in a tournament game that had arrived in southern Italy via Moorish *furusiyya* in Spain. Perhaps the game was distantly descended from those steppe polo games of legend, where Tamurlane used the severed heads of his enemies instead of a ball or dead sheep.*

As in the *hippika gymnasia*, the performers in a carrousel were divided into cartels or teams which often referenced classical antiquity, each headed by a king, prince or duke, and loyally supported by a descending social scale of supporters, pages and musicians. There were synchronized displays of manège riding in quadrilles, parades of huge floats, play-fighting and the martial games of the *furusiyya* and the tournament. The carrousel replaced the tournament *pur* as part of court performance throughout Europe, arranged for triumphs like marriages between dynasties or coronations.

Pluvinel's pupil Louis XIII was betrothed to Princess Anne of Austria and his sister Elisabeth to Philip of Spain, and in 1612 a grand carrousel was organized in the Place Royale in Paris, in which the Palace of Felicity – an enormous wood and plaster construction painted to look like stone and brick – was defended by the five Knights of Glory against the Knights of the Sun, of the Lily, of Fidelity, of the Universe, and then some Romans, for good measure, in mock battles and games.

Tamburs, fifes, cornets, violins and oboes were played by troops of marching musicians. Each cartel of knights entered trailing ornate, lumbering, gold-painted floats, hundreds of footmen, mounted musicians, and dozens and dozens of beautiful horses draped in garments

---

★   And of course today the carrousel has lent its name to the fairground carousel. That gloomy November in Versailles there was a small white carousel with seams of golden light bulbs near the stables, its ivory and gilded gallopers moving slowly up and down yet forward, like well-collected dressage horses.

that matched the colours of their riders' teams, topped with feathers that shivered as they stepped out.

For the 200,000 Parisians who came to cheer, there were a pegasus with wings, real lions, giants, chariots, and a rock some 27 ft high which 'moved of its own accord', belched fire at its summit and had fountains of real water on its slopes. Andromeda was strapped to it, while a huge sea monster lay on guard at her feet.

Pluvinel appeared in a silver cape and mounted on a white-plumed Spanish horse as Valdance la Fidelle among the Knights of the Lily, led by the Duke of Vendôme. The six gentlemen and their pages performed the horse ballet he had choreographed, drawing on the human *grands ballets* of the court, to music composed by Louis's musical tutor Robert Ballard that opened with thundering drum rolls and proceeded to regal hautbois and horns. The knights performed the most advanced movements, like the courbette, as their pages performed the lesser move of the terre à terre – a slow, rocking-horse prance – in a respectful echo of the social order.

The poet Pierre de Ronsard described the complicated geometry of a horse ballet in his *Cartel pour le Combat à Cheval en Forme de Balet*, which probably describes an earlier 1581 performance:

> Now you will see them curvetting as they dance, now retreating,
> moving closer, advancing, moving apart, moving away, coming
> together, meeting in an elongated point and now in a shorter one,
> simulating war in a show of peace, criss-crossing, intersecting,
> in straight or oblique lines, now in a circular figure and now in a
> square one, just like a Labyrinth whose confusing trail leads our
> steps astray by diverse paths.

These riders came, Ronsard said, 'to instruct [the high Prince of France's] people and to render them as easy as their horses are docile under the bit'. The people were horses, the horses people, the political disruption replaced with a harmonious obedience to the measure of the music, a partnership between man and horse, and stately geometry.

At the royal family's command, their nobles had come together to train and obediently learn their moves, they bowed their heads to the throne, their horses bowed their heads to them and danced, and the people of Paris were spellbound in turn.

VI

But the people of Paris, and all those other Europeans of a lower order who lived lives in which the manège was as remote as the gilded galleries of the Tuileries, while hypnotized by royal power, could still see extraordinary performances by horses when the Place Royale was cleared of rolling volcanoes and ostrich plumes. The travelling performers who had made their living vaulting from horses' backs and snatching handkerchiefs from the ground at a gallop were now also reaching new heights of fame and sophistication. They might not match the scale of the royal spectacle, but they were still able to captivate and confound a crowd.

The most famous horse in Renaissance England was a cocky little grey or bay who belonged to a man called William Bankes, who had brought him to London from Staffordshire to make their fortune. Marocco could dance, too – he favoured a jig called 'Canaries' – and more, he could count, play dead, walk on his hind legs, bow for the Queen when told to, and chase his master with his teeth clacking when asked to make a leg for Philip of Spain. He could pick out the virtuous girls in the audience, or the sluts, and once climbed the tower of Old St Paul's Cathedral. Bankes shod him with silver and took him on a tour of Britain.

Marocco was mentioned by John Donne ('the wise politique horse'), by Shakespeare ('The dancing horse' of *Love's Labour's Lost*) and balladeers. In 1595 a comic account of a conversation overheard between Marocco and his master was published as *Maroccus exstaticus*. Here the gelding is a Figaro, the classic comic character of servant wiser than master. 'I know what I saie, and I saie what I know,' he tells Bankes, spicing his speech with 'this Latine I learned when I gambolde at Oxford'. As a horse, he can comment on human folly, and exchanges tart observations about racketeering landlords and immoral, lustful humans.

The real Bankes took Marocco on to Germany, Portugal, Italy and France, where Bankes was accused of sorcery and obliged to give up his training secrets to the Paris authorities – there was no sorcery, Marocco's counting and tricks were guided by tiny cues from his master. Released from Paris, they journeyed to Orleans, where they were again apprehended and accused of witchcraft.★ To save both their skins, Bankes instructed the horse to seek someone with a crucifix on his hat, then told him to kneel before him, then 'rise up againe and kisse it'. Marocco did as he was bid, pressing his lips to the cross, and his master turned to his judges, 'And now gentlemen ... I think my horse hath acquitted both me and himself.'★★

VII

AN INDIAN INTERLUDE

The capriole or goat step was, as the historian Treva Tucker has explained, the most impressive movement of the manège. It was the end point of collection: a sequence that began with a walk with

---

★ Marocco was luckier than another English *cheval savant* which was rumoured to have been burned at the stake by the Inquisition in Lisbon in 1707 after demonstrating its card-playing skills.

★★ Marocco and his heirs, Clever Hans, Beautiful Jim Key and Lukas, are often dismissed as frauds because they could not really 'count', only read subtle cues from their human handlers. However, a 2009 study in which horses were given a choice of buckets into which they'd seen varying numbers of fake apples dropped, indicated that horses, like human babies, can count up to four, or at least knew which bucket held more apples up to that point.

contained power, then a trot concentrated into a passage, and canter to a prancing terre à terre, then the impulsion contained once more until the horse rose in a levade, and then expelled all the energy with the leap and kick of the capriole. Pluvinel said only four horses in France – including La Bonité – could do it. A man's *sprezzatura* was tested to the limit by sitting on the back of a horse who sprang into capriole, kicking 'as though to tear himself in half'.

But no Renaissance horseman attempted to use the goat leap in the battlefield. While that slamming kick might have knocked out a few infantrymen, the steady, rocking preparation required to induce a horse into the classical capriole would have been folly when surrounded by armed foes. When it happened in the carrousels it was only a 'simulating war in a show of peace', but there was a place where a more urgent form of the capriole was used in direct combat, and another tradition of dancing horses both connected to and remote from that of Europe.

A quarter of a century after the death of Xenophon, the armies of Alexander the Great advanced into what is now north-western Pakistan to be confronted by not only cavalry but also war elephants. The Macedonian horses took fright at the armoured pachyderms thundering towards them, and Alexander's all-conquering troops were badly rattled. His precious horse, Bucephalus, was killed. Though his army won, they refused to travel further east. Alexander established a series of satrapies in the territory, and founded a city in memory of Bucephalus – Bucephala.

After Alexander's death in 323 BC, the Jain leader Chandragupta Maurya cut a deal with one of Alexander's former generals, Seleucus, giving the Greeks 500 war elephants in return for Alexander's old territories along the Indus. Chandragupta founded the Maurya Empire with the help of Greek soldiers, and at its peak it enclosed most of the Indian subcontinent.

In Taxila, near Bucephala, Maurya's advisor Chanakya wrote a text called *The Arthashastra* that instructed both kings and their

servants in the correct economic management of an empire. One
section is addressed to 'the superintendent of horses', and it lists
the various manoeuvres considered correct for a warhorse. Although
they are merely named, and there are no instructions for reproducing them, they sound similar to several classical and Renaissance
movements.

The Greeks, who lived in the area before and after Alexander, had
left their influence, easily seen in the use Sanskrit makes of the Greek
word for bridle, khalinos, and like Greek cavalry horses, the Mauryan
destrier of The Arthashastra was required to canter, trot, leap across
ditches or other obstacles and gallop, as well as respond to aids and
cues. The Arthashastra's elliptical descriptions are hard to follow, but it
is at least clear that many different types of canter or gallop – valgana
– were expected from horses, and that these could also be performed
'with the head and ear erect' as slower movements, which sounds
much like the tucked chin and arched neck of collection.

What made the Indian cavalry distinct from that of the Greeks
was their tremendous, poetic vocabulary of leaps, which mainly share
the root pluta, unlike the langhana, which means 'leaping across' and
presumably implied jumping obstacles and ditches. Here are the
leaps:

> Jumping like a monkey (kapipluta), jumping like a frog (bhekaplu-
> ta), sudden jump (ekapluta), jumping with one leg (ekapádapluta),
> leaping like a cuckoo (kokila-samchári) ... and leaping like a crane
> (bakasamchari).

Some of these sound improbable, but given the relatively limited
range of physical movement that horses can achieve, one is tempted
to link some to later Renaissance moves. The European capriole is not
so far removed from the way a frog leaps, with hind legs trailing. A
monkey could be a terre à terre, scutching along the ground from fore
to hind legs. A cuckoo leaps, one supposes, on two legs, like the little
crow of the courbette performed by Pluvinel's Knights of the Lily. And

the baffling 'jumping with one leg' is perhaps a kick, or a high strike, like a Spanish walk or *ciambetta*, which some contemporary horses can perform at a canter.

The *Arthashastra*'s text was lost for hundreds of years, which may be one reason it is seldom mentioned in Eurocentric histories of dressage, although war historians like Ann Hyland give it its due. In later centuries, Indian warhorses were recorded 'leaping' in attacks on elephantry. Sometimes these horses were given elephant masks, in the belief that the elephant would not attack what it thought was its own young. Ann Hyland describes the Delhi Sultanate period, when images of warhorses showed levade, courbette and a bold, upwards and forwards leap called the *oran*, which came from the Hindu word for 'to fly'.

The rider of a horse performing the *oran* could come close enough to an elephant's driver to attack him with a lance, and one famous warhorse, Chetak, at the Battle of Haldighati in 1576, flew at the elephant of Rajah Man Singh, leader of the Mughal army, before striking his front hoofs on the elephant's head as his rider killed the mahout. The elephant scarpered with Man Singh, and Chetak later died of his wounds.

VIII

The carrousel remained a favoured display of royal power until the mid-eighteenth century, and some examples persisted into the nineteenth century, although not at the same level of pomp and expense. Versailles has a huge reproduction of Henri de Gissey's depiction of the 1662 Grand Carrousel – held not far from the site of Pluvinel's old academy on a piece of land now named the Place du Carrousel in its honour. It was choreographed to celebrate the birth of the dauphin and cement Louis XIV's control of the throne after a long regency in which nobles and government had once again wrestled for the upper hand. The parade ground fills the canvas, pushing palace and town aside.

The Sun King used his long reign to attempt to centralize and concentrate power in his own hands, and so, in this didactic display, the five cartels of eleven men represent great empires of the old and new worlds: Romans in scarlet and black (a colour coding for 'the envy of the world'), led by Louis himself; Persians in red and white ('seen and raised above others'), led by his brother; the Turks in blue and black (for duplicity), led by the Prince du Condé – who had led an open revolt against the government; the Duc d'Enghien as king of the 'Indians' in yellow and scarlet (for their riches); and the Duc de Guise as king of the Americans in green and white (for their unspoiled, virgin nature).

Illustrations of the bombastic, allegorical costumes designed by de Gissey still survive. The Duc de Guise rode a chestnut horse whose mane and tail had been woven with dozens of open-jawed cloth snakes, and whose forehead was ornamented with a golden unicorn horn. Its saddlecloth was leopard-skin with dangling lion heads and ribbons, while a dragon reared behind the saddle's cantle, leaping towards the horse's tail. Guise had a golden helm with a dragon on top, and not one or two but three tiers reaching four feet into the air of fulsome black and white plumes. His sleeves were gathered up by snarling animals. The mounted drummers and trumpeters of the Indian cartel had trains of feathers and even stuffed green parrots on their heads. The expense of the costumes – another way to undermine a squabbling aristocracy – must have been astonishing.

The audience for this carrousel was restricted: just 15,000 would watch from specially erected wooden stands, and though they included ambassadors and the Queen of England herself, the towns-people were largely excluded. It was not a public display like Pluvinel's Knights of the Lily performing before 200,000 Parisians. Something was changing.

When Louis moved his court to Versailles in 1682, a 'school of pages' or academy was established in the new Grande Écurie designed

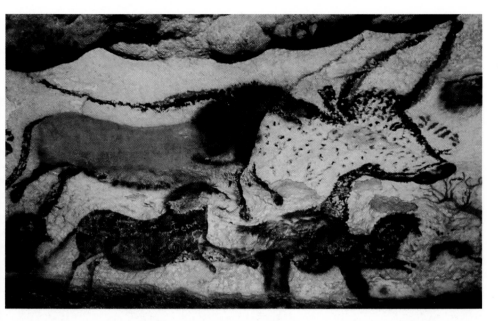

A replica of the cave paintings at Lascaux, featuring a large golden Takhi and two smaller, darker horses.

A 2,300-year-old carpet excavated from a tomb in the Altai mountains shows a magnificently dressed warrior and his leggy chestnut horse.

One of a series of woodcuts by German artist Hans Baldung depicting wild horses in a forest, 1534.

Performer Andrew Ducrow and two equine co-stars in *The Wild Horse and the Savage*, c. 1835.

Mazeppa strapped to the wild horse as it crosses a river on its way to Tartary. Painting by Horace Vernet (1789–1863).

Goering at Carinhall, Schorfheide, with his niece and a shaggy pony, 1938.

A Takhi in deep winter.

Byzantine sarcophagus from the sixth century showing a horse and trainer.

Mamluk cavalry in the *Complete Instructions in the Practices of Military Art*, attributed to Muhammad Ibn Isa Aqsarai, c. 1375–1400.

A skewbald Kladruber called Cerbero performs the capriole. Painting by Johann Georg von Hamilton (1672–1737).

Scenes from the French hippodrama, *Le Cheval du Diable*, c. 1846.

Écuyère Blanche Allarty-Molier and d'Artagnan fly in capriole.
Photograph by J. Delton, 1910.

Bartabas and Le Tintoret hold a photo call for *Le centaure et l'animal* on stage at the Sadler's Wells Theatre, London, 2011.

The Carrière at Versailles, 2011: riders and Criollo horses of the Académie du Spectacle Équestre rehearse Carolyn Carlson and Bartabas' *We Were Horses*.

Jessie Noakes' early twentieth-century print after the medieval *Luttrell Psalter*, with scenes of ploughing, sowing and harrowing.

Old English Black Horse by William Shiels (1785–1857).

by Mansart. The pattern of academies and riding masters established in the Renaissance was still in place, although mounted warfare was becoming still less relevant, which did not stop the number of books produced from rising, nor trammel the horse madness of nobles in general.

Louis appointed François Robichon de la Guérinière, the son of a lawyer from Essay, to run the riding school at the Tuileries, and he would eventually reach the high office of director of the Tuileries Manège under Louis XIV. His keystone text, published between 1729 and 1731, was *Ecole de Cavalerie* (*The School of Horsemanship*). In it can be found the legacy of riding masters who went before him, from Pluvinel to Xenophon, as well as an early defence against the changes that were coming to both military and aristocratic riding.

Guérinière felt already that the golden age of the manège and carrousel was behind France. He distinguished the run-of-the-mill rider from the true horseman: 'practice without true principles is nothing other than routine, the fruit of which is a strained and unsure execution, a false diamond which dazzles semi-connoisseurs'. Ride badly and you would 'mortify the spirit of a noble horse'. One frontispiece to Guérinière's book shows a horse standing proudly at the centre of what appears to be an Enlightenment salon – it is in fact the darkened Tuileries manège – in which sits a crowd of gentlemen in frock coats and wigs. This slender-nosed, intelligent beast is a different creature to the farouche La Bonité. He is both guest of honour at the salon and the subject of scientific examination.

In *Ecole de Cavalerie*, horses are dissected, coached, treasured and enumerated as collectors' items. They are beautiful aesthetic objects with personalities, sometimes 'sly', 'timid', 'slothful' or even 'malicious', but all needing careful handling to give their best. Guérinière evokes the colours of the patrician horses of the écuries in a poetic list that includes five shades of bay from golden to dapple, jet black, rusty black, iron grey, silver grey, dapple grey, flea-bitten grey, cream, pied horses in black, bay or chestnut, and then an exotic series of coats that include tiger, wolf, moor's head roan, porcelain and 'trout'

– black with red and chestnut stippling. Collectors even paid attention to the whorls in the horses' coats, with the most sought after being the 'Roman sword', which laps the roots of the mane all the way along the neck.

As for the movements that horsemen cajoled from these glamorous creatures, Guérinière says that the courbette is used by cavalry officers, but only in parades. Pirouettes could be put to use in rapid turns in combat, voltes for swift movement, the passade* for attack and retreat, passage for leading troops. He does suggest some uses of high-school moves for warhorses, but adds, 'If the airs above ground offer no advantage for war, they offer at least that of imparting to the horse the agility it requires to clear hedges and ditches, which contributes to the safety of the rider.'

But it was this 'clearing of hedges and ditches' that he was ranged against. The cavalries of the future, and the leisure time of the French nobility, would be spent less in making patterns in a sand school on a Spanish horse, and more haring uncollected across the countryside on a tall, lean English thoroughbred. They would make long, fluid leaps mid-gallop over obstacles, not rock steadily till the horse could leap up on the spot and hop or kick out in gambades. If the well-collected horses of the manège went up and down as well as onwards, the thoroughbred just went onwards; its weight did not fall back on rounded quarters, but was slung forward on long, sloping shoulders. Everything changed: the deep saddle became an open, flat plate, the rider rode as a Turkish soldier or Muslim raider, light as a proto-jockey, and the bit was a snaffle rather than a curb. Guérinière admired these English horses, but thought that if they were not trained by his rules, they would be injured as their owners chased them rashly across country.

When an écuyer – a word best translated as horseman, or even, to give it a more courtly ring, equerry – at Versailles, the Marquis de la Bigne, cantered his horse so slowly across the Place d'Armes that it took him a full hour to cross the cobbles from the stables to the

---

★ In the passade the horse canters in one direction then rapidly spins and doubles back on itself. The military usefulness of this move is clear.

chateau,* it was at once an astonishing achievement, and an irrelevant one. Acceleration and exhilaration were now key: the new hussars and dragoons were light cavalry mounted on horses that knew little of haute école. Though they were drilled in manèges they also broke out of the riding halls and across country, swapping *sprezzatura* for bravado. Their style owed more to the steppe and to the Muslim horsemen of the Levant and North Africa than to Pluvinel's academy.

Britain was at the forefront: few Englishmen had ever appreciated what Thomas Bedingfield had called 'sundrie sorts of superfluous dansing and pransing'. Charles I's passion for horse racing and a mania for the new, harum-scarum sport of fox hunting matched the breed created from Arabian, Barb and Turkish trophy horses crossed with those Royal Mares with their faint Spanish ancestry. In the noble equestrian portraits of the eighteenth century, English horses carry lords, but they do not collect themselves in roundness or arch their necks, but stand with open, natural postures, their faces flat or concave, their manes free of ribbons and sequins.

In the nineteenth century, Paris would be swept by Anglomania, when English thoroughbreds replaced the Spanish horse, and the humble French scrub horse, or bidet,** was rechristened the 'poney'. Guérinière's work was preserved by the Spanish Riding School in Vienna and eventually the French Cadre Noir military school at Saumur, but only in the teeth of bitter disputes that still lay ahead.

But the first and greatest rupture to the system of the academies that had brought the carrousel, the manège and the *ballet à cheval*, and the manuals of *reitkunst* and *gracia y hermosura*, the advice from Ancient Greece and the be-plumed and drum-dogged victory parades of so many conquerors and invaders, was, of course, the Revolution, and the ejection of Louis XVI from the palace of Versailles. Even two years

★ Hilda Nelson estimates that the marquis and his horse advanced by 5 cm per stride.

★★ The notorious French bathroom accessory, so baffling to generations of Britons, is named for the French bidet horse because it is straddled. This genteel appropriation has now become the first meaning of the word in France too, as the hero of one 1970s novel fretted, 'Why was the horse that made him bounce around a *bidet*? Bidets were for bathrooms.'

before the Bastille was stormed, Louis had been obliged to shut the Petite Écurie to curtail his expenses. The riding academies were dismantled, the horses and riding masters scattered.

A guillotine was erected on the Place du Carrousel, and Guérinière's old riding school became the home of the new National Assembly. Louis XVI was tried there in 1792. The distinctive structure of a manège can be seen in sketches and engravings of the event: the rectangular shape of the manège floor, two tiers of viewing galleries at the sides, the grander balcony gallery at one end, draped with cloths and rosettes. The allegorical colours of the carrousel were replaced with red, white and blue, which now stood not for some antique imperial power, but liberté, égalité and fraternité.

What the Revolution did not achieve, the Napoleonic Wars nearly completed: the destruction of the Versailles school of classical riding. The Marquis de la Bigne saved the Grande Écurie and turned it into a military riding school, but after a decade it was shut once more.

IX

If the carrousel was gone, its pomp and expense no longer supportable in a nation sick of an overbearing, overspending royal family, its spirit was not forever lost. A new spectacle would rise up that was not only more audacious, but also more accessible: its performances did not wait for royal weddings but were available day after day, for months or even years at a time. And it was open to all who could afford a ticket.

Its first intimations in France came nearly a century before the Revolution, when the cast of a 1682 production of Corneille's Andromède were joined on stage by a real Pegasus, which not only had wings but also 'flew' on wires, his legs paddling furiously beneath him as he whickered and snorted. This astonishing effect was achieved with a saddle reinforced with metal and with the semi-starvation of

Pegasus – as he soared over the boards, he was animated into his performance by the sound of a sieve of oats rattling in the wings. In earlier centuries, horses and donkeys had helped actors to make entrances, or to do some tricks, but this was new: dramatic action – and stunts.

Mazeppa, ou le Cheval tartare, or *Mazeppa, or the Wild Horse of Tartary* were only two examples of the hippodrama, a prolific genre which would have its birth in a London and a Paris where theatre was controlled by the authorities, and only a select few establishments were permitted to perform 'legitimate' or '*grand*' theatre. The rest worked their way around the licensing laws by throwing in acrobats, song, verse and, as it turned out, horses.

Here we rejoin Andrew Ducrow, Philip Astley and their heirs in France, the Franconi family. The tumblers on horseback and the intelligent horse trainers of the streets claimed their moment. Their audience, bulging out of the seams of the rapidly growing cities, brought their new spending power, and the horses and men and women of the road came indoors as the manège riders found themselves out in the cold – though some riding masters and ex-cavalry men found new uses for their skills at the modern hippodrome.

Philip Astley was a former dragoon guard – a veteran of the Seven Years War – who began the modern circus when he gave trick riding shows in 1768 on a field in Lambeth. Instead of performing his feats on an open plan, he used a ring – hence 'circus' – thus having enough centrifugal force to complete more daring vaults and balances. This also had the happy effect of drawing his audience around him, making him the eye of their attention, and creating theatre. He could vault on and off, stand on the backs of galloping horses like the 'man skilled in feats of horsemanship' in *The Iliad*, and snatch a handkerchief from the ground at a gallop.

He kept a horse called Gibraltar that could do everything Marocco could, and more. According to the clown Jacob Decastro, Gibraltar was a Spanish horse – a gift from Astley's general when he left his regiment. This horse, which, among other tricks, could 'take a kettle of boiling water off a flaming fire', acting 'after the manner of a waiter at a tavern or tea gardens', lived to the Methusalean age of forty-two, and was after his death skinned and turned into the drum that provided the thunder sound effects for Astley's productions. Astley's wife was a horseback tumbler, too, and as their popularity grew, they contributed small equestrian presences to some theatrical productions.

Astley added more acts – a strong man, rope dancers, tumblers – and built a new arena near Westminster Bridge. He gave riding lessons and wrote down his wisdom in a book, Astley's System of Equestrian Education. As time went on and his public grew, he built seating, then a roof, and finally created Astley's Royal Grove, which added a stage to the ring, and above whose highest loges rose artful depictions of trees – a whisper both of the first academy of Plato and of the Englishman's patriotic greenwoods. Here the public could see circus feats, amusing skits, 'serious dancing on horseback' and ballet, as well as Astley's 'horsemanship'. Gradually, tableaux and burlettas developed with the barest of plots: a calamity-prone tailor, a highway robbery, the wild horse hunt.

Rivals sprang up, eager to cash in, and others who had served their apprenticeship with Astley blossomed in their own right. Andrew Ducrow, son of a strongman, posed in tableaux as Greek statues so alluring that women broke into his bedroom to leave flowers. In one show Ducrow recreated a Spanish bullfight, in which the bull was played by a 'gentle and beautiful white horse with a bull's skin over his padded neck and body, his head supplied with horns, and his hoofs painted as if cloven, in every respect appearing like a tremendous bull wild and fierce'. Ducrow was mounted on another horse, and thrust spears at the bull-horse until its performance reached a frenzy, and it sank to the ground and died of its 'wounds'. 'After witnessing this performance, no one can withhold his surprise at the perfect knowledge of the business of the scene which this horse evinces,' wrote the critic from the North Wales Chronicle in February 1835. 'There is no deviation from character; he is throughout a bull; his trot, the management of his horns, and the fierce rush of his head all display something more than could be expected even from the most sagacious horse of Mr Ducrow's stud.'

In 1774 Astley had taken his act to a manège on the rue des Vieilles Tuileries, at the command of the Sardinian king. By 1783 he was in Versailles, watching his son John – '*avec des grâces et une vigueur capable d'enchanter le beau sexe*' – perform in the Petite Écurie at the request of Marie Antoinette, who awarded him a gold medal set with diamonds. The French King and Queen granted Astley senior a 'royal privilege' to open the Amphithéâtre d'Astley on the Boulevard du Temple in 1783. He would also enjoy royal patronage in England, but he and his rivals were men of the coming century: enterprising entrepreneurs rather than the courtly riding masters who did as commanded. When, early on, the Parisian authorities told Astley his licence did not cover performances by acrobats, he simply set a stage on the backs of eight horses, and the acrobats performed on that.

The royal blessing from Marie Antoinette was, of course, a disadvantage in 1789, and Astley had to leave Paris smartly, handing over his amphitheatre to an Italian-born bird trainer and bullfight organizer called Antonio Franconi.

A proto hippodrama was performed in 1798: *La Fille Hussard ou le Sergent Suédois* featured Antonio Franconi and son performing a mounted gavotte and minuet, and from there the genre's popularity was assured. The Franconis built the first Cirque Olympique in 1807, Astley having reclaimed his theatre in 1802. As state control of theatre collapsed, the hippodromes were free to produce ever more elaborate and extrapolated productions that evolved from circus to theatre. If the melodramas and *féeries* of the hippodrome were despised by the more serious theatrical establishment, it was probably because the plays were terrible drama, the actors 'bawlers' like John 'Mazeppa' Cartlich and the scripts atrocious, but as spectacle they could not be beaten, and the public knew it.

On 3 February 1846 the Cirque Olympique opened their newest production, promising unmatched spectacle and scenes hard to believe. Theatregoers waited in a building that had much in common with any theatre of the period: they had mingled in the grand entrance hall before making their way up through a warren of staircases and corridors leading to the stalls, circles and boxes. As they fidgeted and murmured among themselves they looked out – from their box, if they could afford it – or up, from the stalls – at chandeliers and a domed, elaborately painted ceiling. Before them the proscenium framed a deep, square curtained stage. The orchestra could be heard tuning up.

Before the stage, where the orchestra pit should have been, was an empty sand circle, neatly raked. And as the curtain began to part, the stage moved slowly, mechanically forward into the arena, and two broad ramps extended to reach the sand, the machinery drowned by the orchestra's stirring strains. Drowned out too, were the sounds of snuffling, of hoofs striking cobbles, and whinnies: beyond the wings stage right was a row of stables. Tonight the people of Paris would ride to an enchanted forest, to Iceland, to Damascus and finally on to hell.

The scene opens with Ulrick, the miller's son, who dreams of being a knight. Here is a gypsy crone called Djina who tells him she can see his future, and that she can offer him his wishes. Foolish Ulrick cannot resist, and Djina – here the action begins! – brings him to the very stables of Satan himself, where the great demon's saddle horse, Zisco, is waiting, 'ocellated with brown spots and slashed with black stripes, like the fur of a panther crossed with that of a tiger'. On his back is a cloth made from a tiger's pelt with twelve stripes.

Zisco can grant twelve wishes for Ulrick, Djina explains, but at a price – five years of Ulrick's life. As each wish is consumed, a tiger stripe will vanish from Zisco's saddlecloth. Ulrick, like every fool with a dusty lamp, is sure that he can win this gamble. First he asks to be a knight, perhaps thinking he can manage on his own from there, and Zisco rears up and tugs on the branch of an oak tree, revealing a standard and suit of armour tucked in the tree's trunk.

But being a knight brings complications, and the temptation to solve them easily means that the stripes peel off Zisco's blanket one after another, like leaves from a tree in autumn. Ulrick's damsel is kidnapped by another knight, and the miller's son turns to Satan's horse for help – Zisco seizes the knight with his teeth and crushes him under his hoofs. Ulrick sails to Iceland to save the son of the Emperor of Constantinople from a pirate, but the pirate lives on a sea-girt rock reached by a hanging bridge that is impossible to cross. Once summoned, fire-snorting Zisco charges across the wobbling planks to snatch up the infant just as the pirate's tower catches fire, and crosses back to the land just before the whole shebang breaks and tumbles into the waves. Polar bears turn on the pirates. The Emperor of Constantinople gratefully awards Ulrick the sultanate of Damascus, but after a brief triumphal interlude, he finds himself pursued through the burning desert by crusaders. One stripe remains. The man who asked for a kingdom asks for a glass of water, and with one blow of his hoof, Zisco opens the flaming mouth of hell, and hapless, foolish Ulrick is borne away to never-ending damnation.

Le Cheval du Diable has not seen limelight since the 1840s, when it kept the audiences at the Cirque Olympique agog for 300 performances. Tastes have moved on, and it is hard to imagine a modern audience could stomach the *féerie* style, in which fantabulous supernatural happenings were taken far more seriously than in a panto. But the nineteenth century was a high point for sensational theatre, before cinema came to knock it on the head, and the masses could flock to see the kind of elaborate, sweeping escapism that had been reserved for the wealthy. Hippodrama offered not the arch allegory of the royal carrousel, but full, immersive escapism for the public. And because the nineteenth century was also the century of the horse, horses had to be part of its fabric, and were, for over a hundred years, before they vanished.

For these showmen, nothing could not be replicated. Where the organizers of the old carrousels turned dancing horses and mythological battles into a spectacle designed to impress their puissance and magnificence upon their subjects, their aristocrats and visiting royalty, Astley, the Franconis and their heirs fought to astonish a crowd so that they would come back over and over, handing over their money for tickets, and – with visions of faerie castles, gun smoke lingering in their nostrils and their hands that had gripped the edge of the stall still vibrating with the thuds of hoofs galloping yards from them – clamouring always for more, more, more sensation!

The Battle of Waterloo with full cavalry charges, volleys of cannon, smoking pistols, a runaway supply wagon on fire and ninety horses? Mr Astley will recreate that for you, or Mr Franconi will reproduce the greatest victories of Napoleon, and the crowd, high on horses and chauvinism, will call out for more. A carrousel of the old days, with every feather and sequin bolder than anything Versailles saw? It's yours. Along with a stag hunt with a real stag, called Coco, Ben-Hur's chariot with four horses galloping furiously on treadmills and dust flying up from jets hidden in the stage, waterfalls, Amazons, jousting, kangaroos, chasms, Rocinante and Dapple flying to heaven on a cloud, blue-eyed cremello ghost horses carrying the spectres of defeated Indians, castles that go up in flames, Lady Godiva, horned horses, the five-year-old 'Elfin Equestrian', Epsom on Derby day, six fairy ponies, one of whom leapt over the others, and 200 Arabian horses re-enacting the Fall of Khartoum. No land too was exotic, no scenario too far-fetched. The lumbering volcanoes and Mount Olympuses of the Renaissance carrousel were outdone night after night.

But while sometimes these horses danced like Zisco when he is welcomed by odalisques in Damascus, their talents, like his, extended far beyond the piaffe. Everyone in the audience knew horses. They lived alongside them, in streets thronging with them. They rode them, drove them, watched them, cheered them on at the racecourse, had perhaps fought in battles alongside them. And at the theatre they saw both horses and horsemanship that they could appreciate for their skill and intelligence. The critics sometimes barely mentioned the human actors. The historian A. H. Saxon, who is responsible for much of what we now know about the hippodrama, says that when one horse was injured, posters would be specially printed to reassure the public that he would return.

X

Haute école was still practised in military academies, although a divide had widened between classicists and modernizers like the Comte D'Aure, formerly of Versailles and now chief trainer at Saumur, who had created a new French dressage drawing on English and Prussian military styles. This was practised outside the indoor manège and encouraged more extended, forward movements as well as collection. Athletic thoroughbreds were favoured over Spanish horses. Its workmanlike qualities meant it was more rapidly taught to the new cavalrymen who had never known the pre-revolutionary academies and were seldom aristocrats.

Speed was also a preoccupation for the foremost classicist, a butcher's son from Versailles called François Baucher – although this was speed in accomplishing what had taken the men of the academies years to achieve. Baucher could take a horse from the knacker's hands and, through his ingenious system of suppling exercises, have it performing the highest school movements in a matter of months. Most famously, he took on the challenge of Lord Seymour to gentle a horse called Géricault, who had a savage reputation throughout Paris, and after three weeks he presented the docile stallion to the public, tamed, collected and obliging to an audience of gratified fans and furious critics. Some accused him of mesmerizing Géricault.

But Baucher was not popular with the purists who sighed back to Guérinière, and who believed the butcher's son made his horses into stiff-legged, high-stepping automatons with their chins pinned to their chests. He also disgusted them by introducing showy new movements, like the 'one tempi' flying changes, in which the cantering horse skips at each stride. In his pursuit of *rassemblage*, Baucher outdid the Marquis de la Bigne by collecting a horse so greatly that it could canter backwards, each hoof set down in turn.

These demonstrations with Géricault and others were performed not in military academies, but in the great circuses of Paris and Europe.

When the hippodrama's popularity stuttered in the mid-nineteenth century, the circus proper began to take over the hippodromes, and one branch of the Franconi family brought haute école to the ring, where it had an audience that included royalty and commoners, and Lautrec and Seurat painted the performers.

Many of the circus's celebrated haute école riders were women, who, mounted side-saddle, were, as circus historian Hughes le Roux put it, 'posed audaciously like a wing'. They were the first women in history to be granted the opportunity to make displays of virtuoso professional dressage and were highly respected for it.* Their routines were every bit as demanding as those of the men, and there's a wonderful photograph of the *écuyère* Blanche Allarty-Molier** riding in full Edwardian rig – rolls of hair under a picture hat, rigid corseted waist, leg-of-mutton sleeves up to her chin – as her horse flies in a full capriole.

Here is the routine of the German horsewoman Jenny de Rhaden, performed on a sloping stage eight metres square at the Folies Bergère:

FIRST HORSE:

Enter ring with a flying leap forwards like the Indian *oran*, followed by a courbette taking Jenny and her mount right up to the edge of the orchestra pit.

Steps to right and left, then voltes, before switching to flying changes at three, two and one tempi.

★ There were – briefly! – women riding in the old School of Versailles. The Cour d'Honneur held a carrousel in 1668 which featured a team of real Amazons – ladies of the court who were led out of the Petite Écurie by Madame la Duchesse de Bourbon as Queen Thalestris, but only the men took part in the mock tournament games.

★★ Blanche also performed the cabrade – a high rear – on a camel. Hilda Nelson has written the best history of these women in *The Ecuyère of the Nineteenth Century in the Circus*.

Pirouette and then the high-stepping Spanish walk, followed by a half pass (in which the horse moves forward and sideways, legs crisscrossing).

SECOND HORSE: Da Capo

Enter at gallop and then pirouette. The horse rears to its full height and walks on its hind legs before making a bow to each of the four corners of the stage.

Four fences are placed in a square in the centre of the ring. Jenny and mount leap each of the jumps.

The horse rears upright once more and Jenny lies back on his quarters, her hair almost trailing the floor, as he takes a few steps.

THIRD HORSE: Czardas, a 'tiger'-spotted stallion

Jenny enters the ring on foot to receive her applause. Czardas gallops in and kneels before her, then lies down. She perches on his side and goes on thanking the audience. The curtain falls.

For both men and women, this high-stakes version of haute école riding was never without risk. Jenny would later go blind when a rearing Da Capo toppled over backwards on to her as she was lying on his quarters. Another *écuyère*, Emilie Loisset, was impaled on the horn of her side-saddle when her horse fell, and died in agony. Baucher never rode in public again after a bizarre accident in which the chandelier at the Cirque Napoleon fell on him as he was mounting his horse, breaking his back and legs.

Haute école was still performed in the circus as the nineteenth century turned to the twentieth, but its popularity was diminishing as new acts proliferated and the population gradually became less horsey. Horses were disappearing from the streets and cars taking over. The liberty horse – an old trick of both the hippodrama and circus

– became more popular, performing rears and circles to cues from a lunge whip held by their trainer. Circuses announced their presence from town to town in parades with gilded juggernaut wagons that resembled those of the Renaissance carrousels, often drawn by cream horses – some of the British royal family's 'creams' even ended up in Sanger's circus. Cinema provided a new source of sensation. In a bid to draw audiences back to the big top, one celebrated écuyère, Solange d'Atalide, was filmed sitting on her horse on the roof of one of the gondolas of the great Ferris wheel in the Vienna Prater as it completed a full circuit in 1914.

Riding academies in Vienna, Saumur, Jerez and Lisbon dug in to preserve and defend the haute école of the eighteenth century, performed on matching horses in antique military uniforms. Military dressage became an international sport as the entire cavalry system gradually broke down and tanks replaced chargers. At the Stockholm Olympic Games of 1912, dressage horses performed no haute école movements at all, although later sporting dressage would return to its classical roots – albeit minus the airs above ground, which became the preserve of the purists of the great academies, with their heavy state subsidies.

XI

But what about the reverse centaur on the dark stage at Sadler's Wells? Where does he come in all this? His name is Bartabas, and his name is Soutine. Soutine is an American Quarter Horse, named for a Belorussian expressionist, and Bartabas is a Frenchman who rechristened himself after a pirate and Barabbas, the prisoner who was spared death on the mount at Golgotha.

I went to Versailles because when I took all those dancing horses from Xenophon's high-mettled parade horse to the Spanish horses of the Sun King with their golden bridles to Zisco the Demon Horse in my hands, collapsed them like a pack of shuffled cards and turned each over, I found Bartabas, and the creams and the Criollos and the Sorraias who were his corps de ballet. The Grande Écurie at Versailles had not seen horses for 173 years until Bartabas brought them there in 2003 and recruited new riders to fill the places of the pages and aristocrats of Louis XIV's time – albeit with a very different philosophy.

The son of an architect and a doctor, he was born Clément Marty in suburban Paris in 1957, and as a teenager bought his first horse with the compensation money he was awarded after a moped accident nearly cost him his feet. He started, like Astley and Franconi, as a street performer in a travelling theatrical troupe that trained rats and crashed anarchically about the streets of Avignon, stealing food from supermarkets. After a brief and failed attempt to become a bullfighter, he founded Théâtre Zingaro, a rackety cabaret gang named for a black Friesian stallion that he thought of as a son. It featured teams of shambolic waiters, a flock of geese waddling chaotically about the stage and Bartabas as a roaring, boastful horse tamer terrified of his charge – the magnificent Zingaro, who chased his would-be master around the ring with bared teeth that closed harmlessly on his sleeve. In 1989 the team's caravans and horseboxes came to rest at their own purpose-built wooden big top on a piece of wasteland in Aubervilliers, a *banlieue* just outside Paris.

Like Baucher and Astley, Bartabas is outside the academy, distrusted by purists, one of whom derisorily called him the Tina Turner of haute école. He's self-taught, has never competed in a horse show, and has even dared to perform non-canonical movements like the backwards gallop achieved by Baucher.

Since that first cabaret, Théâtre Zingaro has produced spectacles and films that combine haute école, vaulting and liberty work in choreography that, although it has distant echoes in the hippodramas and the nineteenth-century circus or the carrousels, is unique. Bartabas has a hotchpotch of non-equestrian influences which he grafts into old and new forms: the films of Fellini and Kurosawa, and the paintings of Picasso. The whole aesthetic of the horse is explored: a pale

Arabian galloping loose; a hunky-munky pinto careering about a ring pulling a gypsy wagon covered in flowers; a thoroughbred in silhouette, positioned like a rocking horse or a Tang dynasty figurine, a lance at its rider's side; a Quarter Horse leaping from side to side in pursuit of a running man, its head low and ears snaky; a heavy grey from the tiltyard performing a backlit piaffe; glossy black Zingaro rolling and rolling in white sand under a spotlight.

To accompany them Bartabas brings, like the hippodrama impresarios and the carrousel choreographers, the exotic: monk musicians from Tibet, Rajasthani drummers, Mexican brass bands, pansori players from Korea and a Moldovan gypsy band. But there are no pasteboard palm trees, or Persians weighed down with plumes, but instead real Cossacks and Kalaripayattu dancers. A trip to one of Zingaro's productions is meant to be more than escapism, a 'sacred' moment in which the spectator experiences 'the feeling of being', or the elusive 'blue note' of jazz that yaws from the tune to strike home. 'La force du cheval est d'être dans le présent et dans l'action,' Bartabas wrote in his manifesto.

But whereas Xenophon prepared horses for his own military glory, and Pluvinel turned dressage into a metaphor for statesmanship, Guérinière an antique art, and Baucher a scientific exercise, for Bartabas the performances he coaxes from his horses fulfil no other function than to celebrate the horse itself. Their beauty is placed in its own service alone. The hippophilia that one suspects is really behind all the Peri Hippikes, the Instructions of the King, the Schools of Horsemanship, is stripped to its heart under Bartabas: unlike Philip Sidney, he is happy to 'wish himself a horse'. He likes to say he doesn't whisper to horses, only listens to them. So as a reversed centaur he and Soutine become one, with the horse's mind leading the human's body.

Bartabas wants horses to take their place alongside other arts, so he leads them up to the table alongside the contemporary dancers, the grand conductors, the Palme d'Or winners at Cannes, sets a bowl of oats down on the starched cloth and watches as everyone freezes and finds themselves in slow contemplation of the muscles in the horse's neck, and the notch above its eyes that fills and empties as the horse chews.

'You must accept that the horse will run your life. He has shaped mine,' wrote Bartabas – who lost his first marriage to his work, lives in a smart wooden caravan behind the theatre at Aubervilliers and exists in a cycle of money not quite made – in his *Manifeste Pour La Vie D' Artiste* (*Manifesto for the Artist's Life*). In the reverse of Pluvinel's 'bringing the horse to reason', he preaches submission to the horse, living by his needs, his rhythms, and taking cues for training from him.

I saw him and Soutine as the reversed centaur at Sadler's Wells in a 2011 show called *Le centaure et l'animal*. It was a joint work with the Japanese dancer Ko Murobushi, whose media is a tortured, writhing dance called the Butoh. As Murobushi, head shaven, body painted silver, twisted and jerked, Bartabas performed four set pieces, each with a different horse. Soutine had his moment as the centaur. Horizonte, a burly, shaven-maned warhorse, piaffed under Bartabas, who was obscured in the white pointed capirote of a penitent in Holy Week. Pollock, the grey Quarter Horse, chased a shadowy runner as Bartabas sat on his back in a cloud of trailing red robes that swirled, just behind Pollock's movement, like ink in water. Le Tintoret collapsed slowly beneath his rider, and they both lay lifeless on the floor, over and over.

Later, I saw his 1993 film, *Mazeppa*, which depicts Bartabas as the Master, an ex-soldier and circus horseman who is a fusion of Franconi and Baucher. The Cirque Olympique is resurrected in wooden form in early nineteenth-century Paris, and filled with a troop of Cossacks, Berbers and loners under the Master. Over the ring looms an immense wooden chandelier like the one that crushed Baucher, and it is lowered to the floor for the company to dine from.

The Master, like Baucher, resists the forward rush of the new Anglophile horsemanship, and the film recreates the Marquis de la Bigne's feat of cantering across the Place d'Armes for one hour, although when the Master and his horse finally reach the end of their slow course, they begin to canter backwards to their starting point: back, back, yearns the Master, against the rush of the new industrial century.

Byron does not feature – this is a very French Mazeppa, so none of the mad lord's slightly camp touches are in play. The artist Géricault haunts the theatre instead, struggling to capture the movement and spirit of horses in paint, and finally being driven mad by the effort: his friends strap him naked to the back of a horse like the Adah Isaacs Menkens and the Cartlichs and all the other hippodrama actors, and set the horse galloping furiously on a treadmill. Art and the visual evocation of it is another part of Bartabas's aesthetic: it's not just the names of the horses that evoke grand masters.

Mazeppa won the technical grand prix at Cannes, which Bartabas added to what would become an eccentric haul of very French honours – not many Chevaliers des Arts et des Lettres have also been made a Chevalier du Mérite Agricole. He has the freedom of a city or two. His familiar features with their scimitar sideburns were drawn by Cabu of *Charlie Hebdo*. He's worked with Pierre Boulez, Philip Glass, Carolyn Carlson, and was teaching Pina Bausch to ride shortly before her death. Hermès made a saddle for him, shaped like a wing, and the tack for all the horses at the academy. The riders at Versailles wear uniforms designed for them by Dries Van Noten: chic, short jackets in muted tones, with dusty silver embroidery on the cuffs and striped sashes at the waist, and capes with fur collars for the winter.

When I arrived in Versailles in November 2014, Bartabas was away on tour with flamenco dancer Andrés Marin in a touring pas de deux called *Golgota*, and the last production of his main equestrian theatre group, Zingaro, had just finished. This *Calacas* had celebrated the Mexican Day of the Dead, and like all Zingaro productions, it was gone, along with the riders, the musicians and some of the horses – never to be produced again. 'The entire Zingaro universe changes,' the academy's secretary, Charlotte de Smet, had told me as she showed me around the back stairways and the red-tiled corridors at Versailles.

The academy was different: the *écuyers* and their horses were the stock company at Versailles. Stock company was perhaps the wrong word – as I explored the corridors, it seemed to me an autonomous monastery of sorts, a small, sealed off kingdom, somehow in the very centre of a touristy town, whose walls were hung with pictures that, like the mirrors in the rehearsal room, reflected only the riders and their horses themselves. If its chief priest was away when I visited, his influence was everywhere.

I suggested to the director at the academy, Marine Poncet, that Zingaro was made in wood, the academy in stone. 'I hope so!' she said. As Charlotte had put it, hand resting on one of the worn bannisters, 'Currently the academy is financially a little fragile.' Bartabas had been invited to bring life to the old Grande Écurie by Versailles, but the academy rented the space from the chateau and were responsible for making up 80 per cent of their expenses in ticket sales and through other commissioned performances. Zingaro had always been independent of the ministry of culture, but the academy was another proposition. Things were not running smoothly.

Since 2003, when the academy was founded, Bartabas had given Versailles shows with steel elephants mounted on wagons rolling around the Basin of Neptune, a pagoda made of light, samurais, whooping archers letting fly at targets and a nightmarish ghost two storeys high that curdled the blood in his own take on Kurosawa's *Throne of Blood*. Versailles wanted periwigs and frock coats. Indeed, in 2013, the palace hired a classical equitation society from Belgium to recreate the 1662 carrousel of Louis XIV in the grounds, without first consulting Bartabas. Earlier, in 2007, a government official who promised the academy funding had suddenly withdrawn a chunk of it, causing Bartabas to tear a radiator off the office wall. 'This life we love, runs against all the values of today,' Bartabas wrote in his manifesto – a book preoccupied with the paradoxes and sacrifices of working with horses. The Franconis and Astleys saw their theatres burned to the ground many times, and eventually the hippodrama became too expensive to produce. Louis XVI was obliged to close the

Petite Écurie and sell some of his thousands of horses. The Zingaristes might be nomads and gypsies, but to run an academy, you need a prince.

There are twelve riders at Versailles – though only eleven during my visit – and they are referred to, as in the old academies, as *écuyers*. The feminine, *écuyères*, is more accurate, as nine of them are women, a far higher count than at the haute école academies at Vienna, Saumur, Jerez and Lisbon, although one which more accurately reflects the wider Western equestrian world. Asked about this, Bartabas had told reporters that more femininity is needed in riding, because it meant not dominating the horse.

The *écuyers* are all French, apart from Anna Kozlovskaya from Russia, whose leatherwork bench I had seen in the dressing room: she grew up in Siberia, where tack was hard to get hold of, so knew how to fix her own. She mended all the harnesses for the academy. The *écuyères* lived in simple mobile homes not far from the chateau, and received a small salary. There were no exams, no qualifications to achieve and little hierarchy – at the top are the *écuyers titulaires*, then the *écuyers confirmés*, the *élèves* or pupils and the young entry – the *aspirants*. Applicants sent a video of themselves riding, and if chosen, attended an audition. The *écuyères* come and they either stay forever, or leave.

They learn fencing, mounted and dismounted, Japanese Samurai or kyudo archery with the long bows I'd seen in the dressing room, Pilates and singing. One, Adrien Samson, had been enchanted by Zingaro's shows as a child and later worked with sport horses before coming to Versailles, where he had been an *écuyer* for four years. He explained the choice of disciplines to me: 'The singing is good for expression, it releases something that would hold you back otherwise. That *exteriorisation* is an important technique for working with the horse.' Pilates was good for core muscles and having an independent seat on the horse. Kyudo was the hardest to master. 'Each discipline brings its interests. We're learning a philosophy of work, too. To learn with Bartabas, is to learn to learn alone!'

Where the old academies of Paris created aristocrats who would master their human subjects as they mastered their horses, the Académie du Spectacle Équestre pursued the blankness of Zen: when the body operates without the conscious mind. It seeks to make riding as instinctive for the human as movement is for the horse. 'We are very busy,' people kept telling me. 'There is so much to do.' But the work was done calmly, with *écuyers* walking from place to place, and going through the rhythms and rites of grooming and checking the horses.

Life was measured out on the sand of the three rings: the Cour d'Honneur, the Carrière, and the Grand Manège. I'd read that at Zingaro, around the Périphérique, everyone passes the day in near silence, communing largely through their seat and hands with their horses. At Versailles, my hosts were chatty and welcoming, but I was aware that my time at the Grande Écurie would be easily erased by the scale of the work they were doing.

They spent twelve hours a day from 7 am to 6 pm, six days a week, in the three courtyards, grooming the same necks, seeing them arch before them over the same raked and re-raked school sand. Every day, they all told me, there is something new: the horse never performs the same movement the same way, relationships between people changed. Some, like Laure Guillaume, Emmanuel Dardenne, Emmanuelle Santini and Anna, had been at Versailles from the beginning, twelve years ago. Laure, who started working at Zingaro as a teenager after a childhood of showjumping on ponies, told me, 'When you've been here twelve years you have a real rapport with your horse. And that sentiment is very important, because the more time you spend with the horse, the more you love it, the more you know it, the more you bring this emotion that is released in the spectacles, because there is love, and sentiment, there.'

Where the uniformed riders and matching horses of Vienna and Saumur preserve technical brilliance, Bartabas's circus in three rings of Zingaro, Académie and solo shows pursued a broader definition of art and creative expression. What Laure had sought at Versailles was 'a chance to transmit Bartabas's equestrian philosophy, of how to use a horse in a spectacle. Haute école can be very technical, and we respect technique above all, but we also try to evoke an emotion, a little artistic sensibility.'

Bartabas's écuyères do not look rigidly ahead like the soldiers at Saumur or the élèves at Vienna, but turn their heads to make eye contact with one another, and hold it like a ballerina 'spotting' her pirouette. They lean in their saddles and trace arabesques with their arms. All is expressive, and not regimented by technical templates. There was a fluidity even to the haute école movements: one day I watched Laure and one of her horses perform a cantered half pass, and it was like seeing a river that suddenly, smoothly, began to run diagonally against its own course.

'Three quarters of people who watch the show are not riders,' Laure went on. 'They don't know what a beautiful piaffe is – and although the show we produce here has an academic basis rather than a theme, it must still touch people. They need to come into our universe, perceive our sensibility and understand our need to transmit and engage.'

Emmanuel Dardenne, or Manu, had been born into classical riding. His father worked with the Portuguese maestro, Nuno Oliveira, and prepared horses for performances and for bullfighting in Carcassonne in the south of France. 'I don't analyse the philosophy behind it,' Manu told me. 'I work with the horses to give pleasure to the public, to work in harmony with a horse, and be as juste as possible. The rest is too complicated for me. They put me on a horse the day I was born, and so I grew up with horses.'

He had a Tiggerish intensity – part clown, part artist – and managed to ask me more questions than I could ask him. After working for his father, he had done military service with the Garde Républicaine, which provides mounted policemen and parade horses for the government in Paris, and then trained horses as a freelance. But he was looking for something closer to the old academies. He called Nuno Oliveira 'le sacre cavalier, my father speaks of him all the time. This type of horseman no longer exists in the south of France, now it's équitation sportif.'

Each *écuyère* or *écuyer* worked at least three regular horses, one from each troop. The Sorraias were the cygnets, demonstrating equitation *aux longues rênes*, a traditional way to start haute école, with the horseman controlling the horse by long reins as he walked behind its rump. The Criollos were used for mounted fencing and anything that involved dashing fluidity. The Lusitanos were the swans, and then there were soloists like Pollock, Le Caravage and Soulages. The *écuyers titulaires* – Manu, Emmanuelle and Laure – worked the young apprentice horses with the shaved manes who had been so eager to talk to me in the stables.

When I visited, the academy was on a tight timetable. The conductor Marc Minkowski had commissioned Bartabas to create a piece for Mozart Week in Salzburg the following January, to a little-known cantata called *Davide Penitente*. The creams and the Criollos would be involved, and the *écuyères* were to sing part of the chorus – it was snatches of this music that I had heard drifting from the rehearsal room across the Carrière. The venue was the Felsenreitschule or stone riding school carved from a cliff in the centre of Salzburg in 1693, when stone for the new cathedral was extracted, and the gallery which had once held spectators watching over the manège below would contain the singers and musicians for this performance. For the first time in over a century, horses would dance on the stage that had once been the manège floor.

The Grande Écurie under the Sun King had its own orchestra, which provided the accompaniment to the carrousels, reviews of troops and fanfares of all kinds and signals for hunting expeditions: a martial soundtrack driven by timpani and drums, overlaid with reedy trumpets, oboes and fifes, brassy horns and throaty bassoons. Like warhorses who were fed when the drums were beaten and guns fired, the horses of the Grande Écurie must have lived with that military music. But *Davide Penitente* was the music of the chamber and the opera house, brought into the stable. The creams learned it through stereo speakers rather than from live musicians.

In between interviews, I sat in on rehearsals as the riders not taking part leaned against the ribbed orange skirting board of the Grand

Manège, one foot propped flat against it, and talked endlessly. *Écuyères aspirantes* or *titulaires*, they had an easy, democratic camaraderie that must have been far removed from the status-conscious hierarchy of Louis's School of Pages. As well as their performance clothes, they had an off-duty uniform also designed by Dries Van Noten, and dressed like dancers in rehearsal: tight jodhpurs, soft, thin boots that clung to calves, aprons like brief skirts, layers of Zingaro sweaters that could be pulled off or on depending on whether one was warming down or warming up. The horses were in half dress against the damp cold too, with charcoal grey blankets with orange logos which were also slipped on or off, as needed.

The riders performed to a CD, with a video camera set up on a tripod on the seating ziggurat where I sat wrapped up in my coat, and they stopped frequently to chew over the finer points of the choreography. I loved the atmosphere of this place in which it was perfectly acceptable to spend hours staring at horses and discussing them. Red and white plastic tape had been used to measure out the dimensions of the Salzburg stage, which seemed much too small for twelve horses, even if they were the Criollos, which would perform a minuet of sorts with the Lusitano Uccello whom I'd met earlier. 'It's a matter of twenty centimetres,' Emmanuelle had commented at lunch as she leafed through sheaves of diagrams and notes that blocked out the choreography. Her curly head nodded as she counted out time to a recording of the music on her phone.

Uccello would be ridden by Manu and Emmanuelle, and he would also duet with Laure on a big cream called Quilate. Out of his stable, Uccello did indeed, as the head groom Philippe had promised me, puffing up his own chest and sticking out his elbows, look more impressive. 'We mainly train our own youngsters, but sometimes we have a chance to buy an older horse who is interesting like Uccello – he might be small but he has a piaffe that is quite brilliant,' Laure had told me, but the little cream stallion was proving problematic. His pas de deux was set to the duet 'Sorgi, o Signore, e spargi' (Arise, O Lord, And Scatter), in which two soprano voices soar against one another. It has a slightly sinister, out-of-kilter edge to it: it opens on a thirteen- rather than a twelve-bar introduction, and the sopranos leap as they beg the Lord to rise and destroy his enemies, reaching piercing notes that unsettle the human listener, let alone the horse. Uccello was afraid of it.

Horses can be cued to respond to music, whether it's gymkhana tunes or a hunting horn, but although Darwin and others have speculated whether horses naturally adjust their paces to a beat, not many experiments have been done concerning their musical preferences, despite the effort made to get them to reproduce their 'dance'. The Renaissance horseman Claude-François Ménestrier suggested that the muscles of men or animals were like strings on a lute, so that 'the one being touched, all the others vibrate'. Thus 'a similar excitation is made in these bodies at the sound of these instruments'. One should, he concluded, understand each horse as an individual in order to select the correct music for it.

One recent study recorded the response of eight stabled horses to different types of music: the horses liked Beethoven and country, but seemed outright disturbed by jazz. A still more bizarre experiment conducted in Iran involved an injured horse being played jazz for twelve hours a day. Another had to listen to classical music. The classical horse recovered in thirty days; the horse subjected to jazz got worse. Maybe the aria sounded like jazz to Uccello. Maybe Uccello preferred real jazz. Who knew? As Laure shrugged when she told me, it's always a mystery. Not even spending twelve hours a day with a horse will enlighten its rider on every count.

Manu and Emmanuelle had meanwhile brought Uccello into the Grand Manège alone to work to the aria. Manu rode the stallion, sitting lightly but firmly in the deep saddle, his reins loose even when they approached the dreaded corner that Uccello had decided contained all that was most terrifying in the music of Mozart. Pretending nothing was wrong at all, Manu circled Uccello at walk and trot, then canter, as the introduction boomed out of the speaker and the first soprano began, 'Riiiiiiise, Lord, and scatter your enemies ...' Manu asked him for a Spanish walk, which Uccello obliged with, crossing the centre of the school one extravagant strut at a time, concentrating intensely. Manu guided him to the bogey corner where he stood, ears pricked, neck arched, as Emmanuelle fed him a treat, and his ears never left the speaker.

But after a few minutes of this gentle routine, the energy that had been backed up in every muscle and channelled into collection became too much and Uccello half reared and made a little dash across the manège with Manu, only to be gently guided back, to stand, nostrils wide, and be fed another treat. Manu then rode him back across the manège at a brisk trot, taking the edge off his energy. Uccello's broad forehead almost crinkled in thought, although he did not dash for the door, or try to unship Manu.

Emmanuelle, her eyes on the cream stallion like a mother watching someone else take her child across an obstacle course, climbed up the seating to the stereo. 'Uccello was badly treated in the past,' she explained to me. 'Someone tried to dominate him and force him to do things, and that's not the way we work here. We need time to build a relationship so he trusts us in unfamiliar situations. Bartabas insists on him being in this performance, but he's still green.' Or, in French, 'He's not ripe yet.'

On bad days, Uccello associated not just the speakers and the corner with fear, but also the entire side wall – the wall which would, in Salzburg, be a cliff face filled with fifty singers, three soloists and an orchestra – moving, breathing, noisemaking musicians. Each time he passed the speakers, he curled one ear towards them; Emmanuelle turned the volume up, as the looped music, just a few minutes long, soared to the leaf chandeliers.

This time Manu slowed the pace and brought the stallion by in a more collected, stately trot. This was successful. On the next pass, the energy was coiled into a passage, and again, Uccello, a picture of focus, passed the speaker. The sopranos shrieked. Manu slowed the tempo again, and asked for a piaffe in the corner, only for Uccello to explode, spin and take off like a bulky deer in a series of leaps, nostrils

shaking deep up his skull. Manu barely moved in the saddle as the stallion bounded, one, two, three, four times, and, circling him, returned
him gently to the corner.

One slightly unstitched trot past the speaker, and within minutes
Manu was riding the stallion on the buckle, his voice as he talked to
the horse just about audible to me where I sat on the ziggurat seating
bank. After ten or fifteen minutes of careful work, a mollified Uccello
could Spanish walk around the terrifying corner. Later, Manu asked
me ironically, 'Tu aimais le sport, eh?' but none of that frustration was
visible in his riding. It can't be – horses, as all those writers of horsemanship manuals knew – learn whatever you teach them, whether
it's passage, fear, bad manners or a courbette. Uccello needed to be
returned to the thing he feared most, but also to feel the same reassur-
ing physical response from his rider each time. Manu couldn't match
his tension, only respond with sympathetic insistence: nothing is
wrong, concentrate on me.

Manu had told me that he came to the academy because he didn't
want to work in the production system he believed had taken over
from Oliveira and the true horsemen. 'You don't respect the horse
enough because you go too fast. Here we listen to the horses, you take
the time to teach them, to change their ideas, to work with them. It's
much closer to the culte of the cheval.'

Although the Salzburg performance was looming, there was
nothing technical in the choreography that Uccello could not do – it
was simply a matter of acclimatizing him to the music. Laure came
in on Quilate, a philosophical cream who stood calmly in the centre
of the manège as his rider pulled her bare hands into her jacket cuffs
against the cold. He reached down to scratch his cannon bone, and
when he was done, Laure picked up her reins and together they
began to move through the figures for the pas de deux with Uccello,
now ridden by Emmanuelle. Quilate was the voice of the first soprano, tracing circles and long curves, before being joined by Uccello,
the second singer, their interactions mimicking the play of the
voices. Uccello's nervous, rogue tension was soothed by Quilate's

steady presence. When the music parted them, Uccello grew more bashful, but he did not spook any more. The other *écuyers* came in to smoke cigarettes by the door and watch, adding interpretations and suggestions.

In the gloom of November, with the bulbs in the leaf chandeliers on a low setting and a weak light pouring in through the great doorway, the cream rounds of Uccello and Quilate's quarters glowed as they passed in and out of light and shade. The raw pine perfume of the seating was stronger than the scent of working horse. Each time the horses passed the plastic tape it shook as the ground and air was disturbed by their energy. When the music was lowered, the soundtrack became the jingling of the creams playing with their metal bits, the creak of saddles, quick percussive pats that the riders snuck on the deep, crinkled, arched necks, or the cluckings and kissings they made to the listening cream ears. The riders' legs were the only part of their bodies that seemed to move.

Dressage is a duet between tension and relaxation, and the curves of the figures traced in the sand were echoed in the curves of the horses themselves as they gathered their bulk and energy into collection: the back arched up slightly to support the rider, the rump and hind legs rounding under the barrel of the body in piaffe. The lower reins of the double bridles hung in loose semi-circles in mirror reflection of the horse's neck arch. Uccello's nostrils made another curve in tension and relaxation as he puffed, having had more work than Quilate. From this roundness Laure and Emmanuelle rose vertically. In good dressage, the seat becomes articulate, speaking to the horse, which appears to have little difficulty understanding the minute shifts of weight – even through a leather and wood saddle – creating that illusion that the rider controls the horse with the barely visible movement of *sprezzatura*.

Everyone seemed stuck on a moment when Uccello cantered in one steeper curve across the school to unite with Quilate. The connection was not being made somehow. There was a discussion over the complex joining-up move, and Manu and an *écuyère* 'galloped' it out on the sand on foot, watched intelligently by Uccello. On the next attempt, Emmanuelle and Uccello made the move correctly, although I couldn't have told you how much was rider and how much horse.

Riding finished, and the unmounted riders drifted away, but Manu, Emmanuelle and Laure were still discussing the choreography when I left. Uccello and Quilate stood with them in their charcoal rugs, at rest, although Uccello's ears were still curious, wondering at the humans, watching me as I slipped out of the main door into the Carrière. What did Uccello think about the things he was asked to do, the connections, the gentle pressure applied and released, as Xenophon had recommended? Does he dance for the sake of dancing, or for Manu and Emannuelle's praise? And what is praise, to a horse? Or music? *Ç'est un mystère.*

XII

The Carrière was lined with low candles burning thick wicks when I went back on the Saturday evening to see the academy's repertoire performance, *The Way of the Horseman*. The seats were sold out. The creaky backstairs I'd padded up and down alone for the previous three days were now full of people bundled in winter coats, making their way up to the warm orange foyer. In the Grand Ménage we sat side by side on the benches on little orange pads as if at picnic tables, waiting for our dinner.

The spectacle opened with three *écuyère*–archers on foot, carrying bows, dressed in long grey robes with leather guards over their chests, hair scraped back, suede gauntlets on their right hands. To the thump of drum and chime of bells, they ceremoniously took up their stance side-on to us, and fitted white-fletched arrows to their bows. The beat slowed the audience down to the pulse of the ritual movements. One after the other, they fired, and their arrows vanished into the darkness at the end of the manège. They bowed and filed out.

The rest was horses. To the stately measures of Bach, we followed Laure and Soulages the black Lusitano through the stages of collection, from walk to trot, and then canter, interspersed with duets, quadrilles and the frantic dashing of the Criollos as their masked riders charged the length of the manège at one another and clashed épées. The words of the Nuno Oliveira were read ceremoniously in the silences as Laure worked Soulages. 'Equitation is a school of abnegation and humility ... It makes men better ... Do not seek public success and the self-satisfaction after the applause, but instead the dialogue tête-à-tête with the horse.' The horses champed on their bits in accompaniment or commentary.

But two thirds of the way through this steady progression of minuets, the manège darkened completely. When the first pinpricks of light glinted in the leaves of the chandeliers, in the reddish half shadow below, the pale forms of the five cream apprentices with their shaven manes appeared, harnessed with nothing but thin tape halters. In the dark at the edges stood five écuyères in long orange skirts. They did not move. The unsettling strains of Stravinsky's The Rite of Spring began to unpeel with sinister discordance from the speakers, and the horses buckled at the knees, sinking like camels into the sand, and rolled and rolled. Their coats clagged with red dirt, obscuring the lines of their muscles, they went down, got up, shook themselves with legs straddled and stiff. They ignored the music. Their own inclination was not to piaffe, but to upend themselves, to wriggle and kick in the dirt.

Their shaved manes gave their heads and necks a strange, elongated and unhorselike look, stripped like dragons or lizards. Once they had rolled, they began to touch noses and leap apart, then nip at each other's necks and twine their heads around each other, lips raised to the sky. One would leap on contact and begin to run, bucking and farting at his pursuer, who set his ears flat back and came chasing after him, head low. Then the pursued would jerk to a halt on his forelegs, and spin to chase his tormenter. One, two, and then three cantered in a rapid circle, momentarily synchronized, and then broke apart, each heading in a different direction or breaking the beat altogether and kicking out with its hinds. Two of the creams groomed each other's necks.

Then, as the brass shrieked and leapt, and the violent strings and the thumping timpani of 'Glorification of the Chosen One' rang out, the five *écuyères* ran into the centre of the arena and began to whirl like dervishes, arms outstretched, orange skirts flying.

Something in the music or the whirling electrified the horses, and they gave up playing and began simultaneously to gallop furiously around the women, all in the same direction, sending the dirt flying, skidding by the low barrier in front of the seating ziggurat, their tails high and ears flat, leaning into the centrifugal force of the ring they had created, while the blur of skirts and galloping bodies and flying tails accelerated one another like a zoetrope. Just as it seemed they might come sailing over the barrier itself, the music cut, the *écuyères* dropped to the floor, and the creams, one by one, slowed to a walk, pricked their ears and became friendly again. They walked to the women, lowered their heads into their laps like unicorns, and were caught.

POWER

Hay is Biofuel

In places where machines can't go, here you will find him too, doing the work of fifty men, carrying his share of man's daily labour ... The only union he knows is with man ... He works with the precision of a machine. A machine that never breaks down.

Pathé Pictorial Looks at Man's Best Friend (1947)

I

The big grey adjusted his weight on his soup-plate hoofs as the planks of the flat-bottomed boat shifted under them. He had perhaps forgotten what it felt like to be on footing that did not sway and lift perpetually, and he no longer minded the sound of the canvas sail cracking and shifting above his head. For six months he had travelled in narrow stalls in the creaking hold of an East India Company ship, rising and falling for 11,000 nautical miles from England to Bombay, around the Cape of Good Hope, across the Indian Ocean. Then he had navigated the swirling waves of the Gulf of Kutch, and finally the wide, brown waters of the Indus. Now the ship plied upstream against the rushing meltwater of the Tibetan plateau as it flowed to the Arabian Sea. They had been caught in a storm so severe that it tore the sail; later, higher up the river, they had stuck fast on a sandbar. They had been fired upon at Oudh, before being welcomed by local troops.

Through all this the big grey had braced and swayed as the river wound through the flat plains of north-west India; he had grown familiar with the damp scent of the river mud thrown up over shoals, and snuffed at the stranger odours from the shore. At night came the soothing fragrance of the hemp and opium that rose from the hookahs of the sailors.

The big grey chewed steadily on the dried grasses in the manger before him, reassured by the four dapple grey mares who chewed alongside him. He listened to the sailors chanting as they hauled the ropes on the sails to the beat of a drum: 'Who has seen the world/The water is sweet./Pull at once,/the port is good.'

From time to time the raft was fanned over to the river's edge, and more grasses taken on, or diplomatic visits undertaken to local rulers who needed to be reassured that the British men on the craft came purely in peace. The countrymen crowded forward to see the big grey stallion and his placid harem, awestruck by the size of them. The local horses were light and small, with straight, undeveloped necks and fine fetlocks. The big grey, even on the short rations and vicissitudes of travel, weighed nearly 2,000 lb; his body, from muscled chest to solid rump, was the size of a hogshead; three spindly cow legs bound together made up one of his cannon bones; his mane was as thick as the pelt on a yak in winter; even his legs had a fringe of long white hair that shivered as he kicked at flies. When he and his mares bent over their food, their necks were like stone archways. Stowed on the deck by them was a crate containing an enormous gilded coach, ready to be drawn by these patient colossi. On up the Indus the horses progressed, stately as kings, as accepting as children, sweltering in the heat.

Occasionally the young Scotsman in charge of the expedition glanced their way, although it was the shore that really interested him: he assessed it, making notes and checking his bound copies of Arrian and Quintus Curtius. This was the territory of Alexander the Great, close to his near defeat at the Battle of the Hydaspes against the elephantry of King Porus, and this young man was his namesake:

Alexander Burnes. He was accompanied by two other British men, a local doctor and servants, but no British soldiers. Already at twenty-six a ten-year veteran of the East India Company, this handsome, arrogant young man was fluent in Hindi and Persian. He had wanted to make this journey for several years, but only now, in January 1831, was it possible at the command of the Governor General of India.

Their boat was making for Lahore on one of the tributaries of the Indus, where, tucked behind two walls and a moat, Ranjit Singh had been sole ruler for over thirty years since he became the first Maharajah of the Punjab. Pugnacious, regal and ruthless, Singh had gone on to carve out an empire for himself, and now King William IV of Britain wanted him as an ally to defend British India against a looming Russian threat north of Afghanistan.

Burnes's party knew that Singh was a horseman – he was reputed to have once launched a war against an Afghan dost to get his hands on a beautiful black Persian mare called Laila – and their gift to him was both a demonstration of generosity and of the British Empire's strength. They had chosen not English thoroughbreds, the most sought-after horse in Europe, but the big grey and his mares – dray horses born to haul barrels of beer from brewery to pub. This plebeian reality was, of course, not quite the spin the Governor General put on them in his accompanying letter, which nestled in a bag of cloth of gold among Burnes's things:

> The King, knowing that your Highness is in possession of the most beautiful horses of the most celebrated breeds of Asia, has thought that it might be agreeable to your Highness to possess some horses of the most remarkable breed of Europe; and, in the wish to gratify your Highness in this matter, has commanded me to select for your Highness some horses of the gigantic breed which is peculiar to England.

The people of Flanders might have quibbled with this, as they had contributed the large black horses that had fed into the branches of the big grey's family tree for hundreds of years, but they might also grudgingly admit that it was the new English farmers who had taken those horses and created bone-heavy prodigies, twice the size of the old Renaissance 'great horses', fed on luxuriant new crops and raised on lately drained turf. The big grey and his mares were British from the marrow out: destriers of abundance and trade, ready power for a century of unequalled economic expansion.

When the boat at last moored near Lahore, the big grey and his mares were led off one by one, stiffly testing their hoofs on the earth, casting about for what might be edible, or the scents of other, strange horses. Sirdar Lenu Sing, garlanded in emeralds and diamonds, met them on the bank with awe, and immediately demanded to see their paces, asking for details to convey to the Maharajah. Burnes wrote later:

> It was not without difficulty that I replied to the numerous questions regarding them; for they believed that the presents of the King of England must be extraordinary in every way; and for the first time, a dray horse was expected to gallop, canter, and perform all the evolutions of the most agile animal. Their astonishment reached its height when the feet of the horses were examined; and a particular request was made of me to permit the despatch of one of the shoes to Lahore, as it was found to weigh 100 rupees, or as much as the four shoes of a horse in this country.

The shoe went on ahead to the palace, and Burnes and his fellow officers proceeded with oriental formality to the walled city. They transferred to more flat-bottomed boats to travel further up the tributary, growing steadily closer to Lahore. At each stage they were showered with purses of thousands of rupees, with sweetmeats, fruits and extravagant lengths of cashmere cloth. Burnes caught his breath when he first glimpsed the Himalayan mountains in the distance.

They rode on elephant-back in silver howdahs, the dray horses plodding peacefully ahead of them, and the coach, drawn by men, at the head of the whole procession. The Maharajah greeted them with a salute of sixty guns firing twenty-one times. It had taken six months to reach him from Kutch. Short but formidably vigorous, scarred from small pox and missing one eye, he was, Burnes said, 'entirely free from pomp and show', and he was in a hurry to see the horses:

> The sight of the horses excited his utmost surprise and wonder, their size and colour pleased him: he said they were little elephants; and, as they passed singly before him, he called out to his different Sirdars and officers, who joined in his admiration.

The Maharajah's response to the Governor General was gracious and hyperbolic; it took up a 5-foot-long scroll housed in a silken bag with pearls on its drawstrings:

> There are in my stables valuable and high-bred horses from the different districts of Hindoostan, from Turkistan, and Persia; but none of them will bear comparison with those presented to me by the King through your Excellency; for these animals, in beauty, stature, and disposition, surpass the horses of every city and every country in the world. On beholding their shoes, the new moon turned pale with envy and nearly disappeared from the sky. Such horses, the eye of the sun has never before beheld in his course through the universe. Unable to bestow upon them in writing the praises that they merit, I am compelled to throw the reins on the neck of the steed of description, and relinquish the pursuit.

For days Burnes and his party were entertained as guests and presented with displays intended to overwhelm them. When Singh ceremoniously presented his troops and his stable to the British a few days later, Burnes was astonished to see the big grey – or was it one of the mares – decked out in cloth of gold, with an ornate howdah strapped to its back, transformed into a royal mount. It was too much for Burnes, after days of high ceremony and hospitality, feasts and delicacies, eye-watering drinks, bolts of cashmere, dancing girls, jewels – including the Koh-i-noor – and horses from Arabia and Persia. He smiled at the dray horse in its finery. Shoes finer than the new moon! Little elephants! Decked in gold! A workhorse in the Maharajah's stable, its hairy hoofs on the same floor as the delicate Turcoman horses who had harnesses fitted with real emeralds.

The Maharajah had not realized what the drays truly were: Trojan horses. They were not just buying diplomatic goodwill, but time for Burnes and his men to make a survey of the Indus in secret. By claiming that the horses would not have survived the long march overland, they had secured their passage on the Indus, and a chance to map it for future British interests. So Burnes smirked.

But in 1831, he was himself too immersed in a world where workhorses underwrote his everyday existence to understand that the Empire was powered by the drays and millions of their lighter, humbler cousins; that Britain owed its industrial and agricultural revolutions not just to its gentlemen engineers and labouring masses, but to the broad chests, tree-like legs and willing nature of its horses. Singh was right to say that his finest Turcoman riding horses were not equal to the mighty dapple-greys.*

Burnes left Lahore for Simla in August, continuing the exploratory mission to Bukhara that would make him famous in England, and bolstering a distinguished career that would end just ten years later in an uprising in Kabul when he was cut to pieces by the Afghans. What happened to the big grey and his mares is unclear. Did they survive the climate and thin fodder of the early nineteenth-century Punjab?

★   The dray horses of Lahore were not the only British work-horses to journey far from England as diplomatic gifts. A heavy horse called Jolly was among five horses presented to the Sultan of Constantinople in 1835. The sultan showed his delight by trying to press some baksheesh on the British consul, who discreetly passed it to Jolly's ostler. The sultan then commissioned two 5,000-guinea snuffboxes with his own portrait on the lid – one for the British king and one for the consul.

Were they bred into the population of local horses, to surface as bone and milky blazes in later generations? Or did they end their days in Lahore as exotics as extraordinary as snow leopards in London?

II

So let us return to London itself, and to the workings of which Burnes seemed so oblivious. This is London in no particular year, where bustles or crinolines might prevail, and top hats be squat or tall, but it is London under Victoria, after the railroad but before the first motor car growled its way on to the street. And it is a country that works to the rhythms of the horse, and whose inhabitants' lives are dominated by the availability of fodder, horseshoes and watering troughs.

Let us take a citizen of this century as our guide to the heart of England in this age of the horse. Our Londoner is a bachelor gentleman – not too rich, certainly not too poor – who has a modest but well-run house, one servant, a respectable source of income and, as yet, no horse of his own.

It is two in the morning, and he is sound asleep in his own bed. Shortly after midnight, he came home in a night cab drawn by a scrawny Irish mare who has known sweeter days, and whose owner dare not send her out on the street in daytime for fear of the driver being waylaid by distressed ladies in large hats. After delivering our guide to his door, the mare has sloped on to seek another fare, only part way into her shift for the night.

As our man sleeps on, the first of the vanners or carters are waking, with bellies rumbling ready for the mix of oats, maize, chopped hay, beans and bran that their yawning grooms will tip into their mangers. The vanners are all good solid colours: greys, bays, browns, chestnuts. They are solid in form, too, ranging from Brobdingnagian animals almost as large as the big grey of Lahore to lighter, faster creatures who nevertheless have thick legs that will drum on 25 miles of city streets

– wood, granite, asphalt – a day for four or five years before bone, ligament or tendon give way. 'What the trains bring the vans must take, what the vans bring the trains must take,' wrote W. J. Gordon, and thus the vanners' days begin early and are spent ferrying everything from pianos to peonies through the streets. In every town and every village in Britain, there are vanners shifting the 81 million tons of goods that the railways put in circulation each year, and the sundry cargoes of ships and barges on top of that. Mr Stephenson's Rocket and its descendants may have rendered the coach horse and the pack-horse train all but obsolete, but they have created a huge demand for work-horses, who also shunt and haul the engines and rolling stock to and from the sidings when the iron horses are indisposed. Several of the old coaching firms have simply transformed themselves into carriers for the new railways.

The vanners live by the hundreds in stables like the ones under Broad Street station, or Camden goods depot, where horses belonging to Pickfords and the vintner Walter Gilbey are quartered in airy, yellow-brick buildings – part barrack, part warehouse – and walk to work through tunnels under the railway lines that connect the road to the canals. By Paddington station there's a three-storey block where broad but steep ramps rise from a brick courtyard to the stables above, and the stables open on to walkways where the horses are groomed if light and weather allow.*

As the horses digest their breakfasts, the grooms fetch the heavy, straw-stuffed collars from the heated room where the previous day's sweat has been dried off and left them 'warm and comfortable as a clean pair of socks'. At 2 am the vanners will emerge – groomed with dandy brushes made of whalebone, hoofs cleaned of straw and dung and checked once more for nails or stones that could ruin them, harnesses adjusted so they do not pinch or gall – and throw their weight into the warm collars. The hardest pull is that which gets the carts in motion from a standstill, and the less the horses have to stop, the easier it is for them. They head to Covent Garden market with

---

★   Still visible as the Mint Wing of St Mary's Hospital, Paddington.

the farm produce, fresh fish and flowers that came in on the railway last night.

On the streets they meet the sweepers, horses drawing a contraption on which a series of brushes rotate on a conveyor belt to scour the road surface. 'Half a million pairs of wheels and a no less number of iron-shod horses' are 'constantly grinding [the roads] to powder', and this must be removed, along with nails, soot, detritus and leaves, all deposited in a sort of hopper on the back of the cart. Other horses draw carts with barrels of water on the back, sprinkling the road behind them to wash away the horse piss and keep the dust down.

If there's snow, then that too must be removed from as many thoroughfares as possible. Later, teams of boys or young women will wait at the street sides with brushes and scoops, darting among the traffic 'in a way that is nothing less than miraculous to the timid on the footway', to remove each dropping. Arthur Munby saw one '[ply] her daring broom under the wheels, which bespatter her with mire as they fly; she dodges under the horses' heads, and is ever ready to conduct the timid lady or nervous old gentleman through the perils of the crossing'. The droppings will later be carted out to the market gardens of the Home Counties in wagons along with the soiled straw and peat from the stables. What the boys miss turns to dry dust that gets into the nap of Londoners' coats, down their throats, in between their hall floorboards and smuts the nose of civic statuary.

At 3 am and 4 am respectively the vestry horse and the coal horse have dined on hay, straw chaff, clover and oats as they savour their soft peat beds before work begins. They are heavyweights, big as drays, 'stout bit[s] of timber' with 'a good pair of breeches', unflappable as they collect their loads. Our Londoner's servant will have a fire ready for him when he wakes because the coal horse will have already made his rounds. The coal itself was got out of the ground at a Welsh pit by squat, part-bred Shires called 'pitters', by Welsh cobs or by smaller ponies drawing tubs from the narrowest seams, their eyes protected with leather shields. Under the English Channel a mile off the Cornish coast ponies ferry copper and tin and sleep in fuggy stables below the sea floor at the Levant Mine. At the surface, more horses draw coal and iron ore along railways from the pit mouth to a depot. The kindling the servant is trying to ignite began its days as timber felled in Derbyshire forests and drawn to sawmills by more horses, who may have also paced the treadmill that drove the saw itself.

Yesterday's rubbish will have been taken away by the vestry horse, both the kitchen waste and bones, rags, dust, cinders, old metal and bottles. After sorting and processing have been done, another horse will cart the waste to the wharf for its journey out of London by barge. In the city's sewage works, horses have worked pumps and harrowed the slurry. When the waterworks were more primitive, they pumped and ferried water to fill private cisterns and took away the mire from cesspools and privies.

At 5 am the brewery Shires and Clydesdales are out on the streets, drawing 8 tons of dray, beer and barrels to replenish the cellars of London's pubs. They won't be out of harness till 7 pm in the evening. They rumble out while the last of the sad night cabs are wending their way home at the end of their shift, and pass the first of the costers' ponies and donkeys, who have picked up fruit and veg from Covent Garden and are dispersing it through the streets of London, to cooks and servants and householders.

The nation's farm horses have digested their breakfasts and stand as they are groomed and harnessed for a day of odd-jobbing about the farm. To whatever needs to be fetched or toted they will lend their massive shoulders. In London, wagons are heading to the docks, where barges of hay arrive on their journey to the mangers of the city's working horses, down the estuary from Essex, from land ploughed, harrowed, sown, cultivated, fertilized, mowed, reaped and 'tedded' or laid out to dry in the sun by machinery drawn by the farm horses. In the West Country, the horses that power treadmills or capstans which drive ferry boats over short crossings are already at work.

At 7 am our man is finally up. He has a lamp lit, filled with oil purchased from a horse cart, and a hearty breakfast, delivered last afternoon by the butcher's boy on a smart hackney that goes like the clappers, back and forth from the shop to customers with baskets of sausage on its flanks. His shirt has come back from the laundry with a horse that used to do six days' cabbing a week and is now down to three, plus this moonlighting for the laundry. His toast is made from bread baked in an oven heated with bricks of 'turf' made of bark ground by a horse turning a mill in Bermondsey.

He drinks his tea, brought from the docks to the tea merchant by vanners, and thinks of the day's work ahead of him, and what envelopes will have arrived on his desk. The Post Office horses have begun their dashing about town after breakfasting in their whitewashed stables on a muesli of hay, clover, beans and straw chopped and mixed by a steam engine. They are also skivvies to the railways, keeping up the rush of incoming and outgoing envelopes and the heavier migration of parcels in their scarlet vans, keeping commerce and families ticking over.

Breakfast done, our man must make his way to work, and if he chooses not to walk, he has three choices: the omnibus, the tram and the cab. By now, the cabs on the street are looking somewhat smarter – the horses are still fresh, not far into their forty-mile day – but they are still the poorest and most angular of the London herd. They last just three years on a six-day week, before slipping down a grade to working fewer days, serving Sundays only or doing the night shift like the scrawny Irish mare of last night. The cab men do not often own their horses, and they are hard pushed to cover the rent they have to pay for both horse and cab: passengers do not like to pay well for a cab, either, so margins are tight. The horses draw two-wheel hansom or four-wheel Clarence cabs at any pace from a walk in crammed streets to a rattling canter when they clear, and fare and driver are anxious to get to their destination. At the cab stand opposite our man's house, they stand with their heads by their knees, and he, thinking ruefully about a pamphlet thrust into his hand by a purposeful young lady last week, detailing the many horrors to which a cab horse was subjected – 'the unceasing sound of the lash in our streets' – decides to forgo adding to this particular horse's burden.

The tram is delayed – notwithstanding the best efforts of the vestry men, the rail is clogged with mud, and the 5.5 tons of tram weigh more to the trammer horse and his partner as they stop and start repeatedly to pick up or set down passengers. They have 13 miles to cover which, though it will be only a matter of hours before they are replaced by a new team, will wear them down faster than any other horse but the cab horse. Four years and they break down, either in harness or after reaching their stall.

So it's the omnibus for our man, pulled by the same awkward-looking, boxy beasts as the tram. Not as light as a cab horse, neither Shire nor cob but somewhere in between. Too heavy-boned to be fine, and too light in the torso to be a dray. Their manes are hogged and their tails docked to keep them clear of the harness and stop them flicking dung into the eyes of other horses or passers-by. They are mares, resting most of their day in stables where they stand between hanging partitions – their collars hanging on hooks overhead – with perhaps a thousand other comrades, with whom they squabble and dispute. They are well kept, with a foreman who knows every one by sight, but two out of three will die on the roads, and they last just five years. Our man hails the omnibus, not thinking that by demanding that it stop he forces the horses to make once more a taxing plunge into their harness, to lift the deadweight of all thirty-nine passengers and the omnibus itself.

So the bussers sigh as their driver flicks them with his whip, and lean once more into their collars. The streets of this age of the horse are hazardous for both animals, like the bussers, and men. Horses are always being injured or killed when coaches or carts hit them, or they themselves spook and bolt still drawing their wagons, and flatten passers-by. They get run through by carriage shafts, or slip on treacherous footing and are unable to rise. In 1834 one bolting dray horse belonging to Elliots of Pimlico fell into a pit near the Treasury in Whitehall and an entire scaffold had to be built to extract it. Another poor cab horse was so startled by the sound of a mechanical 'monster organ' playing on Judd Street that it fled, leapt off the road through a metal railing and plunged into the area of a house on Hunter Street, pulling his cab after him. Four dray horses were needed to pull both out. These horse-filled streets are more dangerous than those of the future, in which the motor car will prevail: there are more accidents and more fatalities for pedestrians, perhaps at least partly because there is barely any traffic management or rules.

The omnibus makes its way carefully among the throng of horses and men, the vanners, the brewers' drays, the coal horses, the costers' donkeys, the vestry horses, the dashing greys of the Fire Brigade, the hearses drawn by glossy black Flanders stallions that know every road to London's cemeteries 'like a book', the doctors' gigs, young horses fresh from the country and being habituated to the sights and sounds of the metropolis by the jobmasters, who will later hire them out, the carriage horses, the riding horses, the trammers, the Post Office teams of four drawing the parcel van, the weary cab horses drawing the hansoms and the four-wheel growlers, the ambulances for humans and for horses, the knacker's wagon – and its later shadow, the cat's meat van.

III

In this city and nation of horses, a poor man or a wealthy man alike marked his progress in life by his capacity to own his own horse, whether it was a pony that lived in a lean-to stable in the yard behind his terraced house or a set of matching blood carriage horses in the mews. People lived alongside horses, in constant contemplation of them, sharing their work and making or breaking their days by them. They had to keep them fit, and if not fully fit, then functional. Their eye maintained the worth of the horse, and the labour it could do. For want of a nail – or because of a stray nail on the road – much could be lost.

There was an entire geography from mews, coach houses, jobmasters' stables, fodder merchants, knackers, harness makers, smiths, wheelwrights, whip makers, loriners, rat catchers, tanners, makers of straw sun hats for horses, strappers who specialized in vigorous grooming, ostlers at inns, riding masters, horsebreakers and common-or-garden grooms. In the same way that the horse left its mark on the titles of the aristocracy of Europe, the *Ritters*, *Cabelleros* and *Chevaliers*, so it marked the surnames of the ordinary man: the Smiths, Grooms, Carters, Wheelers, Wainwrights, Cartwrights, Jaggers,★ Packers and Sadlers.

The journalist Henry Mayhew said that in the mews, the horsemen and their families 'live as much together as the Jews in their compulsory quarters in Rome'. The wandering Cheap Johns who sold gimcrack goods, flimsy knives, nutmeg graters and teapots from horse carts, rinsed out and refilled their horses' feed buckets and washed their own faces in them. Horsemen even risked sharing the same illnesses as their working partners: the respiratory disease glanders was relatively common in horses, and could kill a man in days after a nasty bout of septicaemia.

Horsemen in town and country had their own lore and knowledge and guilds, distinct from those of the men who read Baucher's equitation manuals or pored over the latest veterinary texts. Much of this has inevitably been lost like those 3,000 years of pre-Xenophon teaching, as it was also transmitted by word of mouth and example rather than by paper, and when, long and deep into the twentieth century, the groom, the driver and the ploughman almost vanished, only dedicated oral historians like George Ewart Evans, Bryan Holden and Derek Hollows recorded this world.

There was also tremendous pride and skill involved in the care and handling of horses, and this was appreciated by the same bystanders who rushed to Astley's and Ducrow's performances to admire the horsemanship on show. Farmers in East Anglia pitted their four-square, red Punches against one another in pulling contests, wagering on the

★ Another name for a carter.

result. William Youatt wrote of the county's pride in its 'chesnut' [sic] drays: 'The Suffolk ... would tug at a dead pull until he dropped. It was beautiful to see a team of true Suffolks, at a signal from the driver, and without the whip, [go] down on their knees in a moment, and drag everything before them.' At agricultural shows there were ploughing matches where teams of horses moved along fields with a precision belied by their massive, iron-tipped feet. Miners organized pitmen's derbies where huge crowds turned out to see the lads from competing mines race their ponies bareback around improvised courses, the ponies spruced and groomed to an inch, with bells hanging on their bridles. At city shows like Olympia, costers' donkeys and cab horses alike were polished to appear before the King and the Kaiser in the 'Best and Neatest Licensed Hansom Cab' or 'Best and Neatest Costers' Donkey' classes. The donkeys had wired carnations blooming from their browbands, and ribbons in big bows between their ears.

In 1868, William J. Miles reported seeing a brewery team working in London dropping barrels into a cellar doorway. Without cues from their handler, one horse did the work of 'rais[ing] the butts, and returned and lowered the rope', until everything was stowed away. He then walked over and took his place in formation with the other horses on his team, joined by another who had been pottering loose about the yard.

When the horses were all hitched, Miles followed them out into the streets of London, and was astonished that their handler 'never touched them: between carriages where there hardly seemed room enough to squeeze through, he went without touching, and this, too, by merely waving a bit of whipcord at the end of a long black rod'. He was sure, Miles wrote, that the man must be 'some necromancer waving an enchanted rod'. Another man, selling sprats from a handcart, told Mayhew, 'If I can get anythink to do among horses in the country, I'll never come back.'

The horses themselves were sometimes indulged in their idiosyncracies, especially at the highest end of the working hierarchy. The Henry Meux and Company brewery had a 'very fine horse' which was allowed to come and go from its stable as it pleased, roaming the yard. The brewers also kept pigs, which the horse appeared to dislike, as they ate its fodder. When the horse was given its daily corn it would carry it to a water trough, drop it on the ground and wait for the pigs to come. When the hogs arrived, snuffling hungrily, the horse would seize them by the tail and dunk them in the trough before 'caper[ing] about the yard, seemingly delighted with the frolic', and returning in satisfaction to its stable.

Up the east coast from Kent to Aberdeen reached the Society of Horsemen, who were grooms, drivers, ploughmen and wagoners. Their shared code and knowledge included a variety of magick, charms and recipes for assorted horse ailments, and East Anglian members took a ferocious oath not to share this know-how with 'fool' nor 'madman' nor 'drunkard' nor 'any one who would abuse his master's horses' nor 'any woman at all'. Brothers in the society were sworn to go to another's aid 'within the bounds of three miles except I can give a lawful excuse such as my wife in childbed or my mares in foaling or myself in bad health or in my master's employment'.

Their oath was appropriately equine: 'I to my heart wish and desire that my throat may be cut from ear to ear with a horseman's knife and my body torn to pieces between wild horses and blown by the four winds of heaven to the utmost parts of the earth, my heart torn from my left breast and its blood wrung out and buried in the sands of the sea shore where the sea ebbs and flows thrice every twenty-four hours that my remembrance may be no more heard among true and faithful brethren. So help me God to keep this solom [sic] obligation.'

Horsemen, despite the increasing intrusion of the scientifically inclined veterinarian, kept their own recipes for tonics, draughts and 'balls' – boluses for dosing horses, made with mixtures of powdered spices like liquorice, turmeric, hellebore, cumin or pepper dust. The foreman at one omnibus yard prescribed 'a pint and a half of ale for one horse, mustard for another, a blister for another, poultices for two or three', and 'a drop of whisky for the roan at the far end'. One twentieth-century horseman of the old school kept a notebook of recipes, including one for managing 'a Vicious Horse so as to do anything with him', which included tincture of opium and lunas powder, to be rubbed on the handler's whip and hands and over the horse's nostrils.

The horsemen also had their own methods for horsebreaking, despite the interventions of masters like Baucher and showmen like John Solomon Rarey, who came all the way from Ohio to demonstrate his 'Modern Art of Taming Wild Horses' by which, according to one contemporary, 'ladies of high rank have in the course of ten minutes perfectly subdued and reduced to deathlike calmness fiery blood-horses'. Though Rarey was popular among the drawing-room swells and military men, British farmers and circus horsemen were not fooled, recognizing both his prescription of patience and his trick of strapping up the forelegs of the most troublesome horses to force them to submit.

If there's little recorded for posterity directly by the working horsemen themselves, then there is much from their employers. When the middle classes or aristocracy write their guides to horsekeeping, they are in a state of constant alertness against the swindling groom who might sell their good fodder for poor, or pinch skin when tightening a harness, causing galls. Irish equestrienne Nannie Power O'Donoghue had a dim view of the people she employed to look after her riding horses, for:

> A groom is too often utterly careless … A loose shoe is nothing to him: it does not cause him any inconvenience, not it; then why worry himself? He does not want to bring the horse down to the forge through mud and rain, and stand there awaiting the smith's convenience; not a bit of it. He is much more comfortable lolling against the stable-door and smoking a pipe with Tom, Dick, or Harry.

Authors John Stewart and Anthony Benezet Allen explained that an untrained boy learned by doing and by observing those better than himself, but true expertise to a standard they considered appropriate took years: 'If the boy has any desire to learn, or if any desire can be excited, let him see the stable and the stable-work of a good groom. Show him the horse's skin, how beautiful and pure it is, the stable, how clean and orderly, and the bed, how neatly and comfortably it is made ... In the ordinary course of things the boy may become an expert groom in four or five years.'

Horse dealers were the most distrusted, for they had a grab bag of tricks and deceptions that could hide every fault. 'An Amateur', writing in the Sporting Magazine in September 1839, explained that while many dealers in London 'have as good a character for honesty and fair dealing as men in any other business', there are also 'low dealers' who will hit a horse 'where wheals do not readily shew themselves' so that 'the animal rushes out of the stable, his tail on end, his nostrils dilated, and looking altogether exceedingly plucky – alias extremely frightened'.

Dealers stood horses on a patch of higher ground in the yard to make them seem taller, cracked the whip as they were trotted out so that they leapt and lameness was concealed, and promised that the poor animal 'did fourteen miles within the hour no longer ago than yesterday, with his knee up to his chin ... He's worth a hundred – he is indeed'.

'Such is the force of a horse dealer's eloquence,' said An Amateur, 'such is the easy flow and vividness of his descriptions, and such the beauty and captivating aptness of his similes, that, great as is the fame of Cicero and Demosthenes, were they alive at this day, and to do the utmost their fancy could suggest in praise of a horse, the flowers of their oratory would seem withered and faded when compared to the bright colours in which the lowest of our English copers and horse-chaunters portray their imaginings.'

Teeth were 'bishopped' by carving and burning them so that the horse looked older (no one wanted a horse under five) or younger. Grey hairs were dyed, and horses that had skimpy tails had special wigs: 'Go into the stables at night,' wrote one commentator, 'and then you will perceive the tails hanging on the wall, and the noble beasts, for coolness sake, munching their oats with the stumps of their tails smoothly shaved, almost as smoothly as her ladyship's head, as we perceive when her maid removes the elegantly arranged wig, and hangs up the chignon of such portentous size for the night.'

The hollows that deepen over the eyes of old horses were incised, then inflated with air blown in through a hollow needle or straw, and the puncture hole closed. This rejuvenation lasted long enough to sell the horse, whereupon the hollow cratered again. A bad foot could be numbed with boiling turpentine. An unswollen, unspavined hock beaten till it matched the injured one. Raw, peeled ginger root was crammed under a horse's tail to make it prance. Other horses had their sheaths stuffed with oakum to stop them making indelicate noises when they trotted, or had their anuses cut so that their farts were muted. Small sponges went into the nostrils of a 'roarer' to cover the fault.

Cruelty was not confined to the dealers. As John Stewart and Anthony Benezet Allen explained, many a man was merely getting by, with no real affection for his equine partner. Some working horsemen beat their horses, overloaded them or tasked them with impossible slopes and burdens. Under pressure to earn money themselves, they did not always have the means to be compassionate, and a difficult or reluctant horse could cost them dear. Others were reasonable horsemen, but did not make friends of the creatures they worked with: 'I never was and never will be happy with horses,' one railway carter told historian Bryan Holden of working with horses in the late 1940s. A horse could crush your foot with his own, knock you down, take off down a steep slope with a heavy load before the brake could be thrown, kick, bite or refuse to move. This was not always forgiven, as in Frank Buckland's story of 1850:

A cab-horse in a four-wheeler at the stand by Palace Yard; the cab horse fast asleep with his head hanging down; the cabman also fast asleep, leaning against the horse. The cab horse (I suppose in his sleep) kicks the man; the cabman instantly wakes, and kicks the cab-horse in return. 'You brute,' he said, 'I can kick as well as you can,' and then they went on at it again – the cab-horse and the cabman kick for kick for a minute or so, till at last I settled the dispute by saying, 'Hi! cabby, four-wheeler, go to so-and-so.' 'Right you are,' said the man, and we all three trundled off, as though nothing particular had happened.

Others did their best but were beholden to the horses' owners, the cab firms, the railway companies or the tradesmen. The cab horse was notorious in literature as the most abused beast of the streets, its ribs wretched, its head hanging down to its knees as it rested, but there was sympathy also for his driver, who, according to Anna Sewell's shabby cabman Seedy Sam in Black Beauty, complained bitterly of the terms: 'If a man has to pay eighteen shillings a day for the use of a cab and two horses ... 'tis more than hard work; nine shillings a day to get out of each horse, before you begin to get your own living ... And if the horses don't work we must starve.' His friend adds, 'It is desperate hard, and if a man sometimes does what is wrong, it is no wonder; and if he gets a dram too much, who's to blow him up?'

Horses were fellow labourers as well as mute machines. The horse drawing the reaper had replaced the team of farm workers with scythes. Men, women and horses drew mine carts, and there was only a small leap between the coster who put himself in the traces of his handcart and the one who was able to own a small donkey to do the job instead. Affection, as well as pride, was not unknown. 'He never came home from work until Tabby [his horse] was comfortable for the night,' recalled the niece of one carter at Coventry station goods depot. Another man there would 'bike out into the country, go down in the hedgerow, pull out my knife ... And I'd cut a big, juicy faggot of dandelions. I'd stand by the stable door and shout "CHARLIE! Charlie! Come on, boy!" Up would go his old head, his ears pricked. He knew my voice like a child knows its father's'.

The *Illustrated Police News* was amused by a lad working for the Great Central Railway Company who took Turkish delight a colleague had stolen from his delivery van and fed it to his horse, which he had decorated with a red, white and blue rosette. In the mines, working conditions, dangerous, sweaty, often pitch-dark, were shared between horse and man. One miner in Northumberland during the Second World War called his pony his '*marras*' – a Norse word meaning 'friend': 'They were good company, responsive, and at times, loving ... most of us loved our ponies and treated them as pets. A pit pony would get you back to the stables in the dark. You just held on to its tail or mane and kept walking.' The drivers stole apples to feed them, and once the lads defied their foreman and refused to put the smallest, most overworked pony into harness for a day to give him a break, even though they risked 'stoppage of pay or a dip in the horse trough'.

The horse itself was a byword for human labour. One workman told Henry Mayhew: 'We are used for all the world like cab or omnibus horses. Directly they've had all the work out of us, we are turned off, and I am sure, after my day's work is over, my feelings must be very much the same as one of the London cab horses.' General Booth of the Salvation Army compared Britain's poor to a fallen cab horse in his 'Cab Horse's Charter': 'Every Cab Horse in London has three things; a shelter for the night, food for its stomach, and work allotted to it by which it can earn its corn. These are the two points of the "Cab Horse's Charter". When he is down he is helped up, and while he lives he has food, shelter and work. That, although a humble standard, is at present absolutely unattainable by millions – literally by millions – of our fellow-men and women in this country.' His fellow humanitarians, Booth argued, would not question why a horse came to fall before they got it to its feet, and they should not do so with a fallen human, either.

The reformers who pressed for legislation to improve the lot of the horse also saw a blurring between horse and man, and they recognized that the conditions of working men must be improved in order to help the horse. Many of them had cut their teeth as anti-slavery campaigners, moving on to animal welfare when emancipation was achieved. The *John Bull* magazine complained of the Quaker brewer and founder of the RSPCA Thomas Fowell Buxton that 'Mr Buxton began his career by questioning the title of the proprietor to the slave; which is very much like our doubting he has a right to his own dray-horse'. Anna Sewell's 'autobiography of a horse' was called in America 'the Uncle Tom's Cabin of the Horse'.*

Prizes were given by humanitarian societies eager to promote the good care of animals at the May Day parades in British cities, where work-horses were adorned with flowers, feathers, long raffia tassels, coloured paper darts, ribbons and brasses, hoops of blossoms rising from the hames of their collars like the frames for Catholic icons. In the 1920s, the Pit Pony Protection Society issued pamphlets and began a letter-writing campaign to improve working conditions of animals for whom they believed, as one Miss Gardner wrote to the *Spectator*, 'Life ... is a series of brutal despotic events.'

The Horse Accident Prevention Society campaigned for better road surfaces for horses. The RSPCA provided public troughs, and sought to educate workers on correct harnessing, balancing of loads and handling of horses. The RSPCA paid subscriptions to purchase 'chain' horses that were kept at the bottom of the steepest city hills, and could be hitched alongside the vanners, drays or vestry horses to help them draw their loads up the gradient.

One police constable was responsible for over a thousand convictions for cruelty to horses, and on his retirement a grateful ode was penned on their behalf:

> 'Tis you and only such as you
> Who mark the mute appeal,

---

* Even more inappropriate comparisons between horses and slaves were made by pro-slavery advocates in the eighteenth century, who joked about using horse dealers' tricks to disguise elderly slaves.

Of us poor helpless quadrupeds
When indisposed we feel.

...

So horses, mules, and asses, too,
Their wishes to you give
By neighing 'Honhy, honhy, hon!'
Which means 'Long may you live.'

IV

This high industrial age under Victoria was built on the foundations of another, earlier one, which was the first to bring horses into mass use. It was also the era when the horse itself was developed into the formidable engine of the big grey of Lahore.

Animal power had been put to work since the Bronze Age: oxen and equids replaced man as they dragged weights or stones round and round ring-shaped stone bases, crushing grain, ore or fruits. They turned drums on to which spooled rope or chain, lifted loads from wells or mineshafts, or up slopes to terraces, hauled stone and heavy cargoes on and off wagons or ships. The horse was faster than the ox, and sweated more easily to cool itself than did cattle – and as horses became more affordable, so they increasingly replaced their bovine cousins.

England had begun its long ride to prosperity in the early modern period, and that meant more and progressively cheaper horses. Most were used for riding and for agricultural work, but gradually the coach and cart became more widely used. Since the early 1600s, 'horse gins' or 'engines' had been developed, in which a horse or a team of horses circled in harness, turning a gear mechanism that rotated the drum on which rope twisted, raising coal from deep mines or water from wells.

When wooden ploughs were replaced with more efficient metal contraptions, the ox fell by the wayside, too slow to take advantage of the new ploughs. The uses for horses multiplied: riding, draught, ploughing, milling, mining. This gathering need for horses that had pulling heft rather than agility resulted in increasingly heavy, broad-chested, thick-legged horses. The humanist writer Thomas Blundeville described the carthorses of the sixteenth century as 'great of stature, deepe ribbed, side bellied, and have strong legs, and good hoofes: and therewith will stoope to his worke, and laie sure hold on the ground with his feete, and stoutlie pull at a pink, it maketh no matter how fowle or evil favoure he be'. They were descended from the German, Flemish or Frisian horses brought from northern Europe to carry knights – larger, 'cold-blooded' animals less graceful than the Spanish hot-blood southerners – who had been bred to whatever local stock there was, be it so-and-so's good horse, be it a pony, a Barb or another carthorse, to produce serviceable animals. When later generations of wealthy men brought more great black horses from the continent, they were put in no particularly organized fashion to these 'deepe ribbed' carthorses. The resulting 'Blacks' were not yet the Maharajah's 'horses of superior quality, of singular beauty, of alpine form, and elephantine stature', but the soil was literally being prepared for the big grey of Lahore with the beginnings of the British agricultural revolution.

The countryside was slowly transformed: enclosure meant that larger areas of land were farmed by fewer people, who produced more crops to feed a growing national market. Pasture was turned into arable, and fenland drained into farmland. Fields that had traditionally lain fallow for long periods were now seeded with clover that drew nitrogen back into exhausted soil, and with deep-rooted turnips that restored minerals. Both clover and turnips were fed to livestock when grazing was poor, and the old practice of slaughtering much of the herd before winter fell out of use.

British animals began to wax on this new regimen, and their owners, hoping to make four hoofs yield higher returns for each square inch of grass they trod, began to tinker with their shape by breeding them more selectively. In agricultural illustrations, pigs' faces sank into stacked ruffs of flesh, their tiny trotters overwhelmed by belly; cows resembled barns with pin-sized, horned heads and dewlaps round their knees; sheep were block-shaped, as though ready to be roasted whole and carved directly into efficient cuboid slices, their wool superabundant.

In cattle, pigs and sheep, this profit was measured in sales of meat, hide and wool, but horses, too, underwent improvement for the new economy, which increasingly needed them for draught and farm work. In illustrations, the 'improved horse' is contrasted with its ignoble, somewhat coarse-looking peers. It has a finer head, clean legs, a broader, deeper chest and a neck that rises from that chest almost vertically. Its frame was bred both to haul forward and to thrust its weight back on to solid hindquarters to slow its load on stopping.

This improved horse was exemplified in the 'Bakewell Black' or Shire, as it later came to be known, partly because of one man, a Leicestershire farmer called Robert Bakewell, for whom, as an American agricultural magazine put it in 1842, farm animals were 'as wax in his hands, out of which in good time he could mold [sic] any form that he desired to create'. He bought new mares from Holland and Flanders and bred them to the finest local Blacks he could buy, then bred inwards by putting 'the best to the best' – sires to their foals and dams to their colts, brothers to sisters. He then 'line bred', bringing together closely related cousins.

In a few generations, the ambitious Bakewell had fashioned mighty, patient golems from his drained and irrigated farmland. His stallion, 'K', was described by one admirer as 'the fancied warhorse of German painters' – a reference to Dürer's famous, feather-heeled 'The Great Horse'. He selected only the animals that appeared to pass on the most desirable traits consistently, thus replicating horse after horse.

Bakewell wasn't the only 'improver' – the men of Suffolk had long been busy creating their squat red Punches – but he was the most famous. His model farm at Dishley was visited by royalty and aristocracy from as far afield as Russia, keen to see his Longhorn cattle, Leicester sheep and Blacks. On his walls he hung pickled joints of meat and the skulls of beasts he'd bred: all was measurable, empirical and on display. While his tangible influence has since been disputed, his self-publicity and the enthusiasm of his acolytes – William Pitt wrote that he had a 'restless and aspiring genius' – were the spirit of the new and coming age, bourgeois, not aristocratic, engaged in commerce, not war. Its sons came of age in an era when the nobility had long since lost their control over the economy, ceding to the market at large, and when the individual had plenty of incentives to pad their own nest through innovation and graft. They might not caper and piaffe on Spanish horses, or own as many light-framed racehorses as the aristocrats, but they buttressed their commercial prosperity with the behemoth rumps of Shires and Punches.

When Bakewell hired out his bulls, rams and stallions rather than selling their progeny, he was, as historian Harriet Ritvo perfectly explains, turning genetic potential into money. In the past breeders who sought to improve their herds mated well-built animals to other well-built animals with little regard for their pedigree, and sold as many animals as they could for meat or profit. Bakewell and his fellow agricultural improvers instead created a kind of biological patent intended to deliver what Ritvo describes as 'a template for the continued production of animals of a special type', or 'genetic capital'. The improvers took horses and cattle 'from the natural realm into the realm of technology', as Ritvo puts it, and dubbed themselves creators achieving the feats of Nature or God – all for the sake of honest profit.

By the time the big grey was drifting towards Lahore, the patented English dray horse was considered – by the English at least – to be superior to any working breed in Europe. The descendants of Bakewell's improved horses were a kind of base metal that could be mixed with others to make tough alloys: a squat but heavy 'pitter' for minework; a lighter vanner whose legs were reinforced with Shire bone for hammering on city streets; a jack-of-all-trades carthorse that didn't eat its weight in oats. In Scotland, Bakewell's black horses contributed to the foundation of the Clydesdale breed. British heavy horses were imported to the continent to boost local heavy draughts. Charles Hamilton Smith even reported in 1841 that the Emperor of Morocco had years previously purchased a 'giant black stallion sent from England ... above eighteen hands high ... which was exhibited in London' and had used it to breed 'several splendid black horses ... which were the wonder of his countrymen'.

The uses for the work-horse went on increasing: in tandem with the agricultural revolution ran an industrial one, and the basic horse 'gin' or engine was improved and elaborated to do everything from sawing wood and shelling corn to making cloth. The father of the industrial revolution, Richard Arkwright, had nine horses providing the power that turned a thousand spindles in his first factory, and horses provided the muscles for every stage of wool cloth production from 'scribbling' or combing the rough fleece to carding it, shearing, 'gigging' or lifting the nap of the wool, beating and cleaning the finished cloth.

Horses also lent their strength to the cement mixer of the time, the 'pug mill', an essential part of the brickmaking process. They plodded steadily round and round in stone 'gin-gang' buildings or stalked treadmills, turning gears that drove shafts and the drive belt of the threshing machine that chopped and beat and shook the wheat into straw and grain and chaff. They powered the balers that bound the straw, and turned the mills that ground the grain into flour. In the late seventeenth century the traveller Celia Fiennes passed through Flint side-saddle and saw 'great wheeles that are turned wth horses yt draw up the water and so draine the Mines wch would Else be over flowed so as they Could not dig the Coale'.

Better transport was needed for these goods than the old wagon and packhorses caravans on the mired, rutted roads, but horses provided that, too, as the force behind both the construction and the use of the faster, cheaper and more efficient new canal network that spread across the country, multiplying the navigable waterways and prefiguring the railways. Two canal horses – which were not at first mighty Shires but lighter draughts, or even ponies or donkeys – could draw fifty to eighty short tons – fifty to eighty times what two wagon horses could manage, and 400 times the load of two packhorses. When road construction began in earnest in the late eighteenth century, horses dragged planks that smoothed the roads, and hauled in the rocks and gravel for better footing. Houses, factories and industries were built by horses, and a new landscape – the canals, the turnpike roads and coaching inns – cut out for them. Even the iron tracks of the railway itself were originally created by and for horses. Moving coal from mine to wharf in a constant parade of 2-ton wagons churned up the ground: wood, stone, and eventually iron, were laid over the roads so the cartwheels met less resistance.

Steam power had begun to supplant the horse gin in mine work in 1712, but steam was both dependent on horse power – horses still went down mineshafts to haul back raw materials – and approached from an hippocentric perspective. When James Watt set out to demonstrate the power of his new steam engine in terms that would resonate with the public, he consulted men who worked with horses. In a note from 1782, he writes 'Mr. Wriggley ... says a mill-horse walks in 24 feet diar and makes 2½ turns per minute ... say at the rate of 180lb p. horse.' By his calculations, Wriggley's horse generated 32,400 foot-pounds per minute.*

When Watt began selling steam engines, he approached the breweries. Cooke and Co were the first to purchase one for £200. It replaced four horses. By 1810, Whitbread in London had another which 'pumps the water, wort, and beer, grinds the malt, stirs the mash tubs, and raises the casks out of cellars. It is able to do the work of 70 horses'. Nature was improved upon once more by man, although as steam replaced the horse gin, still more horses were needed, labouring to keep pace with the demands of the railway and the steam-powered factories. And so we reach the scrum of our London gentleman's horse-powered Britain, with its vanners, bussers, cabbers, pitmen's

---

★   Later, more accurate estimations show real horses can only achieve 0.6 of a single unit of Watt's horsepower.

horses, farm horses, cab horses, costers' donkeys, trammers, drays, ferry and railway horses, all leaning their weight into their collars and drawing the nation along.

By 1871, there were as many horses in the city as in the countryside, and by 1901, urban horses outnumbered rural by two thirds to one third. Working horses eventually consumed almost exactly the amount of grain and hay produced by British farmers. Even when steam replacements began to emerge for the horse, no one could believe that it would mean an end. One inventor designed a steam tram whose engine was disguised by a horse's head, neck and chest, with a pilot plough at its feet, a bell between its ears and a booth for the driver where its back should be. It was intended, he said, 'not to frighten the horses', which he clearly could not imagine disappearing from the streets.

But once again, a new source of power accelerated the economy, and horses found themselves outpaced for the final time. In tumbling, clattering acceleration, just as the steady oxen were replaced by the circling horses which were replaced by the hissing, thirsty steam engine, now the steam engine was replaced by the invisible transmission of electricity and the explosive power of the internal combustion engine, and coal, hay and oats with the golden, greasy gush of petrol.

V

In the end it took, as historian Ann Norton Greene puts it, 'three kinds of power to replace horses as prime movers: steam engines for long-distance hauling, electricity for mass transit, [and] automotive power'. The trammers were replaced with electric streetcars in the 1890s. Electrical tramways were cheaper than horses, and both they and automobiles were easier to manage with relatively little training. They did not spook, hurl themselves into basement areas, kick, get ill or go lame. They were faster, and though automobile emissions were

noxious, they did not require scraping off the road surface and dispos-
ing of – instead they dispersed into the atmosphere.

Many feared the new technology. 'Should a cab-horse run away,
or two omnibuses come into collision – there you are,' said the York
Herald, 'but if the auto-mobiles run into each other, where are you?
An explosion and general conflagration, and a deluge of "something humorous, like boiling oil", such, for instance, as occurred in
the tramways accident at Glasgow are the least disagreeable results
which might be reasonably expected.' Others saw a bright side for
horses, who would, they thought, be driven only by those that loved
them, spared from poor treatment by the man who prefers 'to sit over
a petroleum lamp or steam kettle, whose energies he regulates with
pedals and lever'.

In the old pagan D. H. Lawrence's last prose work, the feverish
Apocalypse, he began his reflection on the 'dominant symbol' of the
horse, who 'gives us lordship ... is the beginning even of our godhead
in the flesh', by lamenting, 'Within the last fifty years man has lost the
horse. Now man is lost. Man is lost to life and power – an underling
and a wastrel. While horse thrashed the streets of London, London
lived.' Lawrence, dying of tuberculosis, was writing in 1929 at the mid-
point of two decades in which the number of urban horses plunged
by two thirds. Grooms gave way to chauffeurs and stables and mews
were converted into garages, although they were originally referred to
as 'motor stables'.

Questions were raised in the depressed 1930s about the agricultural jobs and business that would be lost when there was no population
of working horses needing to be cared for and fed, but long before
the Second World War the government had calculated that if Britain
were to be self-reliant in food, it would need to turn to tractors.
Government grants to heavy horse breeders were suspended a month
after hostilities broke out.

But there were also prophets for whom the rush of the petrol age
appeared more fragile than it seemed to the enthusiasts who preferred
to 'sit over a petroleum lamp or steam kettle'. Even in the 1930s, some

delivery or rail firms replaced their new motor vans with newer horses – manpower was cheap enough to make them economical. One goods agent noted, 'Horses – the faithful creatures – are far cheaper for short distance haulage work, and with so much stopping and starting, and their greater ability to manoeuvre, just as speedy, too.'

Sir Walter Gilbey was the grandson of a coachman and the son of the vintner of the same name whose stables at Camden we visited earlier. He inherited both the wine business and enthusiasm for the heavy horse. In the agricultural slump of the 1920s, Gilbey junior had, in the words of Shire historian Keith Chivers, acted as an 'Ezekial, a Jeremiah' prophesying doom as farmland plummeted in value; in the 1930s he had gone on to the London streets to count motorized omnibuses and prove that they, not horse-drawn vehicles, were causing traffic jams. But in 1939 he was 'full of the joy of a man who has waited many years to say, "I told you so,"' when he announced that the second horse age had arrived with the outbreak of the war: 'They cannot get enough horses now,' he cried in delight, 'but once they get them back on the streets they will never let them go again.'

Gilbey was further validated when in 1942 the British government once more began to support the heavy horse-breeding industry for both town and country. In 1943 the War Agricultural Committee exhorted farmers not to 'employ a tractor for a load that a horse can pull'. Soon they were seeking more, smaller collars – the working horses of the war lost condition and didn't fit into the old ones. When bombs fell on the London railway stables, the horses were turned loose in the streets, and, in the chaos of a blitz night, careered on through the streets till they came to Regent's Park, where they were recaptured the next day.

But after the war, and after Gilbey's death in 1945, the urban horse trade, 'dealers, repositories, forage merchants, harness makers – the lot', collapsed. Keith Chivers says that in 1947 '100,000 [horses] were put down, and at least another 100,000 in 1948'. Worse, 'It is pretty safe to say that *40 per cent of these were under three years of age*.'

———

The vanners finally ceded their place in the traces to the automobile, while the milkmen's horses clopped docilely on till the 1960s. The last canal horse worked Regent's Canal into the 1950s, and although in 1948 the British railway system still kept nearly 9,000 horses, by 1967 the last – a Clydesdale called Charlie – had shunted his own horsebox on to the train that carried him off to a retirement of opening fêtes and putting in appearances at industrial heritage centres. He was replaced by a three-wheeled tractor referred to as a 'mechanical horse'. Britain's last pit ponies went to pasture as late as 1999.

The countryside took longer than the cities to strip of horses: it was the mid-1950s before tractors got the upper hand, and hedgerows were ripped up to create huge fields that could be ploughed or sown with maximum efficiency. The number of farm horses in the UK dropped from 1,084,000 to 147,000 between 1944 and 1956, replaced by 370,000 tractors.

People still kept horses on farms in retirement, though the ploughmen themselves were out to pasture. 'I dussn't let the knackers have owd Boxer. The missus and the family would be after me right quick,' one horseman told Ewart Evans. Another was philosophical: 'Horses are dying off on the farms, and they are not being replaced. Before long, if you want to see a farm horse you'll have to visit a zoo; and in about twenty years' time you'll see me a-settin' in there [pointing to his cottage] and doin' my ploughing by radio-control.'

It wasn't the zoo but the heritage centre to which the British heavy horse retired, to give ploughing demonstrations with ribbons in its mane, and to grace agricultural shows decked out in horse brasses. Breweries kept their teams for Lord Mayor's Parades and horse shows. Some Suffolk Punch mares were shipped to the Mona Remount Depot in the Punjab, just 87 miles from Lahore on the other side of the tributary up which the big grey must have travelled, and were used by the Pakistani army to breed formidable russet mules. But the number of the great heavy horses fell steeply, and the Suffolk Punch is currently 'critical', Clydesdales 'vulnerable' and Shires 'at risk' of extinction as the number of mares of breeding age falls.

There are roughly as many pure-bred Punches as there are Przewalski horses today.

In the last thirty years the work-horse has taken on new forms in the UK. An equine *Fighting Temeraire*, he recedes into a misty understanding at the edge of practical and cultural memory, sectioned off into the past. On sentimental tea towels and calendars he, his mate and his ploughman pick their way across a pylon-free landscape that has vanished. On the docks of Liverpool he is cast in bronze, waiting in harness, paid for by the last men to work with horses on the docks. On Queen Elizabeth Street in Southwark he's Jacob, standing on a marble plinth with his mane and feathers blowing about him: he flew in suspended from a helicopter and now stands against the glossy indigo facade of a modern housing development. On the Forth and Clyde canal, two 300-tonne heads of Clydesdales, 30 metres high, bolted together from plates of stainless steel, rear from the quay. In Caerphilly an exuberant pit pony made from 60,000 tonnes of heaped coal shale canters at liberty.

But the old debate as to which was more economical, horse or internal combustion engine, resurfaced from time to time. In the 1980s, heavy horses were still owned by a surprising number of British local authorities, in part because they were cheaper than petrol engines. They mowed football pitches in Dartford, worked in Bolton cemeteries, collected rubbish in Manchester, harrowed public parks in Glasgow, and fourteen Clydesdales replaced the council's lorries in Aberdeen. 'The motor car or lorry is a disruptive force exerted against the vulnerable fabric of town and village,' wrote one journalist. 'There is a case for the equinization of urban transport.'

In the teeth of the 1970s oil crisis, an old Suffolk horseman called Harry Burroughs spoke for many who took up working horses again in that decade, anxious not to lose the skills and knowledge: 'Well … if the oil dried up it won't be a case perhaps of you couldn't do it, you'd have to do it. It would perhaps be going right back not only to ploughing with horses but to ploughing with bullocks. For without the oil what would you do? I sometimes think that it wouldn't be a bad idea if there was some sort of grant just to keep the horse as a stand-by because you never know how the job is going to work out.'

The Orkney poet Edwin Muir saw in the working horse a future saviour. In *The Horses* he describes a Britain sunk in nuclear winter a year after 'the seven days' war that put the world to sleep', where 'we listened to our breathing and were so afraid', as the radios crackle into silence, and warships pass the coast with 'dead bodies piled on the deck'. Tractors are useless, 'dank sea-monsters crouched and waiting'. 'We have gone back/far past our father's land,' suggests the narrator, as the survivors prepare to turn to oxen to plough the poisoned land.

But then the horses come.

> We heard a distant tapping on the road,
> A deepening drumming; it stopped, went on again
> And at the corner changed to hollow thunder.
> We saw the heads
> Like a wild wave charging and were afraid.
> We had sold our horses in our fathers' time
> To buy new tractors. Now they were strange to us
> As fabulous steeds set on an ancient shield.
> Or illustrations in a book of knights.
> We did not dare go near them. Yet they waited,
> Stubborn and shy, as if they had been sent
> By an old command to find our whereabouts
> And that long-lost archaic companionship.
> In the first moment we had never a thought
> That they were creatures to be owned and used.
> Among them were some half a dozen colts
> Dropped in some wilderness of the broken world,
> Yet new as if they had come from their own Eden.
> Since then they have pulled our ploughs and borne our loads
> But that free servitude still can pierce our hearts.
> Our life is changed; their coming our beginning.

Like the knights and their white horses who sleep under the red sandstone Edge at Alderley in Cheshire, waiting for battles that threaten their land, the work-horses wait.

VI

'It's disrespectful to throw away thousands of years of expertise for seventy-five years of fossil fuel. I don't work with horses because I want to go back to the past, I'm doing it because we must preserve this as a valid tool in the toolbox of humanity for the future.'

Jason Rutledge had his baseball cap firmly pulled down over his grey hair, shading his hawkish nose against the July Ohio sun. I'd found him the moment I'd wandered in the main gate at the Mount Hope Auction House in Holmes County, drawn by the familiar brilliant red of his Punch mares, Kate and Sadie, who seemed as astonishing a sight in rural America as the big grey must have been in Lahore. I hadn't realized that of the vanishingly few Punches left, more by far are kept in the USA than the UK – heavy horses might be running to extinction in Britain, but demand is growing in America. That was why, when I'd wanted to see what the future would look like when the horses returned, I went not to a rural heritage centre in the Shires, but to the world's biggest gathering dedicated to farming with horses – the twenty-first running of the Horse Progress Days, held on Independence Day weekend in 2014.

Jason went on, 'The stuff that's going on in the present is what's unsustainable. We know the oil is going to run out. We know all this contributes to our climate change, to carbon, so we've got to develop and perpetuate ways to do a better job for all of humanity, and the horses are wonderful, loyal, kind servants to that end.' In a subconscious illustration of his point, he put one finger on the twisted barley-sugar bit in the mouth of Sadie, and, applying minute pressure, backed her up one stride to stand level with her teammate.

Raised by a grandfather who farmed with horses, Jason learned the job as a young man. He was a serviceman in England in 1969 when he saw a boy walking down a Suffolk lane with a Punch mare called Moulton Princess III. It was his 'Proustian moment ... instant homesickness.' He told me he now bred Suffolk Punches in the Blue Ridge Mountains and ran the Healing Harvest Forest Foundation, using horses to teach sustainable forestry, because 'They're solar-powered. Better for the land, better for the environment.' The working horse has been recast by its twenty-first-century Ezekials as an engine of environmental change with, as the T-shirt of Jason's assistant put it, 'hay as biofuel'.

I lingered to talk and to fool with Sadie's ears while Jason and his assistant finished harnessing her up, before walking on with the promise to watch their pulling demonstration later in the day. Ahead of me lay the open slopes of the showground, dotted with tents and airy temporary barns, and thick with crowds of people through which teams of horses walked or trotted. The pale-green steel auction house stood on the hill above it all, surrounded by barns and stables.

'We call it Horse Progress Days because we're making progress with horses,' one of the organizers, Dale Stoltzfus, told me when I found him in his open-sided box of an office near the main arena. 'This is not about celebrating antiques and demonstrating how they work. This is new equipment that's been manufactured with new components, new types of steel, biothane webbing instead of leather. It's meant to go into the future.' Stoltzfus was born in 1959, the year that the US Department of Agriculture stopped surveying horse numbers on American farms. 'Horse Progress Days has a life of its own,' Dale added, 'that made it exist in spite of the people who were managing it with no records or organization. It had to exist. It couldn't be ignored.'

The gathering prints its own newspaper with details of demonstrations of logging, hay-making, ploughing, market gardening (new that year) and manure-spreading, a parade of breeds, and classes on horse psychology or running a community agriculture project. There was a full programme of seminars at different locations across the showground, with titles like 'Raising rabbits for profit', 'Balancing your soil nutrients', 'Tree cutting and skidding', 'Be your own doctor', and 'What are your produce plants telling you?'. Ever year the turnout grows. Dale reckoned 18,000 people would attend in 2014.

Horse Progress Days moves around the Midwest and northeast, to Indiana, Pennsylvania and Ohio, sticking to areas with large Amish and Mennonite or 'Plain Folk' populations, who dominated the crowd. Holmes County is home to one of the most concentrated Amish populations in America, and the landscape they have created has the quaint, naïve quality of a children's book, with green, rounded slopes cut out against a blue sky, building-block silos, red barns, clapperboard houses and brilliant white fences. There were special lanes on the long, ribbon roads where four-wheeled buggies were drawn by dark brown trotters who pegged along the tarmac like automatons, ears back, hoofs striking in a ceaseless two-four pump. When my hire car's TomTom had finally lost its signal and left the car's avatar floating in mid-air, I'd been directed by stewards on Quarter Horses and fine-skinned mules to park next to rows of tethered buggy horses. The movement is doubling in numbers every twenty or so years, with a new settlement founded every three and a half weeks – by 2050 it's believed there will be a million Amish living in the USA. They are the reason that America's heavy horses, unlike the British Shires, Punches and Clydes, are not endangered, and their use is growing.

The Amish are Anabaptists who left Switzerland in the eighteenth century, reached America and ceased, to some extent, to travel in time, by holding fast in the nineteenth century. They separate themselves from much of the modern world partly from historical habit – they could not practise their faith unmolested in sixteenth-century Switzerland unless they lived in relative isolation – and partly to keep their communities and families close-knit. Physical labour and restricted travel slow down a frenetic world, increasing closeness between individuals and families.

Each small community decides which incoming technologies will be useful – sometimes adopting and later rejecting inventions that turn out to have a harmful effect. That was why I saw both the buggy horses and a yodelling Amish logger riding up and down in his own hydraulic cherry picker: one group of Amish did adopt the car, and once its members took to driving, the group broke down and its members dispersed. The cherry picker, on the other hand, was going to benefit its owner's business without stopping him seeing his neighbours.

Though the Plain Folk farm with horses, they also innovate within those strictures. They were responsible for adapting new farming technology for animal draught, and for the multiplying gins and treadmills and jerry-rigged crop sprayers I would see all over the show site. An Amish-run firm in Ohio called Pioneer Equipment Inc did good business at Horse Progress Days with a fifty-four-page glossy catalogue of farming gear and wagons. Some machines ran partially with diesel engines, but almost all were made to be pulled by equids or oxen. Thanks to the Plain Folk, you can probably complete more farming tasks with horses now than you could before that last US Department of Agriculture horse survey in the 1950s.

For two days I wandered around Mount Hope taking stock of what horse power meant in 2014. I didn't feel like a time traveller in my T-shirt and jeans, though I was in a minority. The Plain Folk women and their daughters wore the same long, simple dresses in blue-toned jewel colours: sapphire, magenta, teal and even deep pink. Their bonnets were crisp and uniform as paper dixie cups. The men and their sons wore plain grey trousers held up with braces, shirts and broad-brimmed straw hats against the sun.

In the ploughing field, I joined the crowd watching the Yoder-Erb Brothers Twelve-Horse Hitch. Enthusiasts have hitched sixty or more heavy horses at a time, but twelve was a record for me, and my introduction to America's draught horse, the 'Belgian' bred in the nineteenth century from the Brabant work-horse. Now unrecognizable from the Old World original, American Belgians are tall as Shires, with less feather, and are always a chestnut blended from tan brick to pale dun, milder than the Suffolk red, with greyed dapples overlaid and a flaxen mane and tail. The massive horses in the twelve-hitch were true to their template: they soared up precipitously like sandstone cliffs from hoofs half sunk in the turf, solid legs turning into broad chests that rose into highset necks and white-blazed heads whose ears peaked 8 ft from the ground.

On a thoroughbred racehorse in peak condition, muscles are carved and defined. These American Belgians had muscle of a piece with their immense bulk, like old-fashioned strongmen with bellies instead of six packs. Their tails were tiny wisps of crinkled tow, docked high up the bone, their white legs were shaved and ended in long, sloping yellowed hoofs, their cream manes trimmed short and even, like the bowl cuts worn by the young Amish men who walked rapidly alongside them, like footmen, watching them anxiously, occasionally taking hold of their bits.

They were hitched four abreast in three rows, with the nearest horses walking on the ploughed strip and the other three on the unbroken grass. The reins of each horse fed from their bits through a ring at the top of the collar and then back through a series of connectors to become just four reins in the hands of the two ploughmen, who stood on a small metal platform, bracing themselves slightly against the power that dragged the plough rapidly along beneath them.

When this juggernaut marched on with its twelve nodding and tossing heads and forty-eight legs it was like standing by as a siege engine passed: the air was filled with the high jingle and clink of the connectors and heel chains, bits champed and mouthed, the work of muscle and mass, the soft rush of the Ohio soil as it was sliced deep, caught and turned over by the plough, leaving a black, shining and broken wake behind like a harbour ferry's. The crowd leaned over the cut turf after the machine had passed, craning their heads to go on watching the horses and see how straight the furrows were.

I did see one tractor at Mount Hope, and a fine, gleaming thing from Massey Ferguson it was too, with a spindly-toothed hay rake and a sleeping man in the driving seat, head propped on his chin, untroubled by the crowd, who pushed on by to get to the horses. With the exception of a small foldable plough pulled by a lone pair of befuddled oxen who were having their first day off the farm in Kalamazoo, everything at Horse Progress Days was hauled by horse, mule or donkey, including the PA systems and commentators, who rode to the demonstration areas on high covered wagons.

Outside one stall, a grouchy Haflinger pony circled round and round, driving the gin that churned ice cream in a can. A Mennonite gentleman demonstrated his horse-powered treadmill, a large, noisy, slanted wooden stairway on which a pinto horse was steadily climbing while looking out over the showground and calling to other horses. 'It can run a wood splitter, a fridge, do the laundry, operate a wheat grinder or charge a battery, and it keeps the horse in good condition,' he told me. They'd been making the treadmills for sale for twelve years and were in the process of improving the design. Orders were stacking up, and a man in England wanted to sell a containerload. 'There's a lot of interest from real horse people,' he added. 'The first pony we used here just jumped right on, and the one before hesitated a little about getting on but as soon as he was up there he was looking around and all right.'

At the top of the field where the farming equipment was laid out on display, I found a pair of American Mammoth Donkeys in dark burnt-coffee brown speckled with white hairs, with leather panels hanging on their sides, on which NOT A MULE was spelled out in metal letters. I got dizzy peering into their enormous ears: it was like looking down the burrow of some small mammal, with thick veins spiralling down into the depths. When I resurfaced, a team of exquisite miniature donkeys with round black eyes, white noses and sides the colour of pigeon feathers was trotting daintily by, drawing a little cart.

I had never seen mules before but here were entire teams standing 17 hands high, with long sagacious ears, mealy-rimmed eyes and milk-dipped muzzles, bred from draught horses and mammoth donkeys like those mules of the Pakistani army. The mule is traditionally the draught beast of the American South, and some farmers prefer it to a horse as it has more tolerance for high heat; it is also considered more intelligent. At Horse Progress Days, they waited impassively with their mates, unperturbed by the buzz of the showground.

That same great, domesticated calm prevailed over every horse, mule and donkey I saw that weekend – the same temperament Bakewell sought, that made it possible to take equids down mines and back teams of Shires up next to a whistling locomotive. How else could all this strength be used in safety? In the barns, horses with names like China Girl, Dick, Dock, Chade and Sonny waited, unable to know when, if ever, their handlers would return, but trusting them to do so. Children ran free among them, stroking big hairy cheeks, unafraid of the elephantine hoofs.

In the produce area I saw a man riding with his small son on the back of another of the big chestnuts, the horse encircled by a light metal rectangular frame suspended from its harness, which supported a crop sprayer set behind the horse's tail. The plastic tanks for the sprayer sat on a little shelf propped on the horse's back behind the rider, swaying with the movement of its quarters. The horse was serene as a planet surrounded by its metal ring, orbiting steadily around the fairground through air that rang out with the sound of boys cracking bullwhips, with humming machinery, gushing hosepipes and the occasional roar of a mule or bellow of oxen.

In the petting farm, children could ride on a 'draught horse carousel' – a large, metal-framed green roundabout with eight arms, to each of which was tied a sleepy heavy horse in a head collar and comfy-looking western saddle. Children climbed on to them by step ladder, and round and round the horses went, a dapple, a chestnut, a pinto, a grey, ears saggy with relaxation, barely noticing the children who wriggled on their backs two at a time.

From one small pen rose like an over-yeasted loaf the largest horse I had ever seen: 'Hershberger's Big Ben', read the sign fixed to the rails with plastic ties, 'Born 5-2003. Height 19H 3". Breed Belgian. Biggest Horse in Holmes County.' Pink skin showed through the shaved white hair on his fetlocks, and his mane was trimmed down to his neck in a creamy white line as distant as a rainbow, dipping down between his ears like a Teddy boy's quiff. A new yellow slip of paper had been plastered on, giving his weight like that of a prize marrow or a baby: '3,006 lbs.'

Gargantuan Ben, all flaxen benevolence, who leaned his colossal head over the rails to take a single slice of apple from a little boy's fingers. He was so big that he looked like a trick of perspective. When an older man held up a waffle cone of hay pellets, Ben slowly, politely, opened his mouth and ate the cone in one. Little girls in pinafores stretched up towards him, placing starfish hands on his pink nose, threatening to lose little fists in his giant nostrils. His head collar dangled from the gate, and the noseband would have reached around my waist.

VII

Kate and Sadie were half asleep when I found them and Jason again at the pulling demonstration. The metal-framed sled or 'stone boat' they would pull was already loaded with 5,500 lb of breeze blocks when I got there, and it stood in the middle of a narrow clearing in the crowd, who gathered close in their bonnets and straw hats, apparently unperturbed by the prospect of being bowled over by two 2,000-lb horses and a 5,500-lb sled. Jason was smoking a torn-off cigarette that he'd saved when interrupted by the previous pull, but when I made my way over he was less talkative than before. He assessed the action and gripped his rollie between narrowed lips.

The demonstration wasn't meant to be a contest like the professional pulling matches held across North America – more a wagerless version of the old farmers' matches mentioned by Youatt, when the Suffolks would drop to the floor to out-haul their owner's rivals. But despite the lack of cash prize, the competition was palpable – there was pride in the horses at stake, and sometimes business, if their owners had bred them and had a few youngsters for sale.

The crowd shaded their eyes and shared maple cotton candy. Jason's red mares were smaller than the other teams: a large pair of skewbald 'spotted draughts', two black Percherons, a couple of mid-sized Belgians and a colossal team of Belgians that included one chunky stallion who was the biggest horse I'd seen on the site after Ben, a dirt bank of compacted, disciplined muscle. A small child could have leapt through his collar without touching the sides.

When it was their turn, the big Belgians were walked out and reversed up to the sled, their evener – the bar dragged behind their heels – was attached by a single hook to the sled, and at a cry from their handler they drew back for a split second, making the chains jingle as they slackened, before hurling themselves into their collars. The sled skidded and screamed as it careered along the white gravel, and the driver ran to one side of it, careful not to be knocked over or have his feet crushed, the muscles on his arms straining as he gripped the reins. The crowd moved back a little as the horses built momentum, hoofs digging into the ground like round spades, quarters churning muscle on muscle, but as soon as their driver shouted 'Woah, woah,' the Belgians stopped and dropped their heads. They were unhitched and walked away, their driver leaning back to balance himself against the force they put on the reins. More breeze blocks were added to the sled.

At 6,500 lb the skewbalds and the Percherons were out. Told to 'git' they heaved, strained and dropped back, the sled unmoved. 'Looks like the four-wheeled drive gear went out on that truck,' said the commentator, before regretting it and adding that these were just farm horses, not professionals. The smaller Belgians hauled the 6,500 lb. The commentator called Jason and he led Kate and Sadie over and got them into position. The mares listened for the clink of chain and hook, or felt the new tension on their traces and leapt, dropping their quarters low as their threw their weight forward, the sled began to move and then to skim along the gravel, and after ten yards with a 'Ho ... ho ...' Jason halted them before they ran into the crowd.

The large Belgian stud kept casting noodly looks at Sadie and Kate, arching his neck. 'He weighs 2,700 pounds,' Jason's helper told me. 'These are meant to be farm horses, not professional pullers,' said Jason. 'Those big Belgians have pulling shoes on – Sadie and Kate are flat shod. Maybe we can get them to distract 'em,' he added with a wink, although the stallion never shifted a hoof towards the mares, as bashful and obedient as a schoolboy.

But hitched to the 6,500-lb sled, the large Belgians hauled as though they had no other role but to heft massive chests against padded leather and wood, and the sled leapt clean off the ground as the stallion's quarters and barrel clashed and flattened like tectonic plates. He was stronger, keener than his partner, as if he thought this haul was a way to impress the mares, or as if he only wanted to please his owners. Jason checked Kate and Sadie's harness repeatedly as the mares dozed in energy-saving mode, still ignoring the Belgian stallion. Whatever adrenal rush had compelled them when they hauled was now dissipated.

At 7,000 lb the small, tawny Belgians had their turn to struggle and drop back. They were out, sweat beginning to dribble from under their collars. I watched as Kate and Sadie took their place again, backing towards the sled, picking their fore hoofs up, now anticipating the pull, keyed like racehorses at the top of a familiar galloping place. They shot forward before the hook was dropped into place, with Jason slewing behind them like a waterskier. He stopped them, backed them up, and this time as the hook dropped they charged, only to fly forward again leaving the sled behind. The hook had snapped. Jason retired them. Kate and Sadie were asleep in seconds behind the crowd, their collars dry.

The big Belgians also broke forward before the replacement hook was dropped into place and had to be brought back before they flattened the crowd like a cyclone through a forest. Once more they backed up, the chains stretched, the hook dropped, their handlers leapt back and with an explosive burst the horses surged as if leaping through waves in deep water. The stud looked like he might haul a small planet, the sun even, and still rush eagerly against that collar.

VIII

'Our motto is possible, practical and profitable,' Dale Stoltzfus had told me. The Horse Progress Days spirit – not the pacific nature of its presiding animals – was that of those nineteenth-century American catalogues dense with patented hay presses, disc harrows, dress patterns and physic cures: it spoke to self-sufficiency, industry, and the carving out of a more self-contained, austere prosperity than that of the aspiring Bakewells and Arkwrights of eighteenth-century England.

Not all of this was due to the Amish – it owed more to a very American mindset than to the eighteenth-century Switzerland the Plain Folk had left behind. 'I'd love to have an agricultural census that included horses, but we're happy to have the government stay out of it,' Dale had added. 'What you see here is only grass roots. There is no government subsidy, the government doesn't even know it exists. And that's sort of nice, because with some things come rules, regulations, impositions, inefficiencies, frustrations.'

I caught a whiff, also, in Jason and others, of the *Whole Earth Catalogue*, that post-hippy bible of off-the-grid living and farming, in which the grandsons of Depression era farmers who left the Dust Bowl for the city returned to the land with notions about organic produce and solar panels. I bought a copy of the *Small Farmer's Journal*, a beautifully produced publication that mixed vintage farm adverts with Kurt Vonnegut quotations: it sold reprints of old farming equipment manuals, chapbooks of readers' poetry, and taught you how to make clotted cream, cook from a winter larder and run a mobile poultry slaughterhouse.

Jason wasn't the only one aware that Horse Progress Days meant independence from the gas that fracking firms were busy blasting out of a slab of crystalline shale that lay under all that good black Ohio soil. A well eighty-odd miles away was shut four months before I arrived in Holmes County after it hit a fault line that no one had known about, and set off scores of earthquakes that rattled nearby towns. Thirty-eight thousand miles of pipeline was planned for the state in the decade to come. Jason's solar-powered horses now seem like a gentler alternative to a growing number of ecologically minded people keen to detach themselves as much as possible from that grid, and it was in small-scale farming that horses were making the greatest inroads into the fossil fuel economy.

In 2000 Michigan State University revived a course for working with draught horses that had been mothballed since 1963, and immediately had thirty enrolments. Another college in Vermont, Sterling, added a draught horse management degree in 2003. Now you can even check if the food you buy has an Equicert mark to prove it's been produced by real horse power. Horse Progress Days was running a three-day workshop for people who wanted to start farming with horses but had no experience. Dale said the numbers of newcomers were growing, but cautioned, 'I describe it as blowing on the embers.' Slow as horses, a movement was developing.

IX

Inside the auction room at Mount Hope, a high wire fence separated livestock – or in this case the speakers – from the audience, and signs on the back wall read 'Troyer Genuine Trail Bologna', 'Yoder Hybrids: Plant with confidence' and 'Bob Puerner – semen and nitrogen sales'. Both stage and seats were full for the International Session – an introduction and a kind of informal progress report from visitors from around the world.

Erhard Schroll of *Starke Pferde* magazine in Germany said that rusty hedgerow ploughs were giving way to Amish-inspired new inventions. Germany had its own version of Horse Progress Days, called PferdeStark or horse power. A businessman called Reiner Weizotski said he was now taking shipping containers of gear back from the States to Europe for a growing market. A French guest said he was working with Breton and Percheron horses on 600 acres of Bordeaux vineyards. When the vineyard owners realized that new vines cultivated by horse power produced grapes two years earlier than those in soil compressed by a tractor, they showed more interest. The only fully horse-powered farmer in the Netherlands talked about the apprentices eager to work on his organic farm with his six horses. The British were importing more machinery for people who already owned a horse and wanted to use it around the property.

From Uganda came Okumu Boniface, who was a guest of an American NGO called Tillers International that maintained old skills like farriery and animal draught work for both Americans and international development workers. They had brought the liver-chestnut oxen from Kalamazoo. Boniface's colleague Bob Okello had worked with the Amish firm Pioneer to create a foldable, portable plough suitable for work with the local Zebu oxen. Boniface counted off the crops Ugandan farmers were now growing on small plots: groundnuts, maize, beans, soya, peas, sim-sim, rice, potatoes, cassava and vegetables. 'Where I am now every household has oxen, so they have food in the house at every time of the year. Sometimes there's just one pair of oxen per village, and there's too much work for those that have the oxen, so fields go unploughed.'

For a huge percentage of the world's population, equid power is not a low-carbon novelty: 33 per cent of the world's 'beasts of burden' are horses. Sixty per cent of all horses in 2011 and 95 per cent of all donkeys were working animals in the developing world, where the money each equid earned was enough to support a family of between five and twenty people. In Ethiopia 98 per cent of rural residents owned at least one donkey, carrying loads of up to 100 kg at a time.

In Tanzania the number of draught animals has doubled in twenty years. Refugees displaced from northern Darfur were asked what they needed to return to their homes, and said firstly security and then donkeys – for with donkeys they could reach water, food, firewood and building materials. This real age of the working equid is a continuum, not a revival.

The second seminar I attended was by an American farmer called Thomas Payne from upstate New York. As Payne walked back and forth under the bologna sausage signs, he told us he had noticed a decade or so ago that, while all the small farms around him which had mechanized had hit the wall, the Amish still kept going. Their farms might be modest, but they were diverse because the Plain Folk deliberately rejected the agri-industrial monocrop movement, using crop rotation systems in smaller fields instead of fossil fuel-based fertilizers on vast prairies. By the mid-1980s they harvested as much corn per acre as monocrop farms in Iowa, for less than a third of the running costs. They had large families in which all members worked from a young age, which boosted the outcome of that careful husbandry, but horses played a role, too.

Payne hadn't put a family of eight to work on his land, but he had, he told us, traded his $25,000 tractor for horses and equipment. 'I saved 85 per cent of fuel costs in the first year,' he said, 'and I've sold more than $25,000 of home-bred horses since then.' Payne handed out to an audience of straw-hatted Amish men a harness maker's advertisement published in the Pacific Rural Press in 1915, which asked, 'Which would you rather have? $250 or $1,850?' A $500 tractor might replace five horses, it went on, but after five years, that tractor would be worth just $250, whereas a stable of three mares and two geldings would be in their prime, the mares having produced twelve new colts in that period, all coming to a total value of $1,850. On Payne's own farm, he explained, 'a $25,000 item, depreciating in value, belching carbon full-tilt-boogie, was displaced ... by horses that have provided income and value of near $100,000, in nine years' time, and 99 per cent carbon-free.'

In the last few decades, various calculations have been made by academics and farmers like Payne comparing the relative efficiency of horse and tractor, see-sawing back and forth between arguments. That a tractor powered by diesel can farm a larger area more rapidly than a horse is incontestable. But at a farm of around 70 acres, or when a greater field of values comes into play, the tractor begins to grind its gears and falter, as Payne had found. As I drove back to my motel that evening, I turned over the statistics in my mind.

Horses can do jobs like logging that are impossible for tractors. Their hoofs, though colossal, are dainty compared to the relentless roll of the tyre tread: that means less soil erosion, less damage to other vegetation and less compaction of the soil itself, which leads to higher yields, as the winemakers in Bordeaux had found.

A tractor is only cheap to run when fossil fuels are cheap: in the oil price hike of 2008, a Tennessee man hitched his mules to the plough and saved $60 a day. And as Harry Burroughs put it in the 1970s, 'For without the oil what would you do?' – one day the last cubic metre of natural gas will have been blasted from the Earth and the last drop of oil suctioned out. A horse powered with hay and oats grown on its own farm, fertilized with the horse's dung and harvested using its energy, is unaffected by an oil price rise or crisis. It produces a third of the methane of a cow, and its 9 tons of annual solid emissions make excellent fertilizer, sparing the need for natural gas or mined phosphate rock or potash.

When the horse produces a foal – and mares can go on working till the advanced stages of pregnancy – it has reproduced itself without recourse to iron ore mining, processing, rubber plantations, silicon chips, plastic plants or mechanics. As Jason Rutledge had more vividly put it, 'You don't find a baby tractor in the woods one morning.' Its offspring can be kept or sold, while an old tractor begins to depreciate on the day of its purchase, even if resold. When a horse dies, it leaves behind compostable material, meat, bone for fertilizer and leather, and not a heap of rusting scrap which must be painstakingly recycled.

In one calculation, a horse-based farm of between 62 and 74 acres made a profit of $21,100 from its horses over a forty-year period, while a tractor-powered farm of the same size had to absorb a $70,000 loss. A meticulous article in a Swedish journal concluded that the energy required in horse farming in 1920s Sweden was 60 per cent locally renewable, whereas modern tractor farming ran at just 9 per cent.

Electrical tractors are at a fairly primitive and puny stage of development compared to the solar engine that is a large Belgian draught horse, and they are manufactured from mined minerals and non-organic materials. Of course, a tractor that runs on biofuel could also, theoretically, help to harvest its own fuel locally, and its production could be powered by rapeseed or sugar beet, but again, it is less efficient than a Suffolk Punch. The tractor requires 2.5 times the bioenergy of a team of horses.

In an article published in 2000 in the *American Journal of Alternative Agriculture*, a force of 23 million horses running and reproducing themselves on alfalfa, corn and roughage were proposed to farm the 360 million acres then in use in America. Just 6 per cent of the farmland – restored and refreshed by good organic shit – would be required for growing that alfalfa, corn and roughage. The tractors, on the other hand, needed 26 per cent of the farmland to be built and to function on biofuel.*

But 23 million horses! Think on it! More than double the current estimated population in the United States – and these calculations tend not to take cities into consideration. How would food reach urban areas from the countryside when the gas ran out? How would conurbations be powered? With horses or electric vehicles? In France and Belgium heavy horses bred in state studs draw wagons that collect household recycling in towns. If horses were once more used for mass urban transport, then still more land would be required to grow food

---

* There is one problem with these calculations for returning to horse-powered farming: they are static. As the population continues to grow, what would happen when more land was needed to feed them and the work-horses?

– remember that the equivalent of Britain's entire wheat output was needed to feed its urban horses in the early twentieth century. What a world would that be, and what a landscape. It would be an America all but beyond living memory – it was 1920 when there were last more horses, donkeys and mules running the country.

This is also where the calculations that are made for a return to real horse power begin to yaw against the current that draws steadily more people from countryside to cities every year. Before the First World War, four in ten Americans worked on the land, and now it's a little under two in a hundred. If the horses returned to draw the combine harvesters for America's wheat fields – as they did in thirty-six-strong teams in the early twentieth century – they would also pull people back on to that land, repopulating an environment that's been scraped and emptied by intensive farming. This, of course, is part of the appeal of the draught horse to the Whole Earthers and the idealists: a new rural landscape refertilized and worked by horses in place of the desertified fields and putrid feedlots of factory farms.

That evening, outside the maple cotton candy cocoon of Mount Hope and the Horse Progress Days, I sat in the diner by my motel and ate a spinach salad that tasted like the leaves were cut from craft paper. The diner was an ersatz farm restaurant which served 'farm-fresh' greige turkey sausages and dyed egg white omelettes under a blown-up photo of a picturesque farm oddly empty of all animals but two riding horses. The local paper, the *Akron Beacon Journal*, was splayed on the table in front of me, full of tales of fracking, empty, foreclosed malls and heroin deaths.

The economic crisis that sent the Tennessee man back to his mules had been slow to lift, and was leaving in its wake an economy transformed by downsizing and new technology. As working-class and middle-class jobs evaporate into hire by the hour, outsourcing, automated check-outs, drones that deliver goods plucked from depot shelves by robots, and consumers 3D-printing their own purchases at home, perhaps the insecurity of a life lived by the weather and the temperament of your work-horses would have more appeal for the

American, who has always relished hauling on his or her own boot-straps, and carving out an autonomous patch in the wilderness.

X

'I grew up in the suburbs of New York. I found environmental history was weighing on me pretty heavily as a kid and a teenager in the '90s, and by my late teens I was really struggling with all the bigger issues and productive means of addressing them. It was more than depressing, it was really crushing for me as a kid. So I sort of departed society briefly.' David Fisher, horse farmer, laughed at this, sitting over lunch at the kitchen table in his Massachusetts farmhouse. He had a long, tanned face under a broad-brimmed straw hat, and a curling beard. One tooth was missing. 'I lived off the grid,' he went on, 'out of my backpack. Bouncing around here and there, from seventeen to twenty-one. I was a ski instructor, worked a trail crew, that kind of thing.'

In the mid-1990s, after seeing 'some pretty frightening farm models out west on my travels – massive monocrops, feedlot beef, fossil fuel fertilizer', he fetched up on an organic farm at a college in New Hampshire and then, via a stretch with an alternative community in the north-west that was endeavouring to live off the grid in homesteads hacked out of the woods, he came back to the Catskills one autumn and could only find a farming gig with a man who kept a team of Norwegian Fjord horses to work a stretch of woodland and a market garden.

'That was just pretty thrilling on a deep level and on a lot of levels. That was a real moment for me. On a very basic sensory level just being around horses was a really nourishing experience, just working with these large warm-blooded furry animals that you can have this incredible partnership with, was unlike anything I'd ever experienced. To use that partnership to accomplish practical work was just great.' David stopped, and his attention drifted as he looked out of the window into the impossibly lush July green of the farm garden outside, and beyond to the pine barn he and his team had built from timber chopped and milled on the farm.

It was the week after Horse Progress Days, and I'd travelled northeast to New England to see what a modern horse-powered farm of the future might look like. Pioneer Valley redoubled Holmes County's pleasant green fields, throwing in wild hedgerows and tracts of inky green forest. Signs in the hedgerows warned about a proposed pipeline that would pass along the top of the valley, bringing fracked gas from Pennsylvania. At night there were huge, booming thunderstorms that refreshed the ground, and I lay awake in my bed in a little wooden house up the hill, wondering what creatures were making the strange noises that caught my city ears.

David now came back to the topic at hand. Lunch was over and he needed to get back to work. 'On a more theoretical level, just the whole idea that we were using living power instead of internal combustion engines to get our work done or fuelling these animals with feed that they produce themselves on the farm just seemed to address a lot of the issues that were kind of critical for me as a child of global warming and pollution and everything that goes along with producing petroleum.'

David's family farm, Natural Roots, ran a Community Supported Agriculture or CSA programme from June till the first frost in October, with the shop open on Tuesdays and Fridays for the subscribers to fill a cloth bag with produce. Natural Roots took in 100 acres of woodland, 20 acres for grass and hay and 7 acres for crops, half of which was either being refertilized by the farm's egg-laying chickens or restocked with nutrients from a cover crop of vetch or purple foxglove.

In the produce season, activity swung back and forth across the valley between a barn that housed the CSA shop and a washroom on one side of the shallow, pebble-bedded river, and the flood plain on the other side, where the vegetables grew, and the laying chickens – splendid creatures with black and white tweed plumage – hunkered under their henhouse out of the sun, or snuggled into dirt hollows to clean themselves or sit and blink.

The other farm buildings were screened from the floodplain and the road by a stretch of woodland and reached by a cool, overhung path with ferns, dock and nettle underfoot, dappled with horse dung in various stages of decomposition from fresh and green-brown to bare dry fibre. There was the newly built golden-coloured barn, and an older barn brought by the landowners from Vermont, which had become home to David, his partner Anna Maclay and their two children, Leora and Gabriel. Beyond the barns stood some sheds that the landowner had stocked with old horse collars, rusting bells, a sled and some farming equipment that was still usable when David got there in 1998 – all picked up when small farms in the area were sold up in the 1980s.

A greenhouse incubated the seedlings before they could be transplanted, and the kitchen garden provided herbs and a tree that scattered long black mulberries profligately on the paths, beds and greenhouse roof. Two sheep sheltered from the sun in an enclosure – food for the farm's work team – and another chicken run held red broilers. Four black and ginger pigs pottered about a partly wooded run at the back of the barn, running obsequiously along the low electric fence when anyone passed, or digging up filigrees of tree roots and making the saplings tremble when they rubbed against them.

In a pasture at the far end of the vegetable fields grazed the five horses that powered the farm: four chestnut Belgians led in seniority by Lady, the grand dame, 'somewhere in her mid-twenties', Pat, an old gelding 'in the ballpark of twenty', headstrong Gus – 'he's a ball of fire most days' – and dim bulb Tim, about twelve and so the same age as Star, the bay half-Percheron, half-Standardbred mare which completed the set.

Horses, sheep, chickens, pigs, vegetables, pastures, hayfields and woods supported five adults – David, Anna, the young twenty-something apprentices Lauren, Nate and Emily – and the children, who ran about about barefoot and brown-skinned between their daily chores and the swimming hollow in the river.

At 6 am the apprentices fetched the horses from the overnight pasture and brought them to the cool dark of the new barn where they sipped daintily from cups that filled automatically with fresh water from the well, and munched on mounds of home-grown hay. Green-brown dung accumulated at the horses' heels as they yawned and chewed, and was later forked over into an enclosure where the long-snouted pigs would root around in it and aerate it before it was carted back out to the fields and spread by the horses.

Lauren showed me a video of the culmination of the hay harvest on her phone: a huge cloud of hay rose up from the wagon on a rope, and flew into the gable doorway under the barn eaves, wisps flying as it was hungrily crammed into the loft. The phone camera panned out to show the distant figure of little Leora leading a horse away behind the barn, providing the traction to haul the hay.

I spent four days tagging along, scribbling, asking questions and occasionally being useful at Natural Roots. I lifted aside trailing melon suckers that were ambitiously striking out towards neighbouring rows. I trellised tomatoes and squashed bright orange, fat Colorado beetles as I found them, snapped off yellowing leaves and sliced green-streaked courgette rods and yellow flying saucer squashes from their vines. After joining in a silent shared blessing, I ate lunch from the farm's fields and chicken coops with the crew and the family at a long wooden table in the garden. Sitting on the grass in the shade near the house, Lauren and I thumbed lettuce seeds into small plugs of soil in growing trays while an over-friendly calico and white cat, driven frantic by the large patch of catnip nearby, tried to climb into my lap. I talked to everyone when work permitted, slowly unravelling the story of the farm.

David started out on his own in the valley in a cabin, growing enough food to eat and a little surplus to sell at farmers' markets while working a part-time job in the town. The farm grew from there, with David borrowing a neighbour's tractor for any heavy work, but, growing steadily more convinced that he must use horses, he bought his first team of Belgians from a dealer in Connecticut for $4,000. 'They were a retired Amish team, they were pretty mellow – very relaxed by that point. They taught me a lot and they were pretty forgiving all round.' It took an investment of $12–14,000 to get the farm started, and there was a grant from the state to build the CSA shop, plus some community help to raise the bridge over the river, but, he said, 'We've never taken a bank loan and we've stayed out of debt by growing steadily.' The farm had been pretty much self-sustaining since the beginning.

As for the horses, 'Well, it's a slippery slope,' he said, wryly. 'You can farm with one horse, but it's a little more limited. So two horses. Double the power. Then one horse is lame for a week, and in our schedule that's a big deal – a week of tillage, planting, seeding. So I got a third horse so I'd have a spare. Then I think, Oh wow, three horses, we can put them on the disk or the plough, so I start relying on three horses. Thought I was going to have to retire one so I got a fourth on loan for a season, and then it turned out the older horse was just anaemic and once he was on iron supplements – boom – he was right back at it. So then I had four horses, which is two teams, and I saw what I could do with that, especially when we started making our own hay. But then you're right back where you started because if one goes lame, you need a spare. So then you have five. We use five enough to justify keeping them. I'm rarely wishing for more.' When the horses' work days were over, they were pensioned at a retirement farm.

The apprentices usually began to work horses during the farm's off season, although they'd been so busy constructing the barn that winter that Lauren hadn't had a chance to start yet. She, Nate and Emily lived in self-contained cabins in the woods behind the barn. At the moment they were moving between farms on apprenticeships of a year or two, as David had done in the 1990s, learning and assessing the land, figuring out what they wanted to do and where. But David had noticed that there was a rising trend for incoming apprentices to already have experience of working with horses.

'The return to horse power is coming,' he mused. 'It's explosive, I would say. There are so many young horse farmers in this region, and in the States that I'm aware of, compared with when I started.' All the current apprentices were planning to farm with horses in the future, just as earlier apprentices had gone on to do in New York, Pennsylvania and Massachusetts. The tendrils were slowly spreading.

Despite the deep, green calm I felt as I helped out with the daintier tasks there were considerable stresses. The hay harvest had been completed just before the Horse Progress Days – a time of high anxiety, when sudden or heavy rainfall could delay the collection of the farm's fuel, or even ruin the crop. Replacement hay would have taken a large chunk out of the budget.

David told me that once on the last day in August a hurricane wiped out all 3.5 acres of crop. Another time, the river burst its banks and took out the bridge that linked the farm to the CSA shop. Anna had to balance the shop accounts and keep over 200 subscribers satisfied, while dealing with other local suppliers who brought beef, ice cream, miso and handmade maple soap to be sold. The horses, too, were capable of contributing to stress: a team waiting outside the CSA barn had once taken off with the wagon, bolted through the ford, across the fields and into the woods, where their neck yoke caught on a tree and brought them to an abrupt halt. (Gus was involved.)

Even with the hay in, this was peak growing season, and activity was constant. The horses came and went from the barn in different combinations: now two horses with the wagon; one with a crop sprayer; four on the disker. The apprentices sped between locations by mountain bike. David rode the high-wheeled cultivator through the vegetables planted for the autumn, its arrow-shaped blades turning over the soil and uprooting the weeds, which withered in the sun in a few hours. Gus and Star drew it, reversing and manoeuvring delicately at the end of each row to get into the right position. Emily, waist-length blonde plait swinging between her shoulder blades, took Pat and Lady out with the small crop sprayer to deal with the Colorado beetles.

I rode with Nate slowly and ceremoniously round the field as we picked up the newly filled plastic crates of vegetables from the end of the rows, throwing damp sacking over each one to keep the contents fresh. Loaded up, the horses walked down through the ford in the river and up the slope to the CSA barn, where they stood and stamped their feet against the insects as the crates were lugged off the wagon and stacked in the washroom.

The produce was blasted clean in big plastic tubs or run through a rattling root vegetable washer which scoured them with rotating bristles. What didn't scrub up had to be carted back down to the pig run, whose inhabitants ran cravenly after the horses along the fence, grunting in excitement as David hurled buckets of daikon and bruised lettuce into their pen. The pigs did not like the daikon.

When we were done, the wooden racks in the shop held a cornucopia: beets were stacked in a cobbled wall of ruby roots, with tangles of pale green garlic scrapes, flat-leaved kale like palms, white Chinese cabbage sliced to reveal a frilled heart, nobby white daikons, earthy bunches of fresh coriander, frisée lettuce beaded with fresh water, and reeds of spring onions. An hour after opening, and much of it was gone or tumbled. Customers sat out front in the sun on benches, picked herbs from the garden or wandered down to the swimming hollow with their children.

On the Friday, I rode with David on the disc harrow, a set of upright discs on a frame weighted down with a Fred Flintstonesque assemblage of rocks strapped on with leather belts. It was used to break down the cover crop before the land could be ploughed, and as the cover crop reached three or four feet high in places, it would take a team of four to flatten it. I stood on the small platform with David, clinging to the metal rail in front – 'If you fall off the back, you're going to get disked' – with four tails before us: blond Gus, marron chestnut Pat, black Star, blonde and chestnut Lady. Even though the evener behind their hocks had been balanced so that Gus took more of the load, he was clearly leaning hard into his collar.

We bumped forward, feeling the resistance of the purple vetch and yellow clover as it was trampled by the horses and then crushed by the discs. The hitching chains clanked against the central pole, the discer rattled and the horses snorted as the tiny bugs flew up from the crushed vetch into their noses. Swallows swooped down through the clouds of insects. In our wake, birds called grackles ran eagerly into the new furrows looking for more bugs.

David kept up a constant chatter with his horses, utterly absorbed by the task. 'Gee Star, gee Lady, good Gus, good boy, Pat,' as the horses moved down the row, then 'Woah,' as we reached the end. 'Haw' to turn to the left, 'Gee' to the right. Driving seemed more of a partnership than riding to me: David's power was limited behind four 2,000-lb horses whose heads were ten feet or so from his hands. They did not have double bridles or nosebands that clamped their mouths shut. No whip or spur was used. The reins were slack. The horses simply knew their job and his voice, and they worked hard. 'Do they know when a field is nearly done?' I asked, as we reached the end of one row and Gus surged into a trot and had to be woah'd back. 'I think so,' said David, eyes on his team, bringing them under control and back on to the field to a new stretch of tall fronds of vetch.

The stubble threw up drops of moisture as we passed over it, and it gleamed on the horses' legs below the hocks. As the team looped up and down the field, sweat began to marble on Gus's hams, but still he surged on eagerly. 'Gus is a mystery but we've worked out a way to work together. He's so keen to pull he just shoots forward. But I've come to appreciate the power,' David had told me earlier. He kept Pat on his toes at least. 'If I put Tim with Pat I get the slowest team on the farm. If I put Pat with Gus I get the fastest. It's like Pat doesn't want to seem old with the younger horse.'

Midway down a row, David called 'Woah', and we came to a crunching, lumpy halt for a few minutes' break and for me to hop down. Water and sweat were dripping from Gus. As I headed on my way, David leaned on to the rail and looked out over his team without saying anything.

Thomas Payne had emailed me after the Horse Progress Days to answer some questions: 'There will be many who consider making a move to horse power insane,' he wrote, 'I say NOT changing direction in a million ways of this sort is more insane.' Would Americans really take to the land to feed the nation with the help of 23 million horses? It feels unlikely – horse-powered farming seems more like a niche carved out in defiance of a bigger system, but there remained the steady, organic growth of the farms, and the spreading roots put out by the new apprentices, restoring the land and feeding small communities. When will the oil run out, and how soon will our solar-powered tractors be here? We, too, might be listening for the distant tapping, the deepening drumming and the hollow thunder of the horses as they come.

At the turn of the afternoon into evening I was still hanging around the barn, chatting with the apprentices as Lady, Tim and Pat, finally unharnessed, ground away noisily on their evening feeds. Anna was closing the shop over the river, and Star, Gus and David were ploughing a section of the produce acreage. Shouts drifted up from the swimming hole. Lauren legged her visiting friend, Daniel, up on to Pat and climbed on to Tim's broad sweaty back, one hand holding Lady's halter rope. Slowly the horses rolled off, their riders' heels dangling loosely at their sides, heading with determined patience towards their pasture beyond the produce fields, unfathomable servants, strangely calm. I walked reluctantly back through the woods, around the fields, waved, and crossed the bridge back to my parked car.

MEAT

Americans Don't Eat Horses

For while patting with pride his soft sleek skin,
We should think of the future steak within.

*Temple Bar – A London Magazine for Town
and Country Readers* (December 1866)

I

The brown mare had given up. She stood on three legs, her near fore
curled under her. The knee was swollen like a boll. Her head was low
and her eyes were dull. She had a white code freeze-marked along the
roots of her mane: W5691. Stickers were glued on either side of her
rump, one with a bar code through the centre, and the other giving her
sale number, 592. Whoever brought her to this place had salvaged her
original head collar and left her with a crude halter made from twisted
blue plastic string.

A raw hole was rubbed in one of her shoulders. The knee nearest me
had a matching puncture. The knee she couldn't bear weight on bore
an open, bloody red wound two or three inches wide, though not fresh
enough to bleed. A shell-like white chip protruded a little from it – broken
bone or a damaged tendon sheath. I didn't like to look closely. She didn't
sniff at a proffered hand, or flinch when we took photos. She was fit for
the wolves, she seemed to say, but she might have to wait for days.

Her pen was like dozens in the American auction barn, a labyrinth
of corridors lined by wooden fences that rose to the middle of my

vision, their planks whittled by the teeth of recurring tides of nervous animals over years. The floor was dirt, with dark urine patches, a scattering of sawdust and old dry dung. Above the pens was a catwalk, where buyers and the curious could look down and assess the flesh below. Some of the pens were barely big enough to contain a single horse, but others were large enclosures, filled with wary or resigned animals. Big fans ran in the walls, driving hot July air above the horses' backs. Overhead the metal roof had once been lined with insulating material that now hung down in ragged strips, and everywhere cobwebs, thickened by dust, straggled opaquely or hung in clots. Not all of the enclosures had hay or water – the mare had none, even if she had looked for it. Many of the horses on her side of the barn were quiet, even when packed two, three or even four into small pens and standing nose to tail.

The brown mare wasn't the only sad story I'd seen: they were scattered throughout the auction barn. Sometimes the story was a fragment, sometimes it was obvious, sometimes it required a little explanation from my guide, an amateur horse welfare investigator who kept tabs on the auction house's business. There were clues: a pony with a bloody nose, another weeping from a nostril, a third with a puckered socket where an eye had once been. There were five old Amish buggy horses who, accustomed to standing for long hours hitched to a rail, were quietly, ravenously, tucking into a rack of hay in their enclosure, their sides corrugated with rib and shadow and their spines rising from their backs like breaching dolphins.

Bald patches where harnesses had rubbed; white scars where nosebands had been left on too long; bloody scratches down a neck; old work-horses with greying sides. Some hoofs had split almost to the coronet, and splayed at the ends; others had the thick, cracked texture of elephant skin rather than the smooth shell-like finish of a healthy hoof. Like rock strata, the hoofs revealed the horses' history.

At the back of the barn one enclosure held perhaps ten horses, including a monumental Belgian whose ribcage protruded like the exposed beams of a cathedral nave, his pelvic bones jutting and hollowed out. His mane looked as though it had been chewed – horses do eat one another's manes and tails when they have been starved. He stood quietly near another horse who attacked the hay rack with gusto, but he did not eat – perhaps his teeth were too overgrown.

Most of the horses were quiet but not exactly peaceful – they were habituated to being taken to unfamiliar places and encountering unfamiliar neighbours, so they were not unsettled, but watchful, with black orbs taking in every move even as they rested hoofs or folded theirs ears softly back. Other horses crackled with nervous animation. A feeble black mule was enamoured of a tough, witchy chestnut draught mare, who swirled in her pen, kicking at other horses. A dark bay mare clung to a thin companion with a ragged mane, hanging her head over her back, and rushing at any horses who came near her with open teeth that ripped at the thin skin over their ribs.

The barn had its own dingy, fearful atmosphere, one carried by the horses rather than the humans who worked there, or came to peer down from the catwalk. These people slapped the horses' rumps, chased them or poked hands through the fences to stroke. In a few days the horses would have cleared through and cattle would take their places, and then the next week the flood of the nervous and the skinny would begin again. Horses seldom passed through the ring twice.

Out front was a soberly painted wooden auction room with steeply raked benches like a nonconformist chapel, smaller than Mount Hope, where just a few weeks before at Horse Progress Days I'd listened to seminars on farming with horses and the relative growth-rates of Bordeaux vines. The brown mare's owner had clearly decided not to spend the thousands of dollars required to fix her, nor the few hundred dollars needed to have her euthanized and taken away by renderers. Instead she had come to the last place where she was worth a few dollars.

Outside in the parking lot, long, gleaming aluminium cattle transporters were parked, ready to ship most of the horses sold that day to Canada or to Mexico, where they would be slaughtered in factory abattoirs, jointed, butchered and shipped to Europe and Asia, reaching the supermarket shelves as smoked cold cuts, steaks, sausage and mince.

Horsemeat has been sacred flesh, unclean flesh, taboo flesh, a cure for poverty and the desperately harvested meat of sieges and warfare. It's been a gimmick, a king's feast and pet food. One billion people in today's world eat horsemeat. It is commonly consumed in China, Russia, Central Asia, Mexico, Holland, Switzerland, Italy, Japan, Belgium and Argentina. Globally, consumption has risen by 27.6 per cent since 1990.

If the work-horse is an industrial and agricultural machine, then meat is its other by-product. Lady, Pat, Gus, Tim and Star at Natural Roots would be retired to pasture when they could no longer work, but when calculations are made for the number of working horses needed to run the modern world, their authors usually assume that the aged will go to the butcher, not the paddock, just as today's Ethiopian working donkeys end up on Chinese tables. In many European countries, government studs pragmatically made over their heavy 'zootechnical heritage' breeds as potential 'chevaline' – which is to horse as beef is to cow and pork to pig – when the work-horse was replaced with the internal combustion engine. If you have livestock, as the saying goes, you have deadstock, and you must find something to do with it. Few horses are raised purely for meat; they usually arrive at this secondary vocation after a complicated life.

In the twenty-first century, hippophagic countries lie cheek by jowl with nations who are repulsed by the thought of equivory: French Canada, Cuba and Mexico by the USA; Belgium, France and the Netherlands by Britain; Japan and Indonesia by Australia. In each case, the country that does not eat horse is happy to ship its own horses to its neighbours for consumption. No other meat animal has been the subject of such debate, and in America horse slaughter has been disputed in Congress and the Senate, used to fight presidential elections, touted as a war winner, an abomination, poison, a sign of unAmericanness, and the saving of horsekind. It has even been the motivation for terrorist attacks.

Federal funding for horsemeat inspection in America was withdrawn in 2005 and, for a while, the country's last remaining abattoirs were permitted to fund their own inspections. In 2007 this privilege was withdrawn and the slaughter of horses was effectively suspended. You can still kill your own horse for meat in many states, but it should not cross state lines, and in some places the sale of horsemeat is illegal. One slaughterhouse survives in New Jersey, processing horses for consumption by zoo animals. Otherwise, disposal is handled in a variety of ways: some horses are buried in landfill or at their owners' homes. Few are cremated. Many are rendered into industrial and agricultural products; owners with the space for it – and the nerve – can compost carcasses.

But America is still actively part of a global horsemeat business that is largely managed by multinational firms in Belgium, France and Holland, and its raw material was the brown mare with the broken knees and the big, starved chestnut that couldn't eat: over 100,000 American horses a year go into the slaughter pipeline. As an agri-industry, it is dwarfed by beef, pork, poultry and lamb; many Americans don't even know it exists. So why did America ban horse slaughter, when the most consistent thing mankind has done with equids over 2.6 million years is to eat them? The horses that dance and that haul are a blink in time, a mere 5,500 years. What makes horses different to the gingery pigs which were destined for the chest freezer at Natural Roots rather than a retirement home? And what brought Americans to this convoluted conclusion, supplying animals for others to kill and eat, as though hippophagy itself were beneath their caste?

II

Horseflesh is dark, with more blood to it than beef or mutton, and, as prehistoric hunters knew and the Botai found when they became horsekeepers, is well supplied with fatty acids and vitamins. The fat is easy to digest, and the meat itself is lean. It has a sweet, gamey flavour. For millions of years, the most complicated issue with hippophagy was catching the horse, but after a millennium of domestication by horse-eating pastoralists on the steppe, horses arrived in the Middle East, and there was an abrupt rupture in their perception and treatment. Their meat no longer figures as human food.

'My flesh [is not] eaten,' says the first talking horse in literature, in fragments of Babylonian cuneiform dating from the seventh century BC. He is a prestigious creature, 'glorious in battle', clothed in expensive copper armour and credited with the 'heart of a lion'. He is not for eating, unlike the ox with whom he contrasts himself, but an expensive foreign import, lifted above the means of ordinary farmers and into the realm of kings and armies as a weapon of war drawing a chariot. The Mitanni horseman Kikkuli said that a trained horse was worth twice as much as an ox used for plough work, and an untrained horse twice as much as a cow. One horse cost the same as a small flock of sheep. This prestige was partly the result of the horse's military role, but also the product of the simple economics of biology and climate. Horses process their food less efficiently than a cud-chewing bovid, and thrive on large quantities of poor-quality roughage like the steppe grasses. In the Middle East their meals had to be supplemented with more expensive barley and other grains.

By the time the Old Testament was written, equids were forbidden flesh. 'Say to the Israelites: "Of all the animals that live on land, these are the ones you may eat: You may eat any animal that has a divided hoof and that chews the cud,"' states Leviticus, and where Leviticus led, Christianity would follow. However, the Bible includes a loophole:

it never explicitly mentions horses, unlike camels, which also are not eaten. Furthermore, after the flood, God tells Noah, 'Every moving thing that liveth shall be meat for you,' which some believe cancels out Leviticus's food taboos – pork and rabbit were happily consumed by later Christians.

In Islam the horse had an exalted role carrying the Prophet and was highly valued in war, as it had been in Kikkuli's time. The Qur'an mentions horses for riding and for display, but is silent on the topic of horsemeat. Some scholars argue that horsemeat is not *haram* but instead *makrooh* – undesirable or even offensive. Others cite a hadith which suggests that hippophagy was permissible: 'On the day of Khaybar we slaughtered horses, mules, and asses. The Messenger of Allah forbade us (to eat) mules and asses, but he did not forbid horse-flesh.' Wild horses were eaten in North Africa in Leo Africanus's sixteenth-century account, but while Muslims in grassy Central Asia still habitually consume horse sausage without a qualm, Arabs today are disgusted by the idea of eating them.

Whoever brought domestic horses to the Italian peninsula and to Greece clearly did not establish an enduring local tradition of hippophagy either. The Romans, who would, no schoolchild will ever forget, eat everything from dormice to ostriches, had a Middle Eastern approach to hippophagy, and would only eat horseflesh in the case of extreme poverty. The meat was purchased not from a butcher but a knacker – in other words, a disposer of carcasses rather than a purveyor of food, suggesting a degree of uncleanliness.

But the territories the Romans conquered as they spread north and west were populated by the descendants of the horse-eaters who had covered cave walls with pictures of the hunt, and by the heirs to the steppe and Caucasian horsemen who had long since travelled west with their horses and settled. Among these barbarians, horses played a symbolic, mythological and sacrificial role. The Huns, the Visigoths, Teutons, Slavs, Vikings, Balts and the Celts and others killed and buried horses with their dead, eating their flesh at funeral feasts. While the Romans were happy to tolerate these practices, their Christian heirs in the empire saw hippophagy as a dividing line between civilized and barbarian, holy and pagan.

Christianity had moved away from the stricter dietary requirements of Leviticus as it broadened its territory, but oddly that old, unclarified restriction on horsemeat emerged as a defining addition – the only food taboo to be endorsed by the church. In 732 AD Pope Gregory III wrote to Saint Boniface about the pagan Germans whom Boniface was struggling to convert: 'You still tell me that some people eat wild horsemeat, and that many also eat domestic horsemeat. From now on, very holy brother, you should not tolerate these practices, but on the contrary completely prohibit them by all means possible, with the assistance of Christ, and impose a deserved penitence, for it is impure and detestable.'

Ten years later, Boniface was complaining to Pope Zachary that his Germans were still eating horse. In the end it took several centuries to establish both the taboo and the faith across Europe, and the association of horse-eating with brutish behaviour was transmitted through folklore, literature or by the Church itself. In the Norwegian chronicle the Heimskringla, King Haakon is offered horse broth by his pagan farmer subjects and when he refuses, they threaten to attack him. Another noble suggests a compromise: Haakon should 'hold his mouth over the handle of the kettle, upon which the fat smoke of the boiled horse-flesh had settled itself'. Haakon covers it with cloth before doing so, and 'neither party was satisfied with this'.

When another Norwegian king – later Saint – Olaf, crushed the Icelanders into Christianity, he cast horse-eating in the same category as the islanders' custom of exposing unwanted children, and any outright pagan worship. He held the sons of four chieftains to ransom, and said that the slaughter of horses for human consumption would be penalized by death or mutilation. The laws of the Grágás proclaimed that, 'People must not eat horses, dogs, foxes, and cats ... If a man eats these animals who are excluded, he is liable to a penalty of lesser outlawry.' You could eat bears (brown and polar), walruses and seals, but if one of your pigs tore into the carcass of your old riding horse, it had to be starved for three months and then fattened for a quarter of a year before it could be eaten, to ensure that not a drop of its meat contained horseflesh. In Russia, prohibitions were still being issued in the twelfth and thirteenth centuries, and the Tartars of the steppe retained their reputation as barbarous horse-eaters long into the nineteenth century.

Travelling in Ireland in the twelfth century, the monk Giraldus Cambrensis reported in gruesome detail on the role of a white mare in the making of a king in Kenil Cunil, in remote Ulster:

> He who is to be inaugurated, not as a prince but as a brute, not as a king but as an outlaw, comes before the people on all fours, confessing himself a beast with no less impudence than imprudence. The [white] mare being immediately killed, and cut into pieces and boiled, a bath is prepared for him in the broth. Sitting in this, he eats of the flesh which is brought to him, the people standing round and partaking of it also. He is also required to drink of the broth in which he is bathed, not drawing it in any vessel, not even in his hand, but lapping it with his mouth.

Some historians link this ritual to an Indo-European horse sacrifice described in the *Rig Veda*, while others believe the account is propaganda painting the Irish as backward heathens, but on this both Giraldus and the revisionist historians agree: horsemeat was figuratively beyond the pale. To eat horse was to be lower than a man, to link oneself as a pagan does, with the animals, lapping at unclean broth like a dog. Haakon, the good king, places a napkin between himself and the faintest essence of horse. Olaf, the saint, punishes the Icelanders for hippophagy. Both sides believe that the food a man eats – especially when it is the substance of animals – becomes part not just of his physical body, but his soul: to the pagans the horse gave strength and fertility, to the Christians it transmitted pollution.

The practice of hunting sanctified some horseflesh for several centuries more: Tarpan were still eaten in some parts of Europe until the 1700s, presumably because as game, the wild horse was in another category to domestic animals. European penitential texts stop short of outright prohibition, but make it clear that hippophagy is not customary. But at the same time that the taboo strengthened, the number of horses in use continued to grow steadily across the centuries, as in Britain during the agricultural and industrial revolutions. Each of these horses left behind its mortal, fleshy remains when its working life was over, and the disposal of this unclean meat became associated with, and left room for, a certain degree of shady practice.

The horse's hide and hair were made use of, but its flesh was generally fed to dogs and cats. Mentions abound of horse corpses left in ditches for dogs, crows and other animals to pick clean. Knackers were still separated from butchers, and even though they disposed of horse and hide without deceit, they were often despised: in 1730 near Inverness a soldier who bought horse carcasses for their hide and hound meat was pursued through the streets by children flinging stones. Being a knacker was considered 'an employment fit only for the hangman'.

Some parts of horse bodies made their way into human dishes, disguised as beef or venison. Horsemeat was undoubtedly also consumed when the death of a working animal created a windfall for hungry peasants who were unable to let so much red meat go to waste. In Italy old work-horses became jerky, and in Spain the horses killed in the bullring could be consumed. Horsemeat also went into the barrels of 'red deer' fed to Spanish sailors on long voyages. Under siege, in famine, the horse was inevitably the resort of the desperate. Soldiers and civilians ate horse when all normality had broken down – although not in every instance. In 1754 a severe famine afflicted Iceland, and local bishops and priests had to encourage their parishioners to eat horse simply to survive. There was no particularly enthusiastic response, and the permission had to be reinforced again in 1757 as the famine went on.

With this taint of desperation came that of social breakdown, of anarchy, the throwing over of old controls, of the Church and other authorities being unseated and their impositions broken down. In 1629 a French groom was executed for eating horsemeat during Lent. In the 100 years before Louis XVI was tried in Guérinière's riding school, illegal trading in chevaline prompted four French royal edicts to reinforce the ban on hippophagy. The country's butchers were worried – under the Ancien Régime they controlled the preparation and distribution of meat. If horsemeat were openly eaten, the lowly knacker would undercut them by selling cheap cuts of work-horses to the poor.

As the eighteenth century drew to a close, the number of work-horses and the number of urban poor continued to grow in France and across Europe, and the piles of mortal, fleshy equine remains did too.

III

We hovered at the barn entrance. Men waited by the door, shooting the breeze. Some were bearded Amish in broad-brimmed and deep-crowned straw hats, others grey-haired 'English' with cowboy hats and plaid shirts covering bellies that jutted over their jeans. People unloaded horses in the car park and brought them to the entryway, where their paperwork was collected and a sticker glued to their rump. My guide 'Jane' pointed out the auction house owner and the various 'kill buyers' – the name American anti-horsemeat campaigners give to what the British would call 'the meat man'.

There are bottom-end horse auctions across America and their consignments reflect local horse use: Quarter Horses and old rodeo animals out west, buggy horses and farm horses in the Amish east, and everywhere ordinary purposeless animals who will be convert-ed into poundage after a trip through the sales ring and across the northern or southern border. I could have gone to Texas to witness these sumps of the American horse industry, or returned to upstate

New York, where on Black Friday 2012 I'd sat by another dreary ring and watched a beautiful – but dead lame – Belgian draught be sold for a pittance.

The number of cattle, sheep and pigs in the country is known to the last unit thanks to federal agricultural surveys, but horses slip – or fall – through the economy, changing their identities and their classification as they do so. A plan to create an official register was mooted in 2010 but never came about, so modern American horses continue to be uncounted and anonymous. A beef steer is a beef steer from the moment of artificial conception till it meets a bolt gun on an abattoir floor, but horses are protean, too mercurial in their human designations to fit into bald United States Department of Agriculture (USDA) statistics. They live far longer than a steer, too. A horse is a racer, a show horse, a mustang, a mistake, a dobbin, a wild hope, a cherished drop of DNA, a farm labourer, and then he changes, becomes a trail horse, a therapy horse, a disappointment, a buggy horse, a saved horse, a lawn ornament, a pet, a parcel of renderings. As they travel through the economic sorting machine, some horses lose what little ID they have – their breed certificates, their names, their histories. They often arrive at their destination blank.

The brown mare with the broken knees still had her name, encrypted in that W5691 brand and easily recovered from the internet that evening. Breezy Knoll Annie was a registered Standardbred born in 2000, descended from the swift Norfolk trotters of eighteenth- and nineteenth-century England that farmers pitted against one another on dirt roads, never allowing them to break into a canter. Crossed with thoroughbreds, they became the modern harness racing horse of the Western world. Annie had raced fourteen times without winning and instead tumbled into the sorting machine, ending up as a buggy horse. There was a patch of road out there that must have claimed those knees. According to the USDA's Animal and Plant Health Inspection Service, or APHIS, she was not fit for transport to slaughter, but the stickers on her quarters indicated that she had been accepted for sale, so here she was again, relabelled at 45 cents a pound.

'To me, horse slaughter is covering up a crime,' said Jane, who had been attending auction houses for over a decade to keep tabs on the horse slaughter business, and who wished to remain anonymous. Even though she'd grown up around horses, Jane had had no idea about the meat business until 1996, when she was working as a trucker and started picking up mentions in CB radio chat near the Canadian border. In the late 1990s she acquired a computer and discovered the internet, and found like-minded people who were mobilizing against the already dwindling number of horse slaughter plants in the USA. Now they tracked the horses that went for meat at different auctions across America. She knew by sight all the buyers, and the drivers who would ship the horses on – often she followed them all the way to Canada.

It took thirteen or fourteen hours' driving in the stuffy aluminium trailers to reach Quebec, where most of the Canadian feedlots and plants were located, and on that journey the horses were packed into trucks with no dividers to keep them from falling against one another, kicking or biting. Horses got trampled and gashed. Jane had seen the trucks rocking as horses fought inside.

'The truck drivers are only meant to drive for ten hours without a break, but the USDA has a limit for keeping livestock in a truck for twenty-eight hours without a break,* said Jane. 'And there's no crosscheck between the USDA and the federal authority. So if they get caught by the Federal Motor Carrier driving more than ten hours without a break, they'll get fined and it goes on the company's record and the driver's record. But the USDA won't do much.' This was why she preferred to report the drivers for breaking traffic regulations than for animal welfare violations.

'I've stood in front of police officers here and asked if they're going to prosecute,' she added, as we peered through the fence at Breezy Knoll Annie. She had once seen a horse with an open fracture at a sales barn. Neither the auction house, the local vet nor the police would do anything, claiming zoning rules – a horse could not be euthanized in the centre of a town. She tried the humane society and the state agricultural authorities. Bucks were passed so quickly that they flew. Later the state agricultural department telephoned her and claimed that the horse's pen had been found open and the horse was missing – Jane, they implied, must have released the horse like mink from a fur farm. 'How far did it get on three legs?' she asked them.

'I know they shipped the horse to slaughter like that. I felt so bad as I thought I could have saved it,' she sighed. Breezy Knoll Annie's condition made her angry, but she knew better than to call the police again – she just documented the wounds with photographs and made a note of the mare's freezebrand and sale number. The networks of campaigners were lobbying for a radical and permanent ban on the horsemeat trade in the country, closing the final loophole so that America could detach itself entirely from the global trade. The steadily accumulated evidence was reinforced with freedom of information requests and compiled into reports by grass-roots organizations like the Equine Welfare Alliance and Animals' Angels.

That year, Animals' Angels published photos of conditions in the feedlots and export pens where horses waited, often without shelter and sometimes for months, before they were taken to the abattoirs. Investigators saw both starved animals and horses overfed to the point of disability and disease, stranded on their sides with their legs sticking straight out from tight, pained bellies. Some horses had visible wounds or even shattered bones. One mare found on a Canadian feedlot in 2012 had a dead foal protruding from her vulva. In 2013, investigators found decomposing horse corpses at another feedlot.

Some campaigners even made it into the slaughterhouses themselves to document what happened inside. Each year brought a new TV documentary or news report, and dingy YouTube videos shot undercover of horses slipping on treacherous floors, or being stunned repeatedly with bolt guns as they went on standing and even whinnying.

Although Canada and Mexico had their own inspections for the abattoirs, and the European Commission periodically audited them too, no one in America appeared to be thoroughly monitoring what was going on at the US end of the production line – the sumps into which all these damaged horses were fed, and from which the pipeline drew. APHIS checked vehicles and paperwork but not horses at the Mexican border. States applied different standards to livestock auctions. If no state livestock inspector or federally accredited vet was present at the sales, the auction house owner decided which horses could or could not be sold, and hoped that any surprise inspection concurred. Given that the owner of the auction house I visited bought and sold horses for meat from his own barn, the nearest there was to a halfway independent official was the vet who ran blood tests for equine infectious anaemia in horses that would be kept in feedlots and had lost their paperwork.

When he arrived, I tried to get his attention as he stood by the chute that would be the exit from the ring. I leaned against the boards.

'There's a horse over there ...'

'OK.'

'She looks like she's got bone sticking out of her leg.'

'OK.'

'She can't put weight on it.'

'OK.'

'I just thought you should know, because it looks bad.'

'OK.'

He didn't meet my eyes or offer to see the mare later. His young female assistant stared at me. The conversation was apparently over.

It wasn't just the visible state of the horses that could be missed or ignored in checks. Any horse treated with one or more of a hundred common veterinary drugs should be temporarily or permanently removed from the meat market. The dewormers, ulcer medications and painkillers of modern horsekeeping are not for second-hand consumption: bute, the 'aspirin for horses', can cause a type of anaemia, and the skin ointment nitrofurazone is a carcinogen. Not too long ago, a significant chunk of the stickered auction horses were thoroughbreds with light aluminium racing plates still on their hoofs – taken direct from some unlucky last-chance race – and dosed for years with anti-inflammatories, diuretics and steroids when they should have had box rest or retirement.

No deaths have been linked to bute in meat, but nor has a safe limit been established by the European Commission, which chooses to ban it outright in horsemeat, imported or home-made. And yet along with names, identities and breed certificates, the veterinary records of horses were sloughed off, and the checks by the Canadian and Mexican authorities didn't seem exhaustive enough to make up the shortfall. The Equine Welfare Alliance had shown me photographs of American equine identification documents or EIDS belonging to slaughter-bound horses in 2010: the forms were blank but for hastily scribbled and dated signatures; often the tick box that stated that the horse had not been treated with medication hadn't even been checked.

In October 2012, a slaughterhouse in Quebec began to turn away kill buyers' trucks after the European Commission, alarmed by drug residue reports, raised their concerns about what was in the horses. The supply line bottlenecked briefly, and both the Commission and the slaughterhouse blamed one another, but when the block was lifted, the Commission decreed that from summer 2013 onwards, lifetime records would be required for horses that ended up as meat in the EU.

At the low-end auction in upstate New York a few months later, I'd seen the kill buyers in the ring anxiously flipping the lips of any horse that looked like a thoroughbred to look for the unique numbers that are tattooed on each registered racehorse's gums. Those horses were not sold. Many anti-slaughter advocates had thought that the European intervention meant the end of the trade in racehorses at least, but by the summer I visited the auction barn, people were wondering where exactly the unwanted racehorses had gone.

Jane thought she knew. 'The racehorses went underground,' she said. 'Now they just go to some middle man who has access to the track. A truck pulls up at night, next morning the horse has gone.' They had vanished deeper into the sorting machine.

IV

By the nineteenth century, the rendering of horse carcasses had become an industrial process that was already part of a global trade. Horseflesh was abundant – it littered the streets in the form of all those broken-down cabbers, bussers, trammers and vanners. The knacker's wagon was a familiar sight, with up to five dead horses heaped on its boards and live horses tied behind it as it processed slowly to its final destination. Some knackers rehabilitated and resold horses at a profit, but often horses not immediately killed were left to starve out their last days pathetically in the yards behind the slaughter-houses, fumbling for vegetable scraps or weeds. In London, according to Henry Mayhew, slaughtermen would 'retire after a few years, and take large farms'.

When the end came for each horse, its mane and tail were hacked off as close to the root as possible, and borne off for upholstery stuffing, barristers' wigs, bow strings or fishing rods. The horse was then poleaxed or had its throat slit or heart stabbed. The blood was collected for fertilizer, for making buttons, for sugar refining, for poulterers. The carcass was hung from beams, and the skin stripped to be tanned into patches for German cavalry breeches, or to make the roof of a London handsome cab, or a driver's whip.

The body was jointed into hindquarters, forequarters, crambones, throat, neck, briskets, back, ribs, kidney pieces, heart, tongue, liver and lights. The meat and offal were boiled for cat and dog food in copper pans 9 ft wide and 4 ft deep – a little raw for cats, and stewed for over an hour as 'tripe' for dogs, two hours at least if the horse was old or diseased. Some flesh was left under hay and straw to generate maggots, which were sold to gamekeepers for rearing pheasants, or as fishing bait, or used as a lure for rats, which were in turn killed and their skins collected.

Tendons became glue. Old nails went back into steel manufacturing. From Berlin the old metal circlets of horseshoes were shipped to China, where they were straightened and sharpened into razors. The hoofs themselves were boiled for glue, combs, toys or Prussian blue. Bones went to fan-makers, turners, cutlers, makers of ivory black and sal ammoniac. More bones – 'racks' – were hacked up and fed into a digester, the grease boiled off for soap and candles, and the oil collected for lubricants for cartwheels, or to supple leather, for polish enamel or glass. Albumen was separated to be made into photographic film. The remainder of the bones also became fertilizer.

This phenomenal efficiency sealed horses into the industrial, modern world: not only did they power it, but their dung, their bones, their blood raised the crops that fed the horses that drew the cabs made with the skin of horses, whose wheels were perhaps greased with horse oil, whose candles in its side lanterns were part horse, whose driver buttoned his great overcoat with horse, and the dog that ran alongside the wheels was powered by horse, too. All was used, nothing left; a small army of workers absorbed the old horse back into the economy in atoms. As an act of recycling, *avant la lettre*, it was complete.

A few years before the French Revolution, in 1786, the physician Mathieu Géraud had suggested that his countrymen's refusal to eat those flesh scraps reserved for dog and cat was a waste of good red meat. This was a new elevation of the thrifty peasant's policy of consuming their old work-horse: an educated man of science endorsed hippophagy on rational grounds as a response to a horse-powered industrial, urban world. But European hippophagy tipped from taboo to acceptability due, once more, to war.

The first European country to legalize the sale of horsemeat for human consumption was Denmark in 1807, after Napoleon's armies laid siege to Copenhagen and its residents were forced to raid the ranks of cab horses and private stables to survive. At the Battle of Eylau in Prussia that year, a doctor in the French forces, Barron Larrey, ate horsemeat and fed it to his hesitant patients in a soup mixed with gunpowder, finding that their wounds healed quickly and that incipient scurvy was cured. A year later, Iceland voted to feed prisoners and paupers on horsemeat. Horses were eaten in the sieges of Hamburg and Genoa. In the terrible retreat from Moscow in 1812, soldiers were observed killing and immediately eating their mounts to survive.

Two years earlier, a senior French judge had established a commission to investigate the potential of hippophagy, and returned a unanimous positive verdict: horsemeat was safe and nutritious – only superstition held people back. The 'superstitious' public were not impressed. Horsemeat might have been eaten by those who lived near knackers' yards, but it was not desirable in the average French larder.

But even if the French were reluctant, in Germany and in Scandinavia, where those last heathen horse-eaters had held out, a series of hippophagic banquets were launched in the 1840s. These were thrown by doctors and by local societies for the protection of animals, who argued that if a horse were used for meat at the end of its life, kind treatment would be guaranteed to it to increase the value and tenderness of its flesh. Those nightmarish walking hides in the knackers' yards would be replaced with well-fed creatures.

From 1841 onwards, German states began to approve the sale of horsemeat for human consumption. Austria and Belgium joined the equivore nations in 1847, and by 1854, Vienna was shifting 32,000 lbs of horsemeat in a fortnight. Switzerland, Norway and Sweden followed. In 1848, Berlin killed 147 old or broken horses for 61,000 lbs of meat, and butchers attacked a restaurant where horsemeat was served. As the revolution of 1848–49 broke out, a veterinary professor argued that if Germans were to eat the sacred food of their ancestors, they would recover the strength of Arminius's warriors, while doctors offered therapeutic horse-broth baths in which children could bathe like tiny pagan kings.

Every one of these countries had a tradition of celebrating the horse as a companion in war, at play and in sport. Each had artists who glorified the nobility and beauty of the animal. Consumption did not go unopposed, but their concept of the horse was flexible enough to embrace its new role as a source of nourishment for the pragmatic and hungry. Two European countries held out: France and Britain, and Géraud's heirs, the men of science who became pro-hippophagic campaigners, were more vigorous in France.

Isidore Geoffroy Saint-Hilaire, head of the natural history museum in Paris and member of the Société Protectrice des Animaux, proposed in 1857 that it was time to take a rational, regulated approach. Do you realize, he asked readers of his 'Letters concerning nutritional matters, and above all, horsemeat', that many working men did not have the luxury of eating meat? How could the Republic tolerate this inequality when the resources were at hand to provide proper nutrition for the labouring classes? France could not rely, like Britain, on its colonies to provide meat. Hippophagy would be not just philanthropic, but also philippic:

> The horse will be treated like oxen, the day when ... he will be called upon to give us his flesh, after having given us his strength: the day when, grown old, he represents nothing more than leather, hair, bone char and a little bone meal, but also, commercially, 224 kilograms of meat. Then we must do something new for him. We let him rest, put him up somewhere, feed him so that he's not a loss; and above all, we heap no more blows on him: because to beat him would be to add to his distress a damage to ourselves: we risk spoiling the merchandise.

At the veterinary school of Alfort in France, at Langham's hotel in London, at the Grand Hotel in Paris, elaborate dinners were thrown to which journalists, scientists and other distinguished men were invited.

At the Grand Hotel, Flaubert was among 129 'of the most distinguished literary and scientific men ... including twenty-four physicians, eight veterinary surgeons, sixteen editors and men of mark' who ate three horses, 'one eleven years old, costing thirty-five francs; a second, eighteen years old, costing twenty francs; and a third twenty-two years old, costing forty francs', although these animals were so emaciated that they did not provide enough meat for the meal.

The French military veterinary surgeon, Emile Decroix, lent his chef, also a former soldier, for the London banquet, and he served up two carthorses and a carriage horse that had, in a more salubrious time of its life, been worth 700 guineas. The elaborate menu of twenty-nine dishes included a Pegasus filet, 'purée de destriers' served with Amontillado, sirloin stuffed with Brussels sprouts 'centaur style', jellied hoof with cherry liqueur and 'boiled withers'. The lobster mayonnaise was made with colourless 'Rosinante oil'. As a finale, an enormous baron of horse weighing 280 lb was borne into the room on the shoulders of four cooks, while a Beefeater in full dress played 'Roast Beef of Old England' on the trumpet.

Photographs of the horses were handed to the guests. The reaction was mixed. 'I do not believe,' wrote one diner, 'that it could have been distinguished in taste from excellent beef.' Others pointed out that if you had an artful French chef, he might as well serve shoe leather in a sauce and it would taste like beef. There was something dishonest about this cheval, compared to honest British beef.

In France, voices were raised against the practice. In his own *Lettres sur l'hippophagie*, Dr Jean Robinet maintained that to eat a horse was to be a barbarian: 'It is not from economic ignorance, nor physical disgust that man has ceased to eat this or that animal, and the horse in particular, but as a measure of how far removed he is from the primitive brutality he refuses to devour for the price of his services the companion of his labour.' Would the horse be followed by the dog, some asked, and even – surely – by cannibalism? The artist Daumier drew a series of hippophagic caricatures, one showing a nightmarish horse, mouth screaming, eyes terrorized, standing on the chest of a horrified man in his bed. Others feared that chevaline would be a new vector for glanders.

Undeterred, Decroix offered a purse of 200 francs to anyone opening a horsemeat restaurant, and 300 for a horse butcher. Samples were given to the poor, and priests were encouraged to set an example by eating it. In the mid-1860s *boucheries chevalines* began to open in Paris, where the municipal council exempted them from sales tax. Strict regulations, as elsewhere in Europe, were enforced: chevaline must be sold separately, in specialist shops that were clearly marked with signs for the literate and illiterate, so that no one could be deceived by this tricksy flesh. A new firm called Rollin and Co. kept stables and pasturage for sixty horses, and used steam-powered machinery to turn them into sausage that was lauded at gastronomic competitions.

And yet many of those intended to benefit from equivory resisted. Servants did not want this inferior meat if their betters did not indulge. Few French cookery books of the period even mention horse. It took the siege of Paris in 1870 to starve the people into acceptance. 'If you had come [to dinner], o beauty I admire,' wrote Victor Hugo to the poet Judith Gautier during the siege, 'I would have killed Pegasus and had him cooked so I might serve you the wing of a horse.'

In Britain, despite Decroix's offer of a medal and 1,100 francs for the first *boucherie chevaline* in London, horsemeat did not really catch on. *Reynolds Magazine* complained in 1866, 'The privileged and wealthy classes are to have a monopoly of beef and mutton; and the poor, politically excommunicated working man is to content himself with the flesh of the horse.' Some of the poorest were grateful, however, although they had probably been eating it anyway. One knacker told a *Pall Mall Gazette* reporter in 1889 that he 'had often seen street gamins buy a slice of dried horseflesh, place it between two bits of bread, and eat it as a sandwich!' But the economics did not quite work in Britain: feed the horses more, and their flesh was not such a bargain any more.

But above all, the new nineteenth-century practice introduced by these distinguished men of science with their common sense, their banquets and experiments, just built on horsemeat's old Roman rep-utation as the food of the poor: stringy, coarse, of dubious provenance and hygiene, salted, stored and barrelled, food for dogs, food for cats, food of the desperate – who would not be eating filet à la Pegasus. The hippophagists' philanthropic argument stigmatized it socially, and their 'philippic' argument only drew attention to the fact that the meat came from the broken and labouring animal.

V

'There are some things in which we do not copy la belle France,' wrote one American newspaper commentator in 1870 in response to the discovery that a Cincinnati butcher was selling sausages stuffed with horsemeat. 'We have not yet felt a disposition … to substitute horse-flesh for beef.'

America contemplated Europe's plunge into hippophagy and con-cluded that, although it could see the rational arguments in favour of serving up a baron of horse, it would rather not partake. European appetite met American distaste, for, while France had to turn to its cab horses and England to its colonies, America had produced from its earliest days an abundance of meat – and with it the inbred expectation that meat would always be cheap and of the best cuts. The open, seem-ingly endlessly bountiful landscape was only temporarily drained by the Civil War, which was followed by the opening of the West for cattle ranching and the flowering of the nation's great romance with beef. Steers were rattled eastwards to industrial packing plants in cities like Chicago and Cincinnati, and from there refrigerated railcars ferried the jointed and frozen beef to the nation's larders.

You could picture a steer grazing on a green, sanitary range out West before it was fetched into town as a nice neat joint, but a horse – as Europe's hippophagy promoters had discovered – was an ever-present, sweating, excreting presence in the city, and it wasn't an appetising spectacle. Forty-one dead horses were removed from the streets of New York each day, along with 40,000 gallons of urine and 4 million lb of manure. 'The horse sweats; the Ox does not; and the meat of the horse seems impregnated with an odor of persipiration,' the writer of the Paris letter to the *New York Times* informed his readers in 1875. As earlier Christians had been repulsed by pagan hippophagy, so Americans felt their gorge rise at the thought of being served one of the carcasses they'd seen on a street corner.

And yet America had a ghoulish fascination for the hippophagic goings-on of the Old World, the grand banquets, the butcher shops and the abattoirs. Saint-Hilaire and Decroix's names required no explanation in the numerous accounts that popped up in American newspapers, whether tiny one-liners on the price of, say, a horsemeat dinner in Vienna, or long, detailed descriptions with pen-and-ink sketches. Ida Tarbell in 1890s post-siege Paris called the 'dreary' abattoir she visited a 'chamber of horrors', confessing, 'I have looked into the revolutionary prisons of Paris and upon the guillotine with less shivering than I felt in this place.' Frank G. Carpenter reported from Berlin in 1893 on the way 'Germans kill, cook and eat the noblest animal known to man. Gallons of soup flowing down throats.' He described the death of a 'magnificent black carriage horse' as 'more like a murder than anything else'. At the horse butcher's nearby, children 'ravenously' devoured bowls of 'horse-bone soup', and the butcher's wife was a grotesque in the accompanying illustrations, with a tiny-eyed conical head, a dark sausage and a sharp knife held in her hands. Carpenter describes her 'cheeks fat and rosy on a diet of horsemeat'. Warming to his monstrous theme, Carpenter gives a recipe for 'mayonnaise of brains' and explains that the shoe leather called porpoise hide was in fact horse.

Not only was it repellent, horsemeat also represented a backwards Old World dogged by war, revolution and social breakdown. The chaos and anarchy of the French revolution, the Napoleonic Wars and Paris siege had lingered in horsemeat like a contamination. American press reports told of riots in Paris in 1870, where a sudden price rise led women to pelt butchers with horsemeat sausage. Groups of Russian students who shared lodgings and communally bought horses to eat were under suspicion as 'so many Nihilists have been found in these communities that the horsemeat eaters are liable to prosecution by the government'. In 1902 students and Tartars in St Petersburg were described as being 'addicted' to horsemeat, and in 1905, in the wake of the Russo-Japanese War, a Kentucky newspaper added, 'There is now almost universal conviction prevailing throughout Europe and America that if ... the rank and file of the Russian army had eaten more rice and less horsemeat, and had drunk more tea and less vodka, they would be much better fighters than they are.'

The new distance between diner and food source also led to suspicion about the quality and content of food in general. Additives to 'freshen' food were developed in the 1870s, making it hard to trust appearances. Several scandals shook public confidence, and the meat industry was especially distrusted as it was believed to both fix prices and to pack sausage with rotten or illicit meat – like the Cincinnati butcher with his 'abomination' of horseflesh sausages.

In this atmosphere, horsemeat was already at a disadvantage, linked as it was to poverty and suffering. 'I had the cholera, nightmare and sea sickness, all at once, and all night,' wrote the *Cincinnati Gazette*'s man in Paris in 1870, after trying horse. 'It makes me belch and heave to write about it, twelve hours after. Every piece of meat I see now transforms into the stringy, coarse, strong, indigestible stuff which – ugh, I shall never allude to it again as long as I live.'

There were rumours from England that horse bone was mixed into waste butter with 'drugs' for colour and flavour, and dried horse liver ground into coffee or mushroom ketchup. In one hysterical newspaper editorial, the writer revealed with disgust that envelope flaps are coated in glue made from 'old nags' that had 'yield[ed] up ... hoofs to the glue boiler'. And, the writer went on, 'should some taint of animal poison lurk amid that "gum", they may soon require other, and black-bordered, envelopes to be licked for them when their mourning cards are sent out'.

Clearly, chevaline was not a meat that belonged on the carefully laid dinner table of a comfortable all-American home as it strove towards prosperity.

VI

But two qualities that were the making of nineteenth-century America – its entrepreneurial pluck and mass immigration – did generate attempts to create a chevaline industry. In veterinary colleges, a Brooklyn 'freak club' and the homes of curious medics, America threw its own hippophagic banquets with a more homely flavour – 'broncho bullion' and ketchup replacing the purée de destrier and Rosinante oil. Meanwhile, newcomers from Germany, France, Austria and Switzerland built abattoirs and set up stall, sometimes without realizing that any form of *boucherie chevaline* was as good as illegal in their chosen state. A 'Patriotic Frenchman' called Henry Bossé opened an immaculate abattoir-factory on Newton Creek in Long Island in 1889 making square, chubby 'saucisson Boulanger', 'saucisson Lorraine' and 150-lb 'saucissons d'art' for export to France and Belgium. His facilities were visited by the local health committee following complaints. Bossé, an exuberant showman, sliced off and ate a cut of raw sirloin before the inspectors, all of whom declined to join him.

He had already had to promise that he would move again in three months. Local farmers reported that their horses would not go 'within a mile' of the small factory, and public opinion held that the board of health 'ought to clean the place out anyhow'. The *New York Sun* reported that demand for sausages at local restaurants had fallen, adding, 'There is no use of talking sausage to a man with a delicate appetite nowadays.'

One 'thrifty young' German immigrant called Henry Bruckman was found slaughtering and smoking horses for his family and neighbours at a quiet canal-side site near Chicago in 1890. Health inspectors doused the meat with kerosene and set it alight, refusing to believe his claims that he wasn't selling the meat under false pretences. His wife was defiant, 'Why, my neighbor used to send up here every morning for a steak, and if you look at her children you will see for yourself that they are fat and healthy.' Bruckman said he was innocent. He was newly arrived, and 'could not understand the laws of a free land which forbade what he thought was a commendable enterprise'.

One way to slake the American entrepreneurial spirit while working around American distaste for horseflesh was to can horses for export to the foreigners who *would* eat them. The best source of this meat was, state veterinarians and ranchers maintained, the 2 million-odd mustangs who competed with beef cattle for grazing in the West, as well as the work-horses raised loose on the western ranges for the cities and farms of America. As horses began to vanish from the streets at the clang of the electric trolley and the first parpings of the motor car, canneries popped up in the north-west, and thousands of range-raised horses and mustangs were herded through their gates. Canners claimed that a horse that fetched $3 on the hoof was worth $30 or $40 canned, although they sometimes had to pay bonds of up to $1,000 to back up their word that their products would not be sold in America.

But the introduction of federal inspections and a special official stamp for horsemeat in 1896 came too late, for the nascent industry was all but dead after a few short years: Europe distrusted American standards and didn't want to eat pickled horsemeat when it could have fresh. Then horse prices lurched up once more when British military agents went West to purchase cavalry mounts for the Second Boer War in South Africa, and the remaining canneries were dealt a further heavy blow.

Would-be horsemeat barons had to be tenacious: the fortunes of the American chevaline industry rose again when the outbreak of the First World War finally drove beef-loving Americans into the arms of chevaline. Meat exports were increased to feed a Europe that no longer scrupled about American hygiene standards, and US Food

Administration director Herbert Hoover cajoled the public into 'meat-less Mondays' in an attempt to avoid rationing at home as the shop counters emptied. 'CHEER UP!' called the *Evening World*. 'Nobody's said anything about Horsemeatless days.'

By 1917 St Louis had its own horsemeat market and was doing brisk business, as 'housewives who tried horsemeat have returned again and again'. America was rioting over rising beef prices, and yet here was a red meat that was cheap, and, now you've had a slice, really not that bad at all. The long-established American expectation of affordable steak was met. The small feral and range horses of the West that could not be shipped to Europe for battle could be eaten to clear the grasslands for the true red meat, beef, and as soon as rationing ceased, it was to beef that America returned with a sigh of relief.

But in the decade following the war, the first industrial-scale processing of horsemeat began with the advent of the new tinned pet food manufacturers, led by America's first true horsemeat baron: Phillip Chapel of Ken-L-Ration, who shipped the best frozen cuts of chevaline to Europe and canned the rest for American canines. Thousands of mustangs were crammed into railway cars, transported across America and processed in his plant in Rockford, Illinois under the eyes of a government inspector and a fluttering Star-spangled Banner. Later, unable to source enough wild horses to meet the demand, Chappel raised what he claimed was the largest herd of horses bred solely for meat in the world on '1,600,000 acres of America's richest range country.' His advertisements boasted that they were 'the healthiest meat animals ... scientifically bred to provide the meat your dog prefers,' by 'the only exclusive makers of dog foods in the world who control the breeding, raising and dressing of their own meat animals'. Despite his attempt to strip away the stigma surrounding 'unclean' horseflesh, American sensibilities were, however, shifting, and the old, prevailing disgust was giving way to a more morally driven disquiet.

In autumn 1925, a series of small fires were started in the plant and in the stockyard where the railway cars rested on sidings. These were easily dealt with, but Chappel began to employ armed guards who patrolled a new, 10-foot-high fence. It wasn't enough – in November, a huge fire broke out, and the guards found a Montana cowboy called Frank Litts among the flames, poised to detonate 75 lb of dynamite in a suitcase. He ran, they gave chase and opened fire, catching him with birdshot. He vanished. Sixteen hours later they found him bleeding in a nearby field. Cornered, he threw red pepper at the guards' eyes before being dragged away. The fire ruined one wing of the plant and did $80,000 of damage.

At his trial, Litts, who twice tried to flee the courtroom, rejected the suggestion that he was insane, insisting that 'the killing of horses is inhumane and contrary to Biblical ordinances'. The ranch he'd worked on in Livingstone, Montana, had sent horses to Rockford, and, so, declaring, 'I would rather see my body or my mother's body ground up and used for fertilizer than to have horses killed like they are here,' he had hitch-hiked to Illinois to save them. He was delivered a short time later to the state hospital for the criminally insane, which he escaped in 1926, making his way to Rockford to try to destroy the plant again. In late 1927, a matron from the city jail spotted him and had him arrested. One hundred and fifty pounds of dynamite painstakingly stolen from the nearby army camp was discovered under a pile of lumber. This time Litts did not get out – re-arrested, he made another jail break in 1931, and was shot in the lung as he scaled a wall. He died in 1938.

VII

'Instead of the public sale of healthy horses, under supervised control and by registered butchers,' Saint-Hilaire had written when he argued for legal hippophagy in France, 'we have the furtive sale of "suspect flesh" in attics, in cellars, by the firstcomer, by smugglers, by prostitutes, by disreputable men with no profession.' And yet, while the

familiar cycle of the legalized horsemeat industry carried on playing out for the rest of the twentieth century in America – when beef
was rationed or expensive, horsemeat functioned as a stopgap – the
stigma of hippophagy and the horsemeat business lingered despite
Chappel's efforts, embedded in the grim tales of foreign practices,
memories of wartime deprivation and hundreds of years of what the
old reformers had called superstition. When rationing outlasted the
Second World War, Republicans nicknamed President Truman 'horsemeat Harry' and told housewives there would be no more hamburger
if he was re-elected.

Worse, the 'disreputable men' carried on their 'furtive sales'
despite ever stricter government regulations. In Texas, a ban on the
sale of horsemeat was pushed through in 1949 after an investigation
revealed that Houston schoolchildren had been eating not just chevaline in the guise of beef, but chevaline laced with sulphite. During the
inquiry, a key witness narrowly missed being shot, two other witnesses had their homes set on fire and one of their maids had acid thrown
at her. In Chicago, a mob boss called Joe Siciliano bombed a restaurant that wouldn't buy the ground 'beef' he sold using old prohibition
networks in the early 1950s, and bribed the head of the Illinois Food
and Dairies division to overlook the 4.5 million lb of mixed chevaline
and beef hamburger that his gang had circulated through the state.
One kingpin in Siciliano's illicit chevaline network disappeared, never
to be seen again, and another was killed in a suspicious car accident.

Stronger now, too, were the feelings of those who shared with
Litts and Robinet the conviction that horses were companions whose
consumption was unthinkable. The dogged campaigning of Nevada
rancher Velma Bronn Johnston – nicknamed Wild Horse Annie – finally brought the beleaguered mustang under federal protection and out
of the slaughterhouse. Animal welfare organisations and politicians
began to demand an end to slaughter when horsemeat consumption
rose again during the oil crisis of the early 1970s, but although the
working horse had been eased out of commission, a new wave of
leisure horses – the trail horse, the amateur's showjumper or barrel

horse – entered into the pipeline. But little by little, America distanced itself from 'Belmont steaks' and horsemeat patties: by the 1980s, beef was so cheap that the prospect of chevaline being anything other than a niche product was remote, and it was even phased out of pet food in the 1990s. Horses were still killed for export though, even as profit margins began to be squeezed, and with them, standards.

On Monday 21 July 1997, a slaughterhouse called Cavel West in Redmond, Oregon exploded and a fire broke out, causing $1 million of damage before local firefighters could extinguish it. Jellied explosives had been poured through holes drilled in the part of the plant containing refrigerators of horsemeat, and 10 gallons more added to a nearby building. Muriatic acid was fed through the air conditioning vents. A spark had begun a fire in the refrigerator unit as electronic explosive devices were being laid, and with a boom and a roar the night sky was illuminated by a fireball, and the horses outside skittered about the holding pens in terror.

That January, the LA Times had run a feature headed 'Trail's End for Horses: Slaughter', which revealed that some 30,000 mustangs had been processed as meat in Cavel West after Bureau of Land Management (BLM) employees falsified their paperwork. The BLM's director had resigned and key witnesses disappeared, stymieing the trial. Now, it seemed, someone had taken action to avenge the horses. The Animal Liberation Front and Equine and Zebra Liberation Network claimed responsibility for the attack: it was their 'vegan jello' that had incinerated 'Cavel West Horse Murdering Plant', they wrote to the press, to 'bring a screeching halt to what countless protests and letter-writing campaigns could never stop ... you know what? The horses don't mind'.

At the end of November that year, the BLM facility in Burns, Oregon was targeted: gate locks were cut one night and 400 horses poured out before fire and explosives destroyed buildings and caused nearly $500,000 worth of damage. In 1998 mustangs were freed from BLM facilities at Rocksprings, Wyoming, where primed explosives failed, and again in 2001 from Litchfield, California. For years, the same 'ecotage' group struck lumber firms, government forestry concerns, meat packers, SUV dealerships and even police stations, and it wasn't until 2004 that thirteen saboteurs were indicted. All but four of the suspects were picked up in 2005, having caused $48 million in damage.

The cowed defendants were given the choice of being charged with terrorist offences or helping the prosecution. One, William Rodgers, committed suicide in a jail in Flagstaff, Arizona in December 2005, leaving a note to say that he had 'fought for all things wild'. 'I am just the most recent casualty of this war,' he wrote. 'But tonight I have made a jailbreak – I am returning home to the Earth, the place of my origins.' Cavel West abattoir never reopened.

Advocates who eschewed the Animal Liberation Front's drastic tactics began to make headway too. In 1998, Cathleen Doyle spearheaded a campaign to ban horse slaughter and make the sale of horses for food a crime in California. 'Prop 6' passed with 60 per cent of the vote. The evidence she had put forward included the videotaped testimony of a kill buyer who recalled collecting horses from 'nickel ads', trapping wild horses illegally and trawling auction houses to fill his contract. Under his stewardship, horses had endured 1,675-mile drives with no stops for watering or rest – sometimes up to 36 hours in overheated trailers. 'The truck just keeps rolling,' he said. One mare kicked him so he jabbed her in the eye to quieten her. Horses that bit had their mouths closed with wire. He bought mustangs that were running a fever so high they plunged their heads into a water trough for relief, and he filled them up with penicillin and shipped them before they could cost him any more. He'd known three firms try to make money from horsemeat in his area, and all were faltering. 'If there was enough money in it there would be more people in it,' he commented. A report commissioned by the USDA blamed the owners who sold injured horses to kill buyers, but also hinted that there was evidence that horses were beaten by slaughter personnel. 'To beat him would be to add to his distress a damage to ourselves: we risk spoiling the merchandise,' Saint-Hilaire had written. And yet the horses went unprotected.

By 2006 there were just three equine abattoirs in the entire country – DeKalb in Illinois, and Fort Worth and Kaufman in Texas – exporting $65 million of horsemeat, and all were owned by Belgian firms. For decades, the slaughterhouses had paid little attention to the anger of locals. DeKalb had regularly exceeded its sewage emission allowances. In Fort Worth, the Beltex plant had blocked the town sewage system, flooding a creek with blood. The plant paid just $5 in federal income tax one year, and the cost of keeping it in the city was far higher than any financial benefit it brought. In Kaufman in the 1980s, the Dallas Crown plant pumped waste into the sewers with such force that residents' baths and toilets filled with horse blood. It blocked city inspectors from entering for nine months, and later ceased to pay its fines altogether. Vultures sat on picket fences.

Kaufman's mayor, Paula Bacon, did not have difficulty raising local support when she lobbied for the full application of the old Texas law of 1949 banning the sale of horsemeat, and both plants had their legitimacy knocked out from under them. The Illinois state legislature banned horse slaughter in 2007 and the Supreme Court refused to hear the DeKalb plant's appeal.

'Horse Slaughtering – the new terrorism?' *Time* magazine had asked in September 2006, as the American Horse Slaughter Prevention Act reached the House of Representatives. 'With all the other problems piling up – soaring energy costs, the war in Iraq, tens of millions of Americans with no healthcare insurance, skyrocketing federal debt,' why, author Douglas Waller asked, was horsemeat the only bill being considered on the house floor? Its Republican sponsor, John Sweeney, nonetheless called horse slaughter, 'one of the most inhumane, brutal and shady practices going on in the United States today'. The bill passed by 263–146 votes, but did not clear the Senate. However, in March 2007 all federal funding for horsemeat inspections was suspended, and the distinctive green hexagonal horsemeat stamp retired, ending the hope of opening new slaughter plants in America.

The sorting machine reconfigured itself, the loophole slipped further along the line, and American horses headed for Mexico and Canada, with their advocates in pursuit, and the horsemeat firms running on ahead.

VIII

If the American horsemeat pipeline to Canada and Mexico were also cut, as the advocates hoped, where would the 100,000 horses saved from slaughter each year live? Who would keep those old buggy horses in hay? Who would pay for the vet to stitch Breezy Knoll Annie's open wounds, for an equine dentist to file the overgrown teeth of the emaciated Belgian draught, and for the farrier to restore those grey, cracked hoofs with vitamins and care? How many 'pasture ornaments' could America take? If fit and healthy horses ended up in kill buyers' trucks, was that not because there was no market for them?

After the federal ban on horse slaughter and before the financial crisis of 2008, horse rescues in Indiana, Virginia and North Carolina were reporting that the number of abused and neglected horses they were taking in had doubled in a year. The horse industry, like the housing sector, had been booming, and had suddenly faltered. When 2008 dawned, the effects of a severe drought were shrivelling the flesh of horses whose owners could not afford leaping hay prices. The number of rescue stables more than doubled from 300 in 2007 to 700 in 2011, but there was no system to regulate them, and no guaranteed funding to keep up the day-in, day-out expenses of all those deworming courses and hoof trimmings. More horses found themselves 'free to a good home' in the classifieds – Craigslist became the new 'nickel ads' – or circling the ring at public auctions.

Strange and macabre Rosinantes materialized across the country. Suburban LA residents woke to find a starved horse lying in the road outside their houses. Thirty or forty carcasses were discovered at the edges of Highway 65 in California in January 2012, dumped alongside defunct refrigerators, old TVs and household trash and picked at by coyotes. Ghostly, shaggy groups of horses with peeling, curling hoofs were found in corrals by law enforcement as their owners pleaded poverty. Two people tried to poleaxe a horse with a sledgehammer. More mustangs died in BLM corrals.

Elsewhere, strangest and most haunting, there were reports of herds of motley horses, old, young, unbroken, lame, thin, with nail holes in their hoof walls and brands on their rumps and necks – sometimes even a wound where someone had tried to cut a brand out of the horse's skin. They roamed wild along the Mexican border in Texas, picking a living from the drought grass and waterholes, or multiplied on land reclaimed from old Kentucky mines. In Nevada, Oregon, Wyoming and Colorado they joined up with mustang herds, got hit by trucks when they wandered on to roads or mauled by mountain lions. In Missouri they mingled with the descendants of horses that had been abandoned in the Depression. Some were believed to be horses shipped by kill buyers but rejected at border feedlots.

About all these apparitions – the botched, the thin, the feral and the wormy – America was engaged in a furious debate. In 2011, the Government Accountability Office (GAO) had published a report suggesting that the ban on slaughter was responsible for all the unwanted horses. The Equine Welfare Alliance produced a deeply researched report picking the GAO judgement apart, blaming instead a six-year rise in the cost of keeping horses, but the USDA still reversed the funding ban, and new plants were proposed in seven states.

The period that followed was full of activism and hysteria on both sides of the debate, some more constructive than others. A member of People for the Ethical Treatment of Animals (PETA) tried to have a creek in Iowa named 'horsemeat forbidden' in Japanese. A man in New Mexico filmed himself ranting about animal rights fanatics as he held his horse, then shot the horse and put the video on YouTube with the title 'Tim Sappington's message to animal activists'. Extreme pro-mustang activists accused the BLM of 'mutilation of mares' for spaying wild horses, or creating 'equine concentration camps' at holding facilities for horses brought in off the range. News broke of a kill buyer who had shipped 1,700 mustangs to slaughter – claiming he was 'doing the American people a favor' as the cost of keeping the corralled wild horses ran into the thousands per head.

It was in the midst of this mania, a week after I visited the upstate New York auction on Black Friday 2012, that I went to the capital to speak to one of the leaders of United Horsemen, the official champions of America's reborn horsemeat industry.

IX

I met Sue Wallis in a busy Starbucks just off a hotel lobby in Washington DC. A Republican representative from Wyoming in her mid-fifties, she was a large woman with straight, drawn-back brown hair and small pieces of turquoise Native American jewellery. She was a Westerner and Republican of a particular gristle, pro-freedom, anti-big government, you leave me alone, I'll leave you alone. Her interests took in raw milk, abortion, medical marijuana and gay marriage – all of which she was in favour of. Sue handed me her card and a United Horsemen pamphlet entitled *The Promise of Cheval*, and we hunched over my dictaphone on the small, sticky table as it darkened outside. Once she throttled up and began to rattle off what felt like some well-rehearsed – but heartfelt – talking points, it was hard to interject.

The problem was, she explained, animal rights groups like the Humane Society of the United States (HSUS) and PETA were out to convince the public 'that all farmers and ranchers and meat processors are evil, uncaring monsters who just want to butcher and abuse horses and every other animal'. The activists did this, she explained, purely for the money. 'They say horses are cut up alive, but nothing could be further from the truth,' she went on, recommending that I watch a documentary by the animal behaviourist Temple Grandin that explained the odd post-death twitchings of the cadavers.

I asked about the technical problems with horse slaughter and drug residue, and she assured me the problem would be solved by the Food Safety and Inspection Service. 'They have committed to us that they will be done with that and we will have inspections by the end of the year.' The new slaughter plant designs would be inspired by the humane innovations of Temple Grandin* and supervised by one of her students. I should not worry, she said, about the stories I'd heard about the horrors of industrial abattoirs, because, 'The United States and the European Union are basically the civilised nations of the world, and here in the US our processing has been governed by one of the highest standards in the world.'

The United Horsemen's plans seemed Utopian rather than market-driven to me. Under their supervision, any owner would be able to register their horse on a nationwide 'Do Not Slaughter' registry, and if a stolen horse appeared at the abattoir, it would be identified and the owner alerted. How this would be achieved when it had never worked before was unclear, as was the funding source for all the administration involved. Unwanted horses could be brought to a triage system or 'Rehabilitation, Training and Education Center' in Hermiston, Oregon, where they would be evaluated, cured of any medical issues at the in-house veterinary surgery and turned into usable animals. Some would work in the therapeutic riding programme for 'people with emotional, physical and mental challenges' that was also located on site. Others would be given to youth programmes. The best would be sold with their new skills. The centre, the United Horsemen promised, would not be linked to the planned horse-processing plant nearby, and yet horses would flow between the two: the most promising in the

---

★ Wallis had earlier gotten into trouble by claiming that Grandin herself would design the plants – something Grandin denied vehemently.

slaughter pipeline would be saved by the qualified horsemen trained at the centre, and the worst horses that arrived at the centre could be trucked a short distance for an efficient end at the plant. It was a sorting machine where every horse met its just, infinitely humane destiny.

But as we spoke of numbers and regulations, Sue became more heated, and her talking points began to slide towards something resembling the patriotic paranoia of the Tea Party – a tendency I'd glimpsed when doing my research before the interview. One United Horsemen YouTube video maintained that the threat from 'radical animal rights activists' wasn't just attacking the meat business, but 'destroying our horseback culture and our American heritage and our freedom to own horses'. Without slaughter, in a HSUS-run dystopia, taxpayers had to pay for horses that had become welfare recipients living out thirty useless years. Another mail-out evoked Germany in the 1930s: 'Like Pastor Niemoller, you will wake one day, or maybe you already have, to find that they have come for your ... business and there is no one left to speak for you.'

As our conversation in the Washington DC Starbucks went on, I felt Sue's rhetoric rise like a helium balloon. Already there was a 70 per cent fall in the number of foals being born, she told me, jabbing a finger into the table as she made each point, and 'generations and generations of valuable genetics' had been lost. 'These people say stop all breeding, stop all slaughter, stop all commerce in horses, but when their old pet dies – and they're going to because we're all mortal – there's nothing to replace it, so what happens in just a generation is we lose our horses.' The opening of the new plants was, she concluded with passion, 'the best way to increase the welfare of the horse and to make sure that our kids and grandkids get to play with horses. That's my mission in life,' she said as she leaned across the table so I could catch each emphatic word, 'and we're getting very close.'

X

I wasn't able to follow up my interview with Sue Wallis, as she died unexpectedly in her home town in 2014, by which time the plans for most of the new plants had had to be abandoned. The mayor and council of Hermiston – which had once hosted a chemical weapons store – refused to even meet United Horsemen to discuss the Rehabilitation, Training and Education Center. Anti-slaughter advocates unearthed an undisclosed burglary conviction on the record of the owner of another proposed plant, and unpaid debts owed by a third. In Missouri a community meeting featured cries of 'Go back to Belgium!' and accusations that horse slaughter was no better than meth labs and puppy mills.

The anti-slaughter campaigners also found themselves balked as they tried to finally end the cross-border trade on food hygiene grounds with the Safeguard American Food Exports Act. The act was introduced to the House of Representatives and Senate but remained mysteriously blocked years later. By 2015 the European Commission had taken the initiative, suspending imports of horsemeat from Mexico and informing Canada that it had 'serious concerns' about 'incomplete, unreliable or false' paperwork, but the meat that the EU rejected found a ready market in Russia and in Asia instead, and the horses continued to rattle to Canada and Mexico in the silvery trucks and trailers.

The European horsemeat industry was reeling from the continent-wide 2013 scandal, which had been, I realized when I looked back at the history of horsemeat, inevitable: the recession had caused a fall in the price of low-end horses and a rise in the cost of beef. All those dodgy tins and suspect sausages of the nineteenth-century had simply transformed into cheap supermarket burgers and microwaveable lasagnes. Once more the legitimate horsemeat business had become Saint-Hilaire's 'furtive sale of suspect flesh'. It seemed to me, sifting through heaps of old newspaper reports, food hygiene manuals, manifestos and legislation going back two centuries, that the cheva-line business was locked into an endlessly repeating pattern: the odd one out in the butcher's canon of beef, pork, lamb and poultry; the cheap, corrupt meat; the philippic solution that led to the sumps of wounded, skinny horses; the philanthropic gift that ended up lining the pockets of the 'smugglers' and 'disreputable men'.

Jane and I were at the barn from early morning till late afternoon on that July day in 2014. When the auction of the horses began, we looked down from the catwalk as the flow of horses in the pens changed, and their agitation rose. They were herded in groups of six or seven down the chute that led to the gate on to the auction room floor, then penned into a holding area, where I saw one pair begin to frantically groom each other for reassurance, chins nested against withers, necks entwined. Each horse raced forward when its time came, ears flicker-ing, looking to the only way out, and the horses behind it were pushed back with boards and shouted at. The sing-song of the auctioneer started up, and the first horse was driven through the gate and out of our sight, as if out of the wings and on to a stage.

In less than a minute, another gate opened and the same horse trotted back out from the sale ring, under the catwalk into another chute lined with large pens. Here it stalled, caught between uncer-tainty and a need to get away from the ring with its sudden audience, before being driven forward by two teenage Amish boys armed with long, flexible whips with flags on the end, who smacked its sides. The boys opened one gate and the horse plunged through it. The gate was shut behind it.

As horse after horse came through, jumpy and rattled, the boys hushed at them, and opened and closed the gates of the pens to sort the horses: two smaller enclosures took horses that were not going for meat, the larger pens were for the kill buyers. These began to fill like ponds overstocked with koi, the horses lashing and circling for space. Up to fifteen horses accumulated in a 20 × 20 ft pen before the horses were all herded back out into the corridor, kicking and squealing at one another, and into a set of larger pens at the back of the barn.

On went the sing-song, and the horses going for $100, $50, filling the kill buyers' pens and then emptying them out. The horses flowed on, into the barn, into the pens, out along the chute and into the ring, back under our feet like a river, and into the pens: kill, kill, non-kill, kill, kill. As horses scrimmaged up against the sides of the pens, a board came loose, and a big draught horse staggered, trying to kick back at a horse that had double-barrelled it. Every now and then one horse would cry out and a chorus of whinnies would answer. The pens filled and drained, filled and drained, each sale over brutally fast.

Breezy Knoll Annie with her broken knees, the starved Belgian, the witchy chestnut mare, the thin mule, the racehorses, the mustangs, the old rodeo broncs, the busted Quarter Horses, the thin buggy horses all streamed into an American auction house on a July day seven years after horse slaughter ceased in the United States, and flowed on into the rill, and then the river of brown, bay, sorrel, paint and grey that flowed out into the global economy, and emerged, finally, from the sorting machine in Belgium, in Japan, in Russia or in China as anonymous dark red discs of chevaline, as cherry-blossom sashimi, as tartare and as beshbarmak stew.

WEALTH

Knight Dreams and Heavenly Horses

The Heavenly Horses are coming,
Coming from the Far West ...
I shall reach the Gates of Heaven
I shall see the Palace of Gold.

Sacrificial hymn, Han Dynasty (101 BC),
translated by Arthur Waley

I

The chilled joints of American buggy horse aren't the only form of horseflesh being transported into China today. Beyond the broken by-products of industrial society there are the other horses, *Equus luxuriosus*, who became all but extinct in twentieth century China, but are now tentatively establishing themselves on the mainland as highly skilled equine migrants. After thirty years of rapid economic development, China has been catapulted from famine and privation to a new, wildly successful capitalism, and as its dollar billionaires and millionaires cast about to spend their wealth, the global equestrian industry has responded by flying in *Equus luxuriosus* in the cargo bays of jets in padded boxes, alongside all that cheap chevaline for noodle stew.

What kind of animal is *Equus luxuriosus*? He does not work like the drays, the buggy horses and the trammers. He's a horse who eats better than his groom and has a family tree as carefully tended as an ancient yew. He knows only custom-made $10,000 calfskin saddles with titanium stirrups, and a $1,300 cashmere-wool rug woven with an 'H' for Hermès. Like the first horses to arrive in Mesopotamia, he is worth the valuable grain expended on him. Like the Spanish horses, the English thoroughbreds and the Moroccan Barbs in the stables at Versailles, his origins are often exotic. His coat of peach blossom, satin or ebony is buffed with soft brushes and lustred with safflower oil, and after he has swum in his pool he is dried under a solarium which plays him music specially composed for equine ears.

In China he populates fragrant equestrian clubs or polo resorts that promise 'a world of utmost exclusivity' where the 'new nobility' pay $165,000 for the privilege of learning to play the sport of kings. His job is to generate status: an infinitely purchasable quality that is amplified by the number of expensive hoofs striking the floor of one's air conditioned stable. The number of rosettes one's daughter wins. The transubstantiation of money into leather-bound, handmade whisky flasks, dry acid champagne, faux-Italianate villas and horseflesh.

In China this is *mianzi*, the reputation and 'face' acquired by 'success and ostentation', and it is to the new wealthy what *sprezzatura* was to the nobles of the Italian Renaissance: in 2013 just 2 per cent of the Chinese population bought a third of the world's luxury goods. This new elite are considered vulgar jokes in the People's Republic: the word *tuhao* is dropped, and it is not meant kindly: *tu* means vulgar and *hao* bullying. Questions are raised in the daily papers as to whether these big-spending tycoons with their Rolls-Royces lined with jade, $2 million Tibetan mastiffs and golden iPhones are good or bad for the economy, especially if they sink much of their yuan into foreign property, overseas casinos, oil paintings and Napa vineyards.

In autumn 2013, on the eve of the Year of the Wood Horse, I wheedled a brief visa from the government of the People's Republic of China and flew to the land of the *tuhao* in Beijing because I knew that in China's long history I would find the perfect illustration of the conundrum that has cropped up in so many societies since the horse was domesticated. Horses that do not work – or what they come to symbolize – are often an illusory, transient embodiment of prosperity, always desired but draining resources in return. *Equus luxuriosus* can be a Veblen good – one that gets its value from the very fact that it is expensive. After all, few horses appreciate value across their life, and all horses grow old and die; unlike the Renoirs and Picassos in Shanghai auction houses, no amount of air conditioned storage will preserve them indefinitely: a $10 million horse will break a leg as easily as a $10 pony. As the saying goes, the quickest way to make $1 million from horses is to start with $10 million.

Even when horses were as the Han general Ma Yuan wrote, the 'foundation of military might, the great resource of the state', the economic complications of laying hold of sufficient numbers to defend and control such a huge population and territory were considerable, and these, too, are part of the story of horses and money. A warhorse, unlike a work-horse, made no direct contribution to the nation's balance sheet, and the maintenance of a cavalry that was not only sufficiently numerous, but also of high enough quality could be a considerable standing expense, even without taking into consideration the horses that drew supply wagons or carried messengers across the empire. This was a common enough conundrum for every nation but the Mongols – Henry VIII was perpetually short of horse power for his campaigns, alternately banning their export or demanding that European rulers provide him with thousands at a time, while in the nineteenth century the Prussians established their own network of state studs in an attempt to be self-sufficient. In China, as we will see, this dilemma was even weightier. Small wonder that from their first arrival, domestic horses in China were treasure houses, repositories of status, tickets to another world, foreign objects of luxury and a political necessity to be sought sometimes at too high a price.

Even hundreds of years later, elite warhorses remained somewhat foreign to China, always in perilous supply, and always western: the very term for warhorse, *rong ma*, translates as 'western foreigner's horse'. And they proved – and still prove in their new embodiment as the racehorses, showjumpers and polo ponies intended to take China to the world stage in equestrian sports – an expensive asset at that, one that Chinese bureaucracy has proved unable to conjure up at will. In settled, agricultural China, roaming, carelessly grazing horses gobbled resources, and the country would ruin itself in silk and tea many times in its efforts to fill its pastures with rong ma and win the long struggle against its enemies on the steppe. Inevitably, the horses themselves became symbols of superiority and difference for their owners as horses, wealth and power intersected throughout Chinese history.

In 2013, I was to find that everything and nothing had changed.

II

When the people of the steppe poured out of their open grasslands in the Bronze Age, they headed west across Europe, recasting the continent's inheritance and filling its graves with chariots and horses, and also east to the northernmost reaches of today's China on the Upper Yellow River, where a group called the Qijia buried their dead with horse jawbones, gold ornaments, copper mirrors and socketed bronze axes from western and southern Siberia. Whether the horses were domesticated and if they were acquired along with the axes and mirrors remains unknowable, but horses next resurface in the tombs of the prosperous and sophisticated Shang, who came from south and east along the Yellow River to conquer the Qijia.

The second Chinese dynasty after the semi-legendary Xia, the Shang built large palaces in walled cities, were brilliant metalworkers, agriculturalists, miners, refiners and transformers of metal ore, but the roaming horsemen at their borders influenced them hugely, and they, too, placed horses in the graves of their aristocracy as an affirmation of power. One Shang tomb at the capital of Yinxu contained no fewer than thirty-seven horses. Other pits held high-wheeled wooden chariots that had been painted with red lacquer, decorated with banners, gold leaf, cowrie shells, bells and trailing ox tails and were used

A London Cab by Charles Cooper Henderson (1803–77).

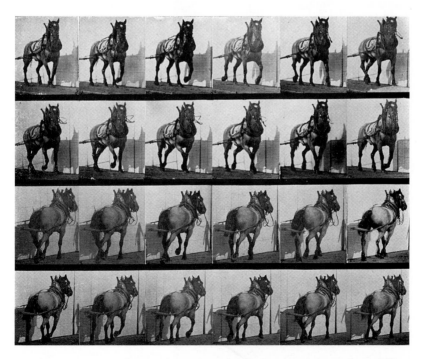

Carthorse sequence from Eadweard Muybridge's 'Animals in Motion', 1880s.

Used Up – a scene at Smithfield market, English School, nineteenth century.

Horses in a holding pen at Cavel West abattoir, Redmond, Oregon, 1996.

Berlin animal welfare campaigners present a buffet of horsemeat, 1903.

US-government-inspected horsemeat goes on sale in Washington DC in 1943.

The famous second-century Chinese bronze of a horse stepping on a swallow, found in a Han dynasty tomb in Gansu in 1969.

A woman of the Tang court plays polo.

Giuseppe Castiglione's 1758 portrait of the Qianlong emperor of the Qing dynasty.

A 1998 re-enactment of Tang-era polo, Xian, China.

Rejoneador Pablo Hermoso de Mendoza fights in *traje corto* in Colombia, 2014.

In the margins of the *Luttrell Psalter*, a horse kicks out at a soldier, 1325–35.

The Battle of San Romano by Paolo Uccello (1397–1475).

German cavalryman wearing a gas mask and carrying a lance, 1917.

Another image from the First World War: *A Horse Ambulance Pulling a Sick Horse out of a Field* by Edwin Noble.

A member of the Caisson Platoon waiting for a funeral service in Arlington Cemetery, Virginia, 2007.

not for plunging directly into combat, but for hunting or perhaps an elevated vantage point for commanders.

These early encounters between the settled, agriculturalist Chinese (the 'sown') and their barbarian, horse-riding neighbours on the steppe set a pattern which would define China for thousands of years. The barbarians and their precious horses would come, repeatedly, from the grasslands north of the Yellow River in what is now Inner Mongolia, from Mongolia itself, and from Tibet in the west. They would come also from Central Asia along the proto-Silk Roads north and south of the 1,000-km-wide Taklamakan Desert in what is now Xinjiang.

Steppe and sown relied on one another for trade, exchanging food, goods and, of course, horses, which became one focus of the struggle between them. The steppe had horses in abundance, the sown needed horses to defend itself from the steppe, and so horses would be the embodiment of both an external threat and of development. They would mean being conquered and being conquerors.

'We will lead a campaign against the Mafang,' read the pictograms etched on a Shang 'oracle bone' found in Yinxu, 'will the Di [a Shang god] support us?' Mafang literally means 'horseland' – the home of foreigners from the northern Shaanxi – and the little drawing for 'horse' with its huge eyes, forked tail and curving ears later evolved into the modern Chinese character. The Shang were perhaps victorious against the Mafang, but from the steppe came a new equestrian innovation, and no oracle bone predicted that at the Battle of Muye in 1046 BC the 170,000 soldiers the Shang sent forth against 48,000 Zhou from the steppe would be crushed. The Zhou had stocked their 300 light sandalwood chariots with archers carrying battleaxes and short swords and sent them directly into the battlefield – transforming them from transportation to weapons of war in their own right. The victory was enough to make the Zhou the third dynasty and initiate the longest reign in Chinese history.

Zhou aristocrats and kings gave chariots with tiger pelt roofs, bronze bells jangling on their yokes and gilt- and scarlet-painted harnesses to warriors pledged to them. Horses were also high tribute: one Zhou king called Mu was presented with a 'Dragon Horse' that could run 1,000 li a day, and had what appeared to be tiny horns above its eyes.* They were the first in China to race their horses, and every spring made offerings to the Ancestor of Horses, and every summer to the First Horse Rider.

But despite their horseyness and their longevity, the Zhou also had to deal with repeated attacks from north and west, and their reign tumbled into chaos after another group of Central Asian nomads drove the royal family east from their homeland in 771 BC, and centuries of warfare followed between competing Chinese states. These were the Spring, Autumn and Warring States periods in which Confucius was born and developed what would become China's guiding philosophy. China resounded to the crash of four-horse war chariots, and measured its rulers' worth in wheels as different states emerged and fought one another – the Chinese word for emperor, huangdi, would later use the character for 'chariot driver'.

The different Chinese states tried to breed their own horses, a task so important that the kings frequently had direct oversight, but raising large numbers of elite horses in central and southern China was challenging: the land either lacked calcium for building strong bones or was too marshy or already well farmed. So trade went on: China needed horses, and the nomads needed farmed food, because the steppe was too arid and windswept for crops, so horses were brought to market by nomadic groups in the autumn and bartered for Chinese goods. This exchange of horses, fodder, farmed goods, weaponry, silks, precious goods and allegiances spilled over sometimes into warfare, for while the Chinese were always greater in number, the nomads used horses – as weapons, as goods – to harry and curtail them.

★   A li has varied over history from 323–645 m, but today is 500 m or half a kilometre. So a '1,000-li horse' should in theory cover 500 km in a day – somewhat unrealistic given that the longest contemporary international endurance races are 160 km. The longest in the world is the 400-km Shahzada race in Australia, but this takes place over five days. Postal riders of the Mongolian era could cover up to 200 km per day on a single horse.

In China as elsewhere, the wealthy preoccupied themselves with the definition of the best horse – how should it be conformed, how kept and how did it move? Bronze statues of ideal horses were made for comparison with real horseflesh. 'A horse's head is king; it should be square. The eyes are the prime ministers; they should be radiant. The spine is the general; it should be strong. The belly and chest are the city walls; they should be extended. The four legs are the local officials; they should be long,' explained Jia Sixie in 544 BC. A *ji* was both a fine horse and a noble person, and a *qianlima* someone who was, figuratively, a horse that could cover 1,000 li in a day like Mu's Dragon Horse. Confucius stressed, 'A thousand-li horse is praised not for its strength, but for its virtue' or obedience, and it took the right man, like the famous horseman Po Le, to recognise a *qianlima* – human or equine.

A new force from the steppe swung the balance once again in the fifth century BC and made the acquisition of horses still more critical: the mounted archers who arrived from the north-west in the fifth century BC as the Chinese fought among themselves. On horseback a warrior was swift in attack and in retreat. He could swivel in his saddle and fire in any direction while his mount galloped on. His armour and equipment were light, and he had been raised to handle and ride horses from the earliest age, and to hunt as soon as he could hold a bow. In a small group of raiders, attacking by day or night, he could sweep away horses, cattle and food, torch settlements and vanish. Massed in ranks, he could make the heavens rain arrows. In the tumultuous periods when no single Chinese dynasty dominated, northern barbarians would swoop down again and again in raids and wars.

Groups like the ruthless Xiongnu of Mongolia, whose foals, historian Sima Qian says, could leap over their dams three days after they were born, could boast hundreds of thousands of archers and had better, stronger, swifter horses than the Chinese – they made little of the mountains separating the steppe from the plains. Hot on their heels were still more brutal and barbaric tribes, who pushed the

Xiongnu deep into northern China and the flat, grazing grasslands there. Often they became integrated into China, and their descendants even founded dynasties of their own. Not until the eighteenth century would the horsemen cease to be a threat.

China, too, had to adopt, assimilate and concede in order to survive. During the period of the Warring States, King Wu-Ling of the Zhao urged his reluctant people to dress like barbarians in trousers and fur caps, practise archery and ride their horses instead of relying on robes and chariots. He led the earliest recorded Chinese army to have divisions of mounted archery, and when he conquered some of these barbarians, he absorbed them into special divisions. Eventually the tumultuous period would end when the Qin, who were known for their horsemanship, living as they did on the edges of civilized China, with one foot in nomad territory, conquered each of the other bellicose Chinese territories and united them under their rule.

Their leader, the brutal warlord Qin Shi Huang, built a more powerful cavalry and expanded Chinese territory deep into the Inner Mongolia grasslands and south to Guangdong and the northernmost part of Vietnam. He drew the states together by constructing an extensive road system for trade and military supply, bringing uniform weights and measures, a standard currency and script. But even the first huangdi could not defeat the Xiongnu, and instead he fused several of the walls in the north against the barbarians.

When he died he was buried in a subterranean reproduction of the imperial capital Xianyang, a booby-trapped, lavish mirror empire so extensive that it has yet to be fully excavated. Horses filled the chambers: closest to Qin Shi Huang's tomb were white-painted bronze horses with half-scale chariots pancaked behind them and harnesses weighted with silver and gold, while 1.5 km away was an army equipped with hundreds of terracotta warhorses painted brown or black, with chests that jutted and manes hogged into a solid crest like a Takhi's, their forelocks split and spitcurled back around each ear. In ninety-eight more pits lay what was left of the imperial stables: the bones of real horses attended by young, beardless terracotta grooms who crouched by their charges. Fragments of leather halters, pottery and basins for feed or water were placed by their heads.

Qin Shi Huang's relatively short reign was followed by further civil war and the arrival of the Han dynasties, who inherited the friable, combustible boundaries haunted by steppe bowmen – now attacking in armies of up to 300,000 at a time. The Han traded silk for Xiongnu horses, but even if the Xiongnu were outnumbered by the Han, they outfought them with superior cavalry and weaponry. They had the grasslands to raise horses, while the Han needed grain to keep a cavalry – and a single horse ate the same ration as a family of six.

The first Han emperor Gaozu established the system of *heqin* or 'peace marriage', in which the Han bartered their princesses and tributes of Chinese wine, grain and silk in exchange for peace and horses. A lonely Han princess dispatched 3,000 miles from her home to marry a nomad ruler in the north-west mourned: 'Alas a yurt is now my home, its walls made of felt,/raw meat is my food here and koumiss my drink/All day my heart aches with pains of home:/Were I only a brown goose and could fly away.' In return, her family received 1,000 horses and promises of allegiance against the Xiongnu, who by this stage had broken the peace they had struck with Gaozu.

But for the Han peasantry, horses were a toll. They could not often afford to use them for draught, and the rolling grasslands they saw overtaken by the army were less productive than the carefully cultivated agricultural plots on which the peasants produced the food that fed the empire. While the aristocracy and wealthy merchants rode by on horses tricked out like those of the nomads with precious stones on their ear fobs and embroidered felt saddlecloths, the peasants raised eyebrows and bent over their hoes or whipped their oxen.

A later folktale is a perfect apologue: a village of peasants live on bark and grubbed up roots after their crops fail, while their rich landlord guards overflowing granaries. A wily peasant caulks one of their skinny, ill-looking work-horses with small silver ingots and a wodge of cotton, and takes it to the landlord. He promises him that the horse can shit silver, pulls out the plug of cotton and catches the ingots with a dish held under its tail, presenting them to the astonished miser.

The landlord demands the horse's price, and the peasant refuses gold, saying he will take only 30 bushels of grain. The landlord thrusts the grain upon him and takes the treasure horse into his home. He waits and waits, and finally is rewarded with a jet of pure, golden urine, cascading all over his woven carpet. The peasants plant the grain, knowing that they will eat again when the crop is raised, and gather more grain to plant the following season. To the peasants who made up most of China, the horse was false wealth, and the land, given over to grain, was more valuable than gold.

III

In the reign of the Han Emperor Wendi, a man who lived at the very edge of China accidentally let slip his horse, which disappeared into barbarian territory. His father counselled patience, and sure enough, the horse returned, galloping together with a much finer animal. Some years later, when Wu Di sat on the throne, another beautiful horse was found at Dunhuang in the Gobi, grazing with a group of Takhi. It was taller than the Takhi and the common Chinese warhorses, and swifter, too.

A soldier set out a dummy holding a bridle by the watering hole until the horses no longer took notice of it, and then hid behind the dummy and quickly wrapped the bridle round the beautiful horse's neck. He sent it to Wu Di's court, saying that he had fished it from the river itself. A nervous, thin-skinned animal, it had legs as straight as 'fruit trees' and it sweated blood when galloped or anxious.* The officials at court knew what they had at once: ambassador Zhang Qian had already reported the existence of these superior, blood-sweating

---

★ Though this detail sounds fantastical, horses afflicted by the parasite *Parafilaria multi papillosa* do appear to sweat blood, and the parasite is common on the steppe.

horses raised on the alfalfa meadows of the Ferghana Valley by the Dayuan, whom some believed to be descended from Alexander the Great's men. This horse was literally heaven-sent from the Grand Unity or highest of gods – a 1,000-li *tianxia ma* or Heavenly Horse 'impatient of all restraint' that 'treads the fleeting clouds … with smooth and easy gait'. It was a 'friend of the Dragon' – the most potent creature in Chinese mythology, the symbol of the emperors, of fecundity and immortality – and Wu Di, who wanted to live forever, was delighted to receive it.

At the advice of court magicians, he had already sent a party at great expense to discover the source of the Yellow River in the Kunlun Mountains west and south of the Taklamakan Desert, where the immortals lived in a golden palace and ate delicacies like dragon liver and bear paws. In emulation, he drank and ate only from gold vessels. Why shouldn't he want to possess the animals that would surely carry him away to everlasting life on the mountains of Kunlun?

Wu Di dispatched an envoy with 1,000 gold pieces and a golden horse of his own. The Dayuan king, playing hard to get, told the envoy that the Han were too far away, and beheaded him, saying he would kill the horses before he handed them over to Wu Di. Wu Di then sent his general Li Guangli across the desert with 20,000 soldiers and 6,000 cavalry, only for them to return empty-handed, starved and sapped by constant attacks from local rulers who objected to the Han armies in their territory.

Wu Di re-equipped Li Guangli with three times as many men and supplies and sent them back out in pursuit of the Heavenly Horses. This time, only half Li Guangli's men made it to the Ferghana Valley, but after forty dogged days of siege, the Dayuan people rose up and killed their leader, thrusting 3,000 horses and thirty Heavenly Horses at the Han general. They also promised that Wu Di would receive twenty Heavenly Horses each year, and gave his men alfalfa seeds to plant and raise. One thousand of the 3,000 horses made it back to Wu Di a year later, after another harrowing journey through the Taklamakan.

Wu Di greeted the surviving Heavenly Horses with a hymn that yearned for endless power and life:

> The Heavenly Horses are coming,
> Coming from the Far West.
> They crossed the Flowing Sands,
> For the barbarians are conquered.
> The Heavenly Horses are coming
> That issued from the waters of a pool.
> Two of them have tiger backs:
> They can transform themselves like spirits.
> The Heavenly Horses are coming
> Across the pastureless wilds
> A thousand leagues at a stretch,
> Following the eastern road.
>
> ...
>
> Open the gates while there is time.
> They will draw me up and carry me
> To the Holy Mountain of Kunlun
> The Heavenly Horses have come
> And the Dragon will follow in their wake.
> I shall reach the Gates of Heaven
> I shall see the Palace of Gold.

Although the tale has the air of a myth, the campaign was real. The true nature of the Heavenly Horses themselves is, however, harder to pin down, because although the most pragmatic explanation would be that they were purchased for Wu Di's cavalries, they are not mentioned in any military chronicles. Perhaps the 1,000 ordinary Ferghana horses were instead the core of a new Chinese cavalry. Sinologist Arthur Waley believes that the *tianxia ma* had a ceremonial role, and that Wu Di really thought they would make him immortal. In Waley's telling, they are perhaps related to the black-headed grey 'treasure horses' that carried the infant Buddha from Lumbini to his father's palace. Others believe they were descended from the sacred white or dun Nisaean horses taken by Alexander the Great as he crossed the Persian empire, and later raised by his Dayuan grandsons and great-grandsons. One popular claim is that they were ancestors of the Akhal-Teke of modern Central Asia, whose hollow hairs refract light and give their coats an unearthly metallic gleam.

While the Heavenly Horses did not help Wu Di reach the golden palace in the Kunlun Mountains to eat bear paws with the other immortals, he did reign for an unprecedented fifty-four years before dying an old and paranoid man, convinced that his courtiers were casting spells against him. He was buried clad in a jade suit stitched together with gold, and guarded by eighty horses.

IV

In the autumn of 2013, I went looking for Heavenly Horses in a Beijing suburb and found myself spoiled for choice. The long trek west across the Taklamakan Desert has been replaced with eager foreigners jetting into the heart of China like the vassals of the past, presenting their finest animals as tribute for the tuhao. The equestrian gold rush triggered by China's economic revolution is beginning to put down roots: in 2012, it was claimed that the Chinese horse industry grew by a fifth each year, and foreign horse copers, brokers and merchants of all stripes have warmly embraced the seemingly fathomless potential of the market. The best of Western horseflesh is on offer to any tuhao who can afford the importation and quarantine fees.

In the echoing halls of an exhibition centre, I drifted along a row of temporary stables as attendees peered in through the metal bars over the doors to see a tall, glossy black Friesian, a clutch of pure-bred Spanish greys – the old royal horses were on the move again, coming to a new plutocracy – and powerful bay warmbloods from northern Europe with frames built for Olympic showjumping. There were leaflets advertising tumbling-maned Gypsy cobs from Ireland, thoroughbreds from Australia, French saddle horses and Arabians from Germany. There was even a 'biotechnology company' in Shandong offering the sperm of a 'blood sweat' 'Ferghana horse' which turned out to be an Akhal-Teke.

'We started low, but already this year we've sold 600 horses,' the lady from the royal Dutch warmblood association told me as I hovered over the leaflets at her booth. 'The Chinese riding clubs order horses by number – twenty-five, say, and all in black, or white – those are the popular colours. Our brand is very popular,' she went on, and by brand she didn't mean the rampant lion and coronet that used to be stamped on to the thighs of the horses.

There are some 300 riding stables in China serving 800,000 hobby riders – fewer than in the UK, which is a twentieth its size – but there is a go-big, *mianzi* extravagance to the new Chinese equestrianism that copies, amplifies and outdoes. It's true that many of the riding clubs are more modest establishments with a few half-Mongolian horses and old Hong Kong racers, but others line their stables with pedigreed foreign horses. In Jiangyin city, the Heilan International Equestrian Club staged a quadrille of thirty Lipizzaners ridden by immaculately made-up young women in scarlet quasi-military uniforms in a manège twice the size of the Winter Riding School in Vienna, with a dozen golden chandeliers overhead and flashing green and purple lights tracking the manège floor. The next year they did it again, but this time the spotlit manège was filled to capacity with scores of white Spanish and Portuguese pure-breds and black Friesians circling and serpentining like an equestrian kaleidoscope.

I stopped to talk to Dr Nils Ismer, whose family had been breeding Arabian horses near Bremen since the 1950s. 'There are maybe 200 or 300 Arabian horses here in China and for sure we're interested in building up new markets,' he said. 'Our main market now is the

Middle East, but I don't see a big growing potential there any more. Here in China we have a virgin market that we need to develop but that has growing potential.'

He thought his Arabians – beautiful, old-fashioned creatures with black-rimmed eyes and the build of their Bedouin warhorse ancestors – would largely end up in the leisure market in China, especially if the Chinese took to endurance riding, which an editor on the country's largest equestrian website had told me was happening. ('It's safer for rich people than showjumping,' she explained.) 'The Chinese have a horse tradition,' Dr Ismer went on, 'but it's been reduced to working horses in the last forty years.'

China does, in fact, have the second highest population of horses in the world after America, but three quarters of those horses are ranged in the north and north-west – often in those old barbarian territories, long since absorbed, like Inner Mongolia, Tibet or Xinjiang. The real Chinese horse culture is one of agricultural work and 'folk racing' of Mongolian or Tibetan ponies over dozens of miles of open steppe, and even horse fights, where men pit stallions against one another for money. These Chinese riders are not counted in the riding club statistics. Chinese breeds like the Xilingol and Jinzhou are mostly crosses of Mongolian or Russian breeds seldom taller than 15 hands, a little improved by the efforts of the Party since the 1950s, but not fit for global competition or luxury status.

Dr Ismer acknowledged the old culture, but the spokeswoman for the Dutch group had spoken of a 'Ground Zero' for her warmbloods. The Xilingol and Jinzhou will not carry a new generation of Chinese horse riders to the Olympics – superior western rong ma were required, and with them, an entirely new infrastructure. There were no specialist equine veterinary clinics in mainland China in 2010. The country's equine vets were all based in Inner Mongolia or Hong Kong, but major clubs also flew in specialists from New Zealand or Europe. The grooms often came from Inner Mongolia, too. Farriery was poor, and fodder had to be imported like so much alfalfa grain. A Dutch embassy report that year claimed that the professional riders employed at many clubs were usually men who had worked for the club owner in another business – often factory work – and who had no riding experience.

An entire stage of modern sporting equine development was missing, root and branch. 'We're selling horses and we're selling knowledge,' said the Dutch spokeswoman. European higher-education institutes and veterinary colleges were making partnerships with Chinese universities. The old British Pony Club manual had even been translated into Chinese.

In the nineteenth century, when the bourgeoisie in Europe and America were flourishing and buying up horses for this new-fangled concept of 'leisure', China was in a steep economic decline that followed centuries of poor growth. Horse racing and the newly anglicized sport of polo returned to China with colonialists, but meant little to the vast majority of Chinese. In the twentieth century, Mao banned horse racing as soon as he came to power, branding it a foreign and immoral sport: the racecourse at Shanghai became the People's Square, rather as Guérinière's manège became the Assemblé Nationale. Polo continued to be played, but the national teams were from Inner Mongolia rather than the Chinese heartland, and other equestrian sports like showjumping were limited largely to the military and expats. In the mid-1970s, the handful of military bases still breeding sport horses were reduced.

Unlike the mega yachts and Rolls-Royces that were trundled into the new China to seduce the tuhao, those twenty-five matching black warmbloods would not be delivered with integrated GPS systems or launch control. As in Renaissance Europe, the skill of the much-admired rider has to match the blood of his horse, and that skill is not purchased with anything other than hours of dedicated practice and tuition. Just as riding was bred into the steppe barbarians, so it is often bred into the aristocratic and the born wealthy: the Georgina Bloombergs, the Athina Onassises and the Zara Phillipes, settled on the saddle pad of a pedigree Shetland shortly after birth and educated through to thoroughbreds. Against this precipitous learning curve, and in competition with other luxury goods and sports like golf, the new Heavenly Horses were on a precarious footing.

The thick, glossy equestrian magazines and pamphlets that piled high in my arms as I wandered from booth to booth were meant to induct the tuhao into foreign equestrian cultures far removed from the stallion fighting and folk racing. A Hampshire boarding school offered places for Chinese children to take A levels and riding lessons while living in a nineteenth-century manor house. Tuhao could play at cowboys on a Montana ranch, or go fox hunting in the Czech Republic. I paused at a monitor showing a fox being ripped apart, and riders in red and green coats leaping banks in Ireland. 'This is the first year we have offered hunting,' the booth's manager told me. 'Chinese people want to know why and what is hunted. Wolves?' He also had a video of a double file of Spanish dressage riders bouncing sedately on white horses in slow motion, with the caption: 'His back, a throne of feathers, will bear you safely at trot or at gallop.'

But puzzlingly, the people drifting delightedly around the booths with me did not appear to be tuhao so much as enchanted commoners, romanced by another vision of the horse on the magazine covers, all knights in armour and side-saddle mademoiselles in velvet habits. They could not afford a throne of feathers, but they still came to see the horses.

A crowd gathered round as a dapple grey called Dark Angel was led from its stall by a man in traditional Andalusian dress. A woman in hot pink and black flamenco ruffles was already mounted on another grey, and they watched as the man made Dark Angel prance in piaffe and perform a levade on the green baize flooring. Cameras clicked, phone screens glowed, showing a dozen tiny dancing Dark Angels, and a young man in glasses bent over his sketch pad to draw the horse with quick strokes.

V

The most natural era of the horse in China fell between 618 and 906 AD, after the Emperor Gaozu established the Tang dynasty from the descendants of Chinese aristocrats who had earlier assimilated north-western nomad groups. The long interaction between nomads and Chinese, with its exchanges and incorporations of knowledge, horses and princesses, continued to meld new ruling classes. The legacy of Wu Di's expeditions to find the Heavenly Horses in Ferghana and the Han efforts to trade and exchange grain and horses blossomed under the Tang, who campaigned successfully in the north-west as the nomads fell back, reopening what would become the Silk Road as they went.

The Tang loved horses and they were successful in raising them because they controlled great swathes of northern grassland for a sustained period: their herds expanded from an initial 5,000 horses to 706,000 just fifty years later in the vast pasturelands that stretched between Tibet and Mongolia. The dynasty tapped their vassals for as many as 50,000 horses in tribute, and required their fighting nobility to breed their own horses, thus increasing the pool that provided everything from war steeds to postal service horses. Officers who failed by just one horse to meet the target set for them on their segment of grasslands were dealt thirty blows with a bamboo stick. Horses were branded according to their role and their strengths: 'flying' or 'dragon' for the best, and 'sent forth' to mark a horse serving in the army or postal system. Artisans and merchants were banned from riding horses, reinforcing the courtly caste.

The vassals' tribute horses came along the Silk Road with their barbarian grooms, four-hoofed embodiments of wealth from Bukhara, Khirgiz, Samarkand and Khotan, exquisite Arabian and Turkmen horses, ram-faced warhorses, travelling with Indian ivory, Persian silver, Baltic amber, gems, furs, Scythian gold and Byzantine glassware. One silk painting shows a grey tribute horse almost swamped by the folds of its embroidered saddlecloth and herded by two 'Tartars' with foolish, hairy faces and horses masked like dragons. The best of these were tapped for the stable of the palace itself, where they were added to herds in corrals named for legendary or prominent horses in history, like the 'Dragon Decoys' or 'Heavenly Park'.

Tributes were not enough. The Tang also exchanged 1 million bolts of silk for 100,000 foreign horses a year, in ruinous, unspooling extravagance. Silkworms were thought to have the same *chi* or 'essence' as horses, and because of this, silkworms were believed to be harmful to horses yet inextricably linked to them. In a fourth-century folktale that became popular in the Tang period, a noble goes to war, leaving behind his daughter and a stallion, whom the girl must care for. The longer the father is away, the more desperate the situation at home becomes, until at last the girl jokes to the stallion, 'If you can find and bring back my father to me, I promise to marry you.' The horse escapes, finds the girl's father and brings him home, to the girl's delight. Hay is piled high in the horse's stable as a reward, but it refuses to eat. If the girl passes its stable, he 'suddenly and passionately rears up and strikes out', until the father asks the girl what has happened and she reveals the whole story. Appalled, he shoots the horse with an arrow, skins it and leaves its hide in the courtyard to tan. The girl mocks the skin, 'You are a beast, and yet you want a human being for your wife?' but the scraped, empty skin begins to move, and then to pursue the girl, engulfing her and carrying her away. Her father hunts for them for days, eventually coming to a mulberry tree where the horsehide, wrapped around his daughter, hangs from the branches: both are turned to silkworms, busily spooling silk into cocoons.

This strange story is harder to parse than the peasants whose horse shat silver. Women were associated with the actual production of silk and its initial discovery; in a more literary legend, the wife of the mythical pre-Shang Yellow Emperor was said to have first unravelled a cocoon after one fell into her tea as she sat under a mulberry tree. She was worshipped as the 'silkworm mother'. The girl stolen by the stallion in the fourth-century folktale became 'Girl with a horse head' or Matouniang, a further reference to the silkworm, which has a long, tapering face with wide-set 'eyes' and a bristling muzzle – not unlike a horse. In a recurring folktale, a Chinese princess sent as a bride to a barbarian prince conceals silkworm eggs and mulberry seeds in her hair and brings sericulture to her new home.* The sinister creature in the story that seizes the young Chinese woman and wraps her in the cocoon is reminiscent of the north-western barbarians who carried off Han and Tang princesses: China still had to give up women for horses, and now it must press its peasant women to make silk they could not eat.

The horse the Tang Dynasty raised on all this silk was longer-legged than the sturdy terracotta horses, its Mongolian and Tibetan blood transfused with Arab, Ferghana and Turkmen refinement. Transmuted into stone, silver and porcelain statuettes, Tang horses are all robust life, standing alert with draped saddlecloths, or gallop-ing with their bellies low. The dance of powerful curves that begins at their Roman noses vaults between their neat, small ears to a high-crested neck, drops to a short, rounded back, rises over an apple rump and finishes with the exclamation of a raised, clubbed tail. They look like they would carry you for days across a mountain range and still have energy to spare. In the words of the eighth-century poet, Du Fu:

> Lean in build, like the point of a lance;
> Two ears sharp as bamboo spikes,
> Four hooves light as though born of the wind.
> Heading away across endless spaces,
> Truly, you may entrust him with your life.

Six of these warhorses were carved in bas-relief on the lime-stone tomb of the Emperor Taizong, where each is named, described and celebrated with its own poem. They were real horses, but they became divine and were worshipped at their own altars outside the

★   But the word for 'cocoon silk' is also a risqué pun on sexual desire.

tomb. Saluzi, called Autumn Dew, has an arrow drawn from its chest by Taizong's general, Qiu Xinggong, who wears a nomad's tunic and trousers, beard and cap, quiver at his side. The horse pulls back slightly, but his ears are pricked as he braces for the pain. Of Autumn Dew, the emperor's poet writes, 'It was feared along the region of the three rivers;/It struck awe into the enemy on all battlefields.' Qing Zhui the piebald was 'Light-footed, a streak of lightning,/It was full of natural spirits', and Bai Di Wu the 'white-hoofed crow' 'could run with the wind', carrying his rider to victory so easily that 'with one look I brought peace to Shua'.

Tang horses did not just go to war and gallop postal routes. Although it is unclear when exactly the game of polo reached China, the Tangs were its most enthusiastic players. Tang polo was variously a war game, a court rite, a sport and a drunken party for aristocrats, commanders and emperors. It came from somewhere on the Silk Road and steppe – the oldest polo goals may have been unearthed in Iran, but it's not beyond reason to speculate that many horse-bound nomadic peoples played mounted games in pursuit of a hollow wooden ball, a dead goat (as in buzkashi) or other token to keep their warriors fine-tuned or their herdsmen entertained.*

In the Tang game, sixteen players took aim at peculiar goals a third of a metre wide and ten or twenty times that height. Emperor Taizong built goals with golden dragons on the lintels, and had gongs beaten when a ball went home. One of the Emperor Dezong's generals had a party trick of stacking twelve coins on the pitch, then galloping at them and whacking them high into the air, one at each pass. Emperor Hui-tsung made four generals play off to decide who would head the garrison at Sichuan. One later Tang emperor demolished the altar where he had been consecrated into power so that stick and ball could be played on the land instead.

Horses also provided court entertainment beyond the polo field. Long before Europe first held its carrousels with Pegasi and quadrille

---

★ In 2015, stuffed, sheepskin balls and L-shaped sticks dating from the eighth century BC were found in Xinjiang.

horses adorned in sequinned snakes and cloth of gold, envoys from Tibet were greeted by Tang horses with unicorn horns, the wings of phoenixes and trappings made from precious metals. Emperor Hsüan-Tsung had a troupe of a hundred dancing horses who had honorary ranks, like 'Imperial Pet' or 'Household Favourite'. They were harnessed with bridles tacked with silver and gold and had pearls and jade beads plaited into their manes and tails. Poets called them dragons and Heavenly Horses who, 'though full of high mettle ... do not advance, but stamp with a thousand hoofs'.

Their most famous dance was the 'tilted cup'. On a flower-painted platform that reached three storeys in height, the horses 'whirled round and round at dazzling speed', 'the tossing of their heads and the lashing of their tails ... in perfect time with the music'. At the end of the twenty movements of the dance, the horses picked up sloping cups in their mouths, 'drained' them, and lay down 'drunk'. Another performing horse was even hefted up on a divan held aloft by a strongman.*

But this horsey dynasty did not mark a final indistinguishable blurring and mutual assimilation of Chinese and nomad cultures. The Tang dynasty was following the course of all the earlier ruling clans, slowly sinking into decadence, corruption and intrigue, and the foreign influence of the nomads was noted and resented. One Tang prince, Li Chengqian, had his own felt yurt decorated with wolf's head banners, where he would speak Turkic languages and roast sheep he'd stolen. The women of the court dressed their hair like Uighurs from Xinjiang and played polo with dash, wearing nomad trousers stuffed into broad-calved boots. The poet Yuan Chen opined:

---

* Perhaps a Guoxia or 'under the fruit tree horse', a tiny breed smaller than a Shetland originally from Korea. According to historian Edward Schafer, they were 'gaily decorated' and 'carried the gilded youth of T'ang to drinking parties in the gardens of the capital during the height of the spring flower-viewing season'. Some believe they were also ornaments of the Han court, where they drew the vehicles for court ladies.

Ever since the western horsemen began raising
     smut and dust,
Fur and fleece, rank and rancid, have filled
     Hsien and Lo.
Women make themselves western matrons
     by the study of western make-up;
Entertainers present western tunes, in their
     devotion to western music.

Polo became a theatre for the clash between northern, nomadi-cally inclined Chinese and settled Confucian or Taoist courtiers. One Taoist priest complained, 'Polo hurts the vitality of the players and it also hurts that of the horses.'* Worse, polo seems to have become a tool of courtly intrigue: several individuals were conveniently removed from the wider play of court or army life by being run over by their enemies' polo horses, and Emperor Muzong was knocked out of con-tention by a mid-game stroke.

By 751 AD at the Battle of Talas, the Tang were losing control of parts of the Silk Road to new mounted warriors – 'black-robed Arabs' with lances and bows, who came from the Iraqi Abbasid Caliphate and were abetted by the Tibetans. The dynasty was also still fighting a running, intermittent war with northern nomads like the Jins, who, though often defeated in battle, were never suppressed in war, draw-ing on their own impressive trading links and the vast numbers of horses they could raise.

When the end came, it began in the north. General An Lushan, an associate of Emperor Xuanzong's favourite concubine, rose up and declared himself emperor in his own right in 755. An Lushan had

★ The Taoists seem generally uncomfortable with the use of horses. One fourth-century BC text complains that Po Le has ruined horses by clipping them, keeping them in stables and forcing them to race. The I Ching concurs: 'When there is Tao in the empire/The galloping steeds are turned back to fertilize the ground by their droppings./When there is not Tao in the empire/War horses will be reared even on the sacred mounds below the city walls.'

been in control of a chunk of land north of the Yellow River and had over 100,000 troops; he had been smuggling away choice horses for years. The fighting lasted seven years and drove the emperor from his own court – where he had once provided the ingrate An Lushan with a beautiful home stocked with sandalwood couches and golden ornaments. The Tang had lost their territorial gains in Central Asia and Tibet and the mulberry orchards around Hopei where many of the empire's silk cocoons were gathered. They lost much grazing for their horses, too.

Now the horses on which the 'might of the state' rested became more expensive. The Uighur of Xinjiang, emboldened, asked for thirty-eight pieces of silk for a single horse. The Tangs turned beggars, running up debts as they desperately tried to restock their breeding farms – in 762, they owed nearly 2 million pieces of silk. Almost 15 per cent of the central government's income was blown on horses in 842, enabling the Tang to attack the Uighur successfully, although this solved neither the shortage of horses nor the hole in the treasury. Tang rulers claimed a full fifth of their wealthiest merchants' worth in silk, and attacked the Buddhist monastery system as corrupt and decadent – with a weather eye on their land and treasures. Sapped and shaken, the dynasty collapsed, and once more a period of competing, warring clans followed.

VI

The only way of accounting for the unusually beautiful October day was that it was a fortnight after Golden Week, when factories across the country had closed and 1.3 billion Chinese found themselves on vacation. The clean air of the shutdown lingered like holiday goodwill. Over Chaoyang Park on the fourth of seven Beijing ring roads, the skies were deep blue and there was a fresh, brisk breeze that bent the tops of the silver birches lining the entry roads.

I parked myself with my interpreter in some temporary stands in front of the Olympic Beach Volleyball venue, an amphitheatre that had seated 12,000 in those brief weeks of glory in 2008 and had since been quietly falling apart. But, like the Beijing skies, it was looking – ostensibly at least – fairly good five years later, its peeling sides stripped and re-covered with a three-storey-high advertisement for a diamond- and ruby-studded Rolex. Behind the judges' booth there was still a set of Olympic rings clustered round an empty torch holder. In another arena before us, the smaller contests supporting the Fédération Equestre Internationale's Chinese showjumping league were playing out, with riders in slimline scarlet and black hunting jackets and vented plastic helmets – a post-lycra take on the hunt uniforms that evolved in the Shires of England – rising and falling as their horses cleared the striped poles. Behind us a huge screen relayed their rounds, explained by a state TV sports channel presenter in red trousers and trendy square glasses. Serene muzak played in the background, interrupted by a jazzy fanfare if a horse and rider jumped clear.

'There are five reasons in total why I want to be a showjumper,' Yuan Maodong told me when he joined us. 'The first one is my big brother is the manager of a horse training area, and I was affected by my brother so I just want to ride horses. Second, when I was a kid, my family raised horses, but they were only small, not for riding.' Yuan was from a rural part of Shandong province, south-east of Beijing. 'Third, I want to represent China in the Olympic Games. Fourth reason, when I was a kid I dreamed of being a knight – I had the knight dream.' Yuan Maodong nodded and smiled as his words were conveyed to me by the interpreter. 'Fifth reason, I come from a poor family so if I don't ride I have to choose another job.'

When he was not competing, Maodong was training. 'You need a team of people to help you train the horse,' he explained, 'and then you have to have your own horse to fight in the game, and last you have to have sponsorship.' His sponsor was Mr Wu, a burly, affable businessman who ran three riding clubs in the Beijing area, and had purchased Maodong's dark bay French saddle horse, Bukaji. I'd just seen them jump a smart clear round and win their class. At the prize-giving ceremony, as three slender maidens in white fluted gowns and fur bolero jackets stood by on a red carpet, Bukaji had done his party trick – a Wild West rear. Yuan Maodong grinned in delight, pointing to Bukaji and demanding applause for the horse before their victory lap. The rosettes were deep and petalled like chrysanthemums, the trophy a silver rearing horse.

Yuan Maodong was lucky in Mr Wu – there were no established pots of money for launching a showjumping career in China. The first Chinese showjumper to qualify for the Olympics was the 'peasant champion' Li Zhengqiang from Guangdong in the far south of China, who sold his quarrying business to found a riding club which he named Camelot. Unlike the organizations that supervised high-medal sports like swimming and gymnastics, the government-run Chinese Equestrian Association provided no funding for its 2008 team, so Zhengqiang's hometown had to step in with cash to supplement the loans he took out to purchase Jumpy des Fontaines, a French stallion worth 1 million EUR. Another of China's Olympic showjumpers cobbled together money from commercial sponsors. The third was funded by a wealthy businessman father. The team all finished far down the ranks, and the sole dressage qualifier, a woman from Xinjiang, came forty-first out of forty-six. The three-day eventing qualifier, an old Etonian born in London and based in England, was eliminated in the cross-country round.

The younger riders were taking their turn in the ring now, and Mr Wu watched with satisfaction, hands resting on his thighs. He had ridden for a while before he had a bad fall, and told me he started his first club in 2003 with friends who loved horse racing. 'The gambling part was banished, but people can watch this showjumping just for fun and that's OK. Horse riding has become popular in China in the last ten years,' he told me. 'It's only just started. My clubs have about 4,000 members and most of them are Chinese – 50 per cent are teenagers, and some of them are dreaming of becoming professional riders. They start out doing riding, dancing and skating, and gradually they drop everything but the horse riding.' Of the rest, most were like Yuan Maodong, in their thirties. 'There are more girls than boys. I think it's because teenage girls are smarter than teenage boys. It's more graceful for girls to learn to ride. Boys are naughty at that age. You can't make them focus.'

Because there was nobody with the experience or skill to teach riding on the Mainland, Mr Wu employed Dutch instructors, part of a new migrant population of peripatetic trainers from old European equestrian families who were splitting seasons between East and West – the kind of horse professional whose lean, bowed legs have been whittled by three generations of high-level riding, their faces elegantly corrugated by squinting at crop after crop of expensive sport horse yearlings each spring, or looking on as a clutch of their own children hoovers up the pony championships.

All this came at a price: Mr Wu's Knight Union clubs had membership fees of between 11,800 and 29,800 RMB a year; the income of an average Chinese family in 2013 was 13,000 RMB. I guessed that its members were probably the residents of the service apartments for 'young and rich' that I'd seen advertised on Beijing billboards.

'It's every boy's dream to become a knight,' Mr Wu told me, as Yuan Maodong excused himself and made his way to the temporary stabling to see Bukaji. 'That's why my clubs are called the Knight Union Equestrian Clubs – because of the knight spirit. Our knight is like a European knight – he rides a horse, and is brave, selfless and devoted.'

Mr Wu used the term qi-shi, which referred to a European knight, rather than the Chinese da-xie or xie-ke, meaning warlord, or hsia, knight-errant or outlaw. Like the chivalric knights of the West, China's warlords and knights were often far nobler in verse than the reality, and the hsia are also somewhat rough around the edges – they are bandits, not aristocrats, noble in character, not bloodline. They have little religious affinity, and a dismissive or bawdy rather than worshipful attitude to their ladies. They were not Confucians either, and they tended to come from the north, but they were Robin Hood figures who helped the poor – in literature at least – and chose 'death over dishonour'. They have a different aspect of 'face' to mianzi: lian, or personal integrity.

Their horses were essential parts of their panoply, as for any chevalier or Ritter: the poet Yü Hsin describes how 'wandering knights join their horses together,/their golden saddles covered with willow twigs ... Gay with wine, the men are half drunk;/Wet with sweat, the horses are still proud'. The Han warlord Xiang Yu had a black horse named Wu Zhui or 'dark fur' which carried him at his defeat at the Battle of Gaixia, where, surrounded by his enemies' troops, he wrote in despair ' ... the times are against me,/And [Wu Zhui] runs no more;/ When [Wu Zhui] runs no more,/What then can I do?'

The knights' popularity seems to have arrived long after they disappeared from the historical record: the famous classical poems about them were originally written in the Han period, but generally draw on real figures alive at the time of the Warring States, and they are dug up from history when needed and poured into new forms – folktale, opera or novel. In tuhao China the knight spirit was presented as a purchasable side order to polo or showjumping, bestowing mianzi on post-Maoist entrepreneurs as they joined the global elite with their sport horses and luxury goods. But the component parts of a 'knight dream' that was more lian than mianzi were assembled under the Song dynasty that followed the Tang, when the hsia also took on a symbolic, nostalgic role in times of upheaval and uncertainty.

Weary of constant attacks, the Song enlisted the Jin or 'Gold' people from the most north-easterly part of what is now Inner Mongolia to free China from the threat of the Khitan Empire of Mongolia and Siberia. But once victorious, the Jin swept on to Beijing and then the Yellow River valley where the Shang had forged China. They took easy control of all the northern grazing lands, helping themselves to 10,000 Song horses. At the Song capital of Kaifeng, they captured Emperor Hui-tsung and his family, and carried them back north. Hui-tsung died of distress not long after, but his son, Ch'in-tsung, is said by one chronicler to have been sent on to the polo pitch with the last Khitan emperor to knock some balls about, and then ambushed by bowmen on horseback. Ch'in-tsung's body was left on the field, and horses ridden over it.

In the south, the Song retrenched and dug in: they still presided over the majority of the Chinese population, and its enviably rich agricultural land, but the pastures on which the Tang had raised their herds were lost, and the horses produced in the south were of poorer quality and scarce. Only five out of six of the Song cavalry had a horse to go to war on in 1061, and one viceroy in Sichuan pointed out, 'The [Jin] enemy is strong because it has many horses, and we are weak because we have none. Any three-foot-high toddler knows the difference between strength and weakness.'

The Song court tended to ride sedan chairs rather than horses, and eschew the dangers of the polo field. Its women had their feet bound and stuffed into tiny slippers rather than the broad boots that Tang women wore when they dressed their hair like Uighurs and galloped about the polo pitch astride. The emperor Hsiao-tsung, who had once been left hanging from a beam when his pony took off into a nearby building, was persuaded off the field by his councillors, who held that polo was 'of no help to military training ... anything could happen with a quick swing of the stick or a sudden stop of the horse'.

The Song's bureaucratic solution to the horse shortage was heavy-handed and ultimately disastrous not just for the Song cavalries, but also for many peasants. China might not have horses, but it did have tea, and nations rich in horses sought bricks of densely packed black and green tea leaves. While the main Silk Road now lay beyond the control of the Song, the Tea Horse caravan road that ran through the spectacular mountains of south-western China to the Kingdom of Tibet was still within their territory.

Tibet – with its 'wild beasts' of men who 'smelled so bad you could not approach them' – was the only feasible source of horseflesh and must now be wooed, and offered tea in exchange for horses. The

Sichuan Tea and Horse Agency rapidly established a total monopoly over the internal tea market, buying from growers at rates the agency chose, and selling some to merchants or shopkeepers, but keeping most for the Tibetans. The Tibetans, however much they wanted tea for trade, were no fools. For these black and green bricks they exchanged poor-quality horses, or else fewer horses than the Song bureaucrats anticipated. They starved then overfed the horses, so that they were lively when the inspectors arrived to view them. Song tea growers were squeezed harder to meet demand, and taxes piled on to them and to merchants.

Even when the Song horse buyers laid hands on enough head of horses, they were poor at handling them, and the route back to their homelands traversed treacherous mountain country. One in five of the hard-won animals were lost. The horses damaged hoofs on the rock and had little grazing as they made their way through the gorges. Transported by river raft, they were pitched out into rapids. They became sick, and no one knew how to care for them. In 1173 two transports of horses collided with one another at night, and fought: 'Half the horses lay dead, their flanks ripped open and their intestines flowing as thought they had run into spears or lances.' The tea growers had rioted, and then the price of tea collapsed and silk and brocade had to be sent out once more in exchange for weak, then further weakened horses, and then silver itself.

In 1234, the Mongolians under Genghis Khan made their move. First they overran the Jin – one Jin general was too busy playing polo to rouse his army – overtaking the north of the empire, and after a few decades of uneasy coexistence and then outright war, it was Genghis's grandson, Kublai, who united southern and northern China under the Mongolian yoke, and crushed the Song. Beijing – flattened and rebuilt by Genghis – was his capital, and Shangdu or Xanadu in Inner Mongolia his summer home.

For the next century the Mongolians ruled as the Yuan dynasty: the nomad takeover was complete. Their garrisons consumed all that the Chinese peasantry could offer, and the Chinese found themselves ranked below not just Mongolians but also other Central Asian barbarians. The Yuans requisitioned large numbers of horses – far more than China could easily produce. If Marco Polo is to be believed, Kublai's personal herds included 10,000 white horses, the milk of which was reserved for his family only. The Song were forbidden to breed their own horses, and military training and hunting were ruled out. Only Chinese generals working within the Yuan army or officials in the court maintained familiarity with these practices.

It was under the half-defeated Song that the tales of the *hsia* knights errant from the Warring States period had became popular, as the Chinese comforted themselves with notions of knights who resisted both a corrupt government and un-Chinese invaders from the steppe. Now, under the later Yuan, two of what would be known as the Four Great Classical Novels were developed from older poems, folktales and legends into the knight dream. Both of them retreat in time to earlier ages where China was free of the Mongolians.

Water Margin or *Outlaws of the Marsh* is based on legends surrounding the real outlaw, Song Jiang, who ran the now-loathed early Song government ragged until he was captured. The fictional Song Jiang, by contrast, is the 'protector of justice' with 'eyes like a phoenix' and both a cultivated man and a skilled fighter, taking a stand against the ruling class, who, as *Water Margin*'s readers knew, had later delivered China into the hands of the Mongolians. *Romance of the Three Kingdoms* is set in the aftermath of the Han dynasty's collapse, when three new dynasties fought for sixty years. It focuses on the hero Guan Yu as the warlords of the three kingdoms of Wei, Shu Han and Wu clash, regroup, ally themselves and tear one another apart again. Guan Yu is paired with a semi-legendary warhorse called Red Hare, whom he prizes above the *mianzi*-giving jade, brocade cloth, mansion and ten beautiful concubines that his host, the warlord Cao Cao, tries to win him over with. The novels are long, sprawling tales of shifting allegiances, set-piece battles and heroism – largely conducted on horseback.

Both were also widely read even under Mao, who wrote poetry referring to *Three Kingdoms* and put Song Jiang and the bandits of *Water*

Margin on the side of the peasant against the feudal rulers★ – the fact that Chinese knights errant were more Robin Hood than Sir Lancelot made them more ideologically acceptable to the Party. Guan Yu's refusal to be corrupted by Cao Cao's lavish gifts was also an exemplary demonstration of *lian* over *mianzi* – he prizes Red Hare above the temptations of luxury and the flesh because Red Hare will carry him swiftly to the rescue of his 'oath brother', Liu Bei.

When Mr Wu and Yuan Maodong told me they'd had the 'knight dream' they were talking of both European knights and the novelized, Party-approved version of the real knights of China. The image of Chinese people on horseback that most Chinese TV viewers saw was not the red-jacketed showjumper, but the heroes of the rambling, blockbuster TV series that the central government paid to have made from both *Water Margin* and *Romance of Three Kingdoms* in the 1990s. *Three Kingdoms* had a cast of 400,000, cost 150 million yuan, featured 100 on-screen battles and ran to eighty-four hour-long episodes which have been re-spun repeatedly since. Fifty horses were purchased from New Zealand for the cavalry scenes, as the local horses, though authentic in size, were believed too small. In Beijing alone, more than half the TV-watching population had seen every single episode.

When the Guan Yu of the television series is presented with Red Hare – played by a fine chestnut thoroughbred – the wily Cao Cao says, 'A horse is the soul of a warrior,' adding to Guan Yu, 'The steed can only be matched by you.' *Three Kingdoms* has also been requisitioned for video games, in which a snorting Red Hare stomps on his pixellated enemies, his eyes glowing scarlet through a flame-shaped champron.

At the horse fair in Beijing I'd seen young men clustered excitedly around a stand called Nomadic Wind, where a Mongolian was playing a horsehead fiddle. They snapped photos of a suit of leather armour on a dummy, a bow and arrows stacked against the wall, and a wooden shield carved with a deer with enormous antlers. The world expert in mounted archery, a Hungarian named Lajos Kassai, had visited China

★ Though Mao did later attack Song Jiang for selling out and working for the emperor.

in 2012, and people were starting to get interested.* He'd even taken on a Chinese student. If Mr Wu wanted to open more Knight Union clubs and reach beyond the tuhao, I would have suggested digging up the showjumping arenas and setting up targets and tents for archery camps where Beijingers used to twiddling their thumbs on game consoles could string a recurve bow and make the dust fly on their own Red Hares.

VII

The next day I passed north of Beijing to the Great Wall on a multi-lane highway rumbling with dust-covered juggernauts, speeding out of the gravitational pull of the seven ring roads and the high-rise clumps that reached on and on. My driver, Mark, was happy to be out of the city too. 'The feng shui for this place will be good because it's in the northwest,' he said. 'That's where the good, cleansing wind comes from. There's a lot of poverty south of Beijing.' A rampart of rocky slopes rose straight from the plain, littered with huge yellow boulders, and the neat, grey crenellations of the restored Wall rose and fell along the peaks and gorges as vertiginously as a roller coaster.

We were heading beyond the Wall to the flat grasslands that became Chinese once more under the Ming who followed the Mongolian Yuan. As a new dynasty struggling to secure their rule, the Ming modelled themselves on the Tang. They, too, had to deal with ambitious, horsey neighbours, the Mongolians, Koreans, Timurids and the Jurchen – descendants of the Jin who had overrun and humiliated the Song. The Ming built much of the Great Wall as we know it today, using billions of bricks and paving stones over the old ramped earth fortifications begun by first emperor Qin Shi Huang. Like earlier dynasties, they drew on nomadic culture for their court ceremonies, and there were

---

★ Kassai was in China to train Matt Damon who, for some inexplicable reason, was about to play a mounted archer active during the construction of the Great Wall.

still Yuan Mongolian personnel in the Ming court's retinue. At the Duanwu Festival at the summer solstice, polo was played and competitors took turns to 'shoot willow' – firing arrows at a strip of willow they circled on horseback. The emperor presented horses and silk to his barbarian guests, army men and courtiers.

But trade with Mongolia still fell flat in the early seventeenth century, and the 100,000 imported *rong ma* needed to maintain the Ming cavalry were hard to acquire even when the government monopoly on the tea-horse trade remained in operation. Emperor Xuande stressed that horses were 'first priority' and when, as his men assured him, he had sufficient cavalry, he still 'did not dare be the slightest bit lackadaisical'.

Horseyness was also seen by some palace insiders as an unhealthy influence perfumed by a Mongolian 'stench of mutton' as barbaric as the Tibetan stink. Chinese aristocrats warned that eunuchs who hunted with the later Emperor Zhengde did not respect the etiquette of the palace or know their place. There was also an extravagance to horsemanship that troubled some: Zhengde's large hunting caravans of courtiers expected food for themselves, their hounds, hawks and horses to come from the local peasantry wherever they paused. The emperor Jingtai was told by a minister that 'pearls, jades, hounds, and horses, rare birds and exotic beasts' would prevent his heart from being 'pure and without desire'.

The Jesuit priest Matteo Ricci noted that the Ming cavalry horses were so 'lacking in martial spirit that they are put to rout even by the neighing of the Tartar's steed'. When the Ming wrestled with peasant uprisings and the disastrous effects of famine, flood and border disputes in the mid-seventeenth century, the descendants of the Jurchens, the Manchus, swooped south from their base in Manchuria and, after protracted skirmishing, took power, becoming the Qing dynasty.

On the other side of the Wall, the murk and concrete of Beijing gave way suddenly to the farmland of Yanqing county. The harvest was being gathered in. Wire cylinders the size of haystacks stood in the yards of farmhouses, part-filled with corncobs. In the fields, dry brown stalks withered in the golden autumn light as tethered, rangy-looking horses picked among them for stray kernels and grass. The roads were lined with willows or silver birches whose leaves were turning yellow, and here and there reed-fringed lakes could be seen through the copses. Every now and then we passed someone in a face mask sweeping the road with a besom.

A sign promised the 'Kangxi Grasslands' – two stylized yurts and a man on horseback. This turned out to be a holiday camp where Beijingers could ride horses, drink mare's milk and sing Mongolian songs in Mongolian gers, playing at warriors in the spirit of Kangxi, the second Qing emperor and the finest ruler of the last dynasty. He was raised as a nomad and spent three months of every year hunting north of the Wall in the grasslands I was crossing. In his sixty-one-year reign he brought stability to the vast and always fractious territory of the Chinese empire, securing, among other things, the tea-horse trade to Tibet. It was he who commissioned the famous prayer not for Heavenly Horses and immortality, but, more pragmatically, for a healthy herd:

> Oh Lord of Heaven, Oh Mongol leaders, Manchu princes, we pray to you for our swift horses. Through your power may their legs lift high, their manes toss; may they swallow the winds as they race, and grow ever sleeker as they drink in the mists; may they have fodder to eat, and be healthy and strong; may they have roots to nibble, and reach a great age. Guard them from ditches, from the precipices over which they might fall; keep them far from thieves. Oh gods, guard them; oh spirits, help them!

The Qing absorbed some Han military officials into their Eight Banners or administrative divisions, marrying them to Manchu women, but the bulk of the Han Chinese population were treated as lesser citizens, with the men forced to adopt the Manchu hairstyle of a shaved forehead and long pigtail. The Qing also restricted the breeding of horses for two and a half centuries during the same period that horse ownership broadened in Europe: Han bannermen were permitted to raise just seven horses to a Manchu's forty-five, and ordinary Han men were banned from breeding horses outright. After strengthening their hold on China, the Qing moved to the north-west, recapturing Xinjiang and laying the geographical footprint of today's China before they reached their peak in the late eighteenth century, and as the saying has it, found that, 'You may conquer an empire on horseback, but you cannot rule it on horseback.' They followed the familiar slow slide into corruption as a new set of foreigners – the British Empire – began to insinuate and then bully their way into China.

For a century and a half the Qing had a strong cavalry, but when they finally lost power in 1914, despite the huge increase in the Chinese population, they had only 6,000 more horses than they had kept under Kangxi. The story was old: military horses had to be raised in the north, but if the Manchus requisitioned land for studs, they evicted local peasant farmers, who then scraped together lower yields and therefore lower taxes on their new settlements. It became more economical for the bannermen to let the peasants flood the horse pastures and grow crops, so the bannermen lied to the central government about the size of their horse herds. By the 1850s, the rebels who arrived from Nian in eastern and central China could mobilize more cavalry than the Qing themselves, and their uprising in the provinces of Shandong, Henan, Jiangsu and Anhui caught the Manchu armies in a horse-poor state.

When the Qing gave way to the Republic in the early twentieth century, Han Chinese in Jingzhou taunted the fallen Manchus, 'Now you are no longer riding horses, we masters are riding horses; if you are riding again, we masters are going to beat you up.'

The soft, golden yellow ranch house of the Sunny Times Polo Club had an interior poised somewhere between Yanqing county and a worn but well-heeled country house in Dorset. Over the fireplace were three crossed swords on a shield, and below them the mantelpiece was crammed with horse figurines and trophies, a St Regis polo ball sitting on a tripod of miniature mallets, a bronze of a cowboy, a model horse flocked in camel-ish fur, a Greek horse and a lively Tang-style porcelain statuette. The walls were hung with a series of English hunting paintings fusty with hedgerows, red coats and top hats. The room lacked only a Labrador, or perhaps one of those chow chow 'lion dogs' I'd seen puffing in Beijing in their thick fluffy coats.

An older lady poured tea into a cube-shaped glass before me on the long wooden table before scurrying off, with a smile, into an odoriferous kitchen just off the hall. A team of students from Beijing were rattling about, comparing schedules and working through checklists, dashing up the wooden staircase to the first floor and returning moments later. It was the fifth running of the club's international tournament, and my host, the club's owner Xia Yang, was the absent focus of all this bustle, the man for whom the steaming teapot and the checklists were filled.

There were just four teams playing: All-Blacks from New Zealand, the British Exiles, the Royal Polo Club of Barcelona and the Chinese-Argentinian team of Yang, his coach and three moody-looking Argentinians, one of whom had played on a World Cup-winning team. The opening rounds would take place over the next two days in the grounds of the ranch house, and the finalists would meet on the Saturday at the new pitch a kilometre away.

Behind me, three huge floral sofas held an elegant Chinese translator and the young and plummy representatives of British Polo Day, a sort of travelling sporting troupe borne aloft by luxury brands that sponsored tournaments in exotic locations like Singapore, Mexico or Morocco. Modern polo eats wealth – a team of four riders and more horses can be stratospherically expensive to maintain – so the British Polo Day camels arrived in China laden not just with their Exiles team, but with promotional bumf featuring Land Rovers accessorized by spaniels, Ettinger leather flasks with plumed royal insignia, coral Hackett polo shirts worn by blue-eyed, curly-haired young men,

Taittinger crisping with bubbles, gold-embossed stationery and Park Lane hotels.

Xia Yang arrived in his white polo breeches and a blue US Polo Federation sweater, surrounded by a school of personal assistants and student interpreters. He seemed to smile perpetually, delighted by the activity around him, his glasses sitting on a broad, freckled nose under cropped hair. 'Sunny Times is not about luxury, it's about comfort, it's meant to be like an English home. People should come here to relax,' he told me. He had made his millions as an architect then a property developer, and now also worked as a fellow at a 'research centre for luxury goods and services'. The 1.4 million yuan he had put into Sunny Times was pigeonfeed compared to the capital behind some of the wildly lavish polo clubs I'd read about, but here, as I soon realized, the stereotype of the *tuhao* was combatted, not wooed.

From behind a door next to the kitchen came hoarse whinnies and the scrape of hoof on concrete floor – 'I love horses so much I wanted them to be part of my home,' he went on, and so they were – the stable door opened directly into the living room. Xia Yang was as smitten as all those old cavalry officers and aristocrats for whom the horse was 'the beast of most bewtie, faithfulnesse, courage'. The house and its grounds were dotted with his sculptures of them.

The translators relayed his story in the third person: 'He could not afford his own horse until 1996, and this was a Yili' – a Xinjiang breed infused with thoroughbred and Russian blood – 'that broke the record in the national race in China six months later.' Horse racing was briefly legal in China at the end of the 1990s, and five courses were built in a flurry of enthusiasm, offering punters a chance to play a 'guessing game' on the results rather than betting outright, but a crackdown followed, and the CEO of one course was jailed for corruption. But in the meantime Xia Yang had seen a vision: Prince Charles playing polo against the Sultan of Brunei. The knights on their golden saddled horses and the Tang figurines came to his mind.

Not only, he realized, did Chinese horses suit polo better than racing, but the game could have a civilizing effect on the *tuhao* – he later told one reporter, 'In the west, polo is said to be an aristocrat's sport ... We don't have aristocrats in China, but we do have a lot of people who have got very rich very quickly. I want to encourage them to behave like gentlemen, and playing polo is part of that.' I'd read in one of the polo magazines lying around that he'd first wanted to ride after seeing an image of a Tibetan on horseback in traditional dress. When I pushed him about the knight spirit, he laughed, 'We also have the sense of being a hero when we ride a horse – the knight spirit. We had this kind of spirit in ancient times, many heroes and some generals rode horses, so people want to be like them. They admire them. It makes them feel more powerful.'

He went to Australia to learn the game, and in 2006 founded Sunny Times. 'People started to be interested instantly. We have mothers, fathers, kids from three to sixty-five who want to play polo or just ride a horse. This is the most professional club in China, and all people in China who want to play polo come here.'

We walked around the grounds of the club accompanied by the little knot of interpreters. There were 160 horses, the students explained, and we passed large dirt-floored pens, one containing that year's foals, the other the yearlings. They pushed curiously towards us before pulling back in shyness and retreating to tubs of hay which had been brought, the students told me, in some awe, from north-west China. The horses were not imported *Equus luxuriosus* but a home-bred brew of local breeds, Russian and even Arabian blood, and in one round pen was an especially fine-looking stallion, who shrilled to the mares who remained out of his sight but not scent somewhere on the remaining 2,000 acres. Another corral held a fuzzy collection of British native ponies for the junior members of the club. I saw the vegetable garden where white cabbage and brown-sheathed onions were grown in the shade of a yellowing set of racing starting stalls, long unused, and the cool white stables where the day's mounts dozed. Yang's favourite, Honey, a diminutive grey with a bristling, cropped mane and Mongolian head, snuffed at my outstretched hand and moved closer, making the translators giggle nervously.

'He likes you very much.' Yang added with pride, 'He's very fast, loves polo.'

I was offered the chance to ride to the new pitch – 'It can hold 10,000 spectators. This one just 1,000' – but had to decline as I didn't have the right clothing. My time with the man was up, and Mr Yang excused himself with a handshake as he had to go and practice.

At ease in the leisurely atmosphere of the club, I spent the afternoon lazing in the sun on the wooden stands that ran the length of the old pitch, watching the games unfurl and tuning in and out of the conversations of the teams and organizers around me. The British Polo Day people discussed 'market shares'. I learned that the new pitch had only been turfed three weeks ago. 'That's China!' someone said – a pitch from nowhere! Dancing unicorns and dragon horses! The silent grooms from Inner Mongolia waited with fresh ponies that pounded at the ground impatiently, their tails clubbed like old Tang horses' and legs wrapped in bright yellow, red, white or black bandages to match the players' colours. That afternoon, dawdling unaccompanied down a path to talk to some isolated horses in a small enclosure, I found the grooms' quarters: a neat little bungalow with large windows, packed with five single, iron-framed beds. The lanky young British men towered over the Mongolians as they collected their mounts; they wore sashes over their white shirts, and helmets that resembled solar topis.

When the ponies drew near, the gameplay seemed a little arcane: three riders on a single team would mob the ball at once, the ponies grouping, swarming awkwardly and then scrimmaging. A lot of inconsequential tapping about of the ball would follow, the horses' legs snatching up and manoeuvring like a flock of flamingos jostling for space. Cries of 'Ball!', 'Pallota!', 'You're not there, I tell you to get there, you get there!' and guttural Spanish curses rang out. The horses jumped to get out of the way or to place themselves next to the ball, snorting and pinning their ears back as other ponies crowded them, and then suddenly, out of all the blocking and footsie, a player would balance the ball on a mallet and then whack it into the air, and the whole field would be off, galloping flat out to the far end of the pitch, scattering like a retreating cavalry, intent on the tiny, round white hare.

As they disappeared towards the smog that seemed to make a distant wall around the farm, the audience's attention drifted and the scrimmaging seemed far removed. The wind rustled the dying leaves on the trees, and the horses in the pens and paddocks began an echoing rondeau of whinnies to one another, led by the stallion in his little enclosure. A wind turbine span in the distance, and geese honked as they streamed by overhead, leaving the grasslands for the winter. I strolled around the horse lines and watched a ladybird climbing up the yellow leg wraps of a pony, and saw one of the grooms kick a horse that had laid down with its saddle on. The horse ignored the groom. We would reconvene on Saturday for the upshot of two days of tapping mallets and dozing ponies: the final between China-Argentina and New Zealand.

Earlier Xia Yang had shown me his sculpting studio – a white plastic tent at the head of the polo pitch, whose curtains he could pin back so that he could see the horses in the fields if he needed to check their anatomy. He had been working on his latest piece for three months in preparation for the tournament final. It had, the students had told me, been cast in bronze for presentation that Saturday at the final to the embassy of the subject's owner, Sheikh Mohammed Bin Rashid Al Maktoum, the Emir of Dubai, vice president and prime minister of the United Arab Emirates, architect of the 'desert miracle' of golf courses, skyscrapers and the $1.25 billion Meydan Racecourse, and owner of the Godolphin racing team.

In the centre of the tent stood an all but life-size, unfinished sculpture of a thoroughbred called Jalil, lean as the point of a lance, sloping of shoulder, with two shed halves of moulds lying on either side of him, as if he had just burst from them like a beautiful, muscular silk moth.

VIII

I put my dictaphone on the desk with an apology, and Mr Li Nianxi of the Chinese Equestrian Association turned on his own recording device. Li's office, in an anonymous campus on the western side of the fifth Beijing ring road, was dominated by a huge reproduction of the classical Chinese character for a horse that I recognized from the old Shang pictogram. 'Can you see the real horse in it?' he asked me. I pointed to the sweeping brushstroke at its base. 'Is that the tail?' 'I suppose so!'

Li must have been in his fifties, with greying hair, frameless glasses and an Adidas tracksuit worn over his buttoned-up shirt. His desk held a small figurine of a prancing bay horse and a bottle of mineral water from New Zealand. He was, by and large, chatty, amused and forthcoming. I hoped to learn from him just how the communist government aligned its own knight spirit with that of the exclusive polo clubs and million-euro Heavenly Horses. I wanted to ask about one mystery in particular, as I had been blocked when I had tried to approach other organisations about it. As horse racing declined globally – punters now having the choice to gamble on anything and everything on the internet – and thoroughbreds were turned into saucisson, the brash press releases and puff pieces I'd read before I arrived in China were fond of proclaiming that the 'traditional power bases' of racing in the West would give way to a resurgent Asia and Australasia, spearheaded by Sheikh Mohammed's Dubai. Racing was to be shipped into the land of the golden yuan as easily as leather flasks, vintage red wines and Rolls-Royces.

If Lipizzaners could dance by the Yangtze and polo ponies live in air conditioned stables in Tianjin, why hadn't the most ancient of horse sports returned to the mainland yet, I wanted to know. But first I asked Li Nianxi about the government's support for the Olympic disciplines.

'The horses that were originally used for fighting in the Civil War were used in more peaceful ways, in farm work, or like in Europe, for equestrian competition,' Li explained. Polo and dressage were handled by the National Sports Association in the new China, and, as in Europe, the original competitors were military men. He painted a picture of a steady but gratifying rise in popularity in the last decade, especially in showjumping, while other equestrian sports were gaining. 'At last year's national sports competition the stadium was almost full for the dressage.'

I worked my way around the stereotypes of Chinese sports. Would there be special intensive academies hothousing young riders like gymnasts or swimmers? Mr Li chuckled, and said no such thing would happen, but they were putting in place a national competition structure through which young riders could progress. 'Maybe in the future we will have something like the Pony Club or educational programmes. This field is developing very fast, just like the Chinese economy.'

The Saudi royal family was said to have given their national showjumping team millions to buy 'the fifty greatest horses in the world'. Did the Chinese government invest in horses? At this he laughed again. 'The CEA doesn't buy horses. Most are privately owned. But with the development of the nation in ability and financial support from businessmen we can import more horses from Europe, so it's getting better.

'If demand domestically is growing even faster, it will be necessary for us to breed our own horses in the future,' Mr Li went on, although by 'us' he did not mean the Chinese government. Dr Ismer had told me that a firm in Shandong was already copying the template of nineteenth- and twentieth-century Prussian studs to create a home-grown Chinese warmblood by mating imported sport horses to the Buohai, a heavy hybrid of Mongolian, Soviet and Belgian coldbloods.

Behind me on the bookcase was a framed photograph of a familiar horse face, a patrician bay with a white blaze and an expensive leather halter: Sadler's Wells, an Irish-owned champion racehorse which had gone on to be one of the most successful stallions in history. Li hadn't mentioned horse racing at all, but he was delighted to quiz me about the photo. 'Do you know who that is?' He was pleased, too, when I correctly named the stallion, so I decided to push on.

I had read plenty of newspaper articles about a move to make horse racing legal on the mainland once more, I said, but was confused about what was actually happening. These stories all promised the dramatic opening up of the largest untapped horse racing market in the world, but nothing ever actually seemed to happen. Could he explain? Li's body language shifted. He began to look out of the window. He gave a long, solemn reply, but at the end of it my interpreter turned to me, smiled, and explained brightly in English, 'He didn't really answer the question. Shall I ask again?' This time Li was hesitant for the first time, and answered with his hand over his mouth.

'There will be more horse racing with horses that are imported. The public is willing to watch.' And that was that.

Certainly, the Chinese public were willing to watch horse racing, but that wasn't all they wanted to do at a racecourse – and that was a problem for the government. That the Chinese love to gamble is as popular an assumption as saying they love food or the colour red. Gambling on horse races is legal in Hong Kong and Macau, but it's a narrow needle eye controlled entirely by the city's government without the intervention of independent bookies. The Hong Kong Jockey Club pays its taxes and turns over to charity profits that run into billions of dollars a year – and that gambling money constitutes 10 per cent of the city government's entire income. On the mainland, since Mao's ban on 'immoral' gambling those who wish to gamble must choose between paltry state-run sports lotteries, jetting in *tuhao*-class to casinos in Macau, Singapore or the Philippines, or simply gambling illegally.

At the horse fair I'd found a stand for the *Racing Post*, where an enthusiastic young man who looked like the computer nerd in an action film – spikey hair, black trench coat – had assured me that the forthcoming Chinese edition would be perfectly legal. They would print a disclaimer telling readers to contact their local government for details about the law, he said, but they would, of course, merely be providing information for keen followers of the sport or Chinese owners who had horses based abroad.

But some investors hoped that Chinese owners would be able to see their horses run on the mainland, and they were prepared to pay a high price. Fantastic sums have been sunk in racecourses at Wuhan, Nanjing, Chengdu, Tongzhou and Tianjin. At the Tianjin Equine Culture City, the press releases promised, stands in the shape of a plate glass phoenix would rise from the turf, and up to 4,000 horses – home-bred, home-raised and sold at the complex's own auction house – would be provisioned with fodder grown and processed on the grounds, and treated by vets and horse technicians trained in an on-site college. Thousands would be employed in a cluster of hotels, a riding club, riding show facilities and a luxury housing complex: it would be a Hoover Dam of a project, a Yangtze river bridge, a Great Wall, an emperor's underworld tomb achieved in Five Year Plans. The cost leapt between $2 and $4 billion.

The Tianjin phoenix was a partnership between a Malaysian billionaire, the China Horse Club, Sheikh Mohammed's luxury race-course development firm in Dubai and the local government: the Tianjin State Farms Agribusiness Group. Later they would be joined by Sadler's Wells owners, Coolmore. Sheikh Mohammed was sponsoring a jockey-training programme and Flying Start programme – a kind of MBA in horse racing – which was pushing through its first Chinese graduates, two of whom would find work at the China Horse Club. The phoenix's sweeping wings were supposed to preside over the opening meet while I was in Beijing, and yet I couldn't get a response from the club.

Against these tens of millions and billions of dollars and press launches and websites with golden crests and promises of apotheosis into the global elite was set a steady, ominous accumulation of sudden reversals of fortune. The racecourses at Guangzhou and Tongzhou were shut for flexing the laws on gambling a little too far. A hundred or so horses at Tongzhou were euthanized. Four years later, the villagers who had leased the land to the racecourse had blocked the gates to the stables where nearly 2,000 thoroughbreds were still stabled. The land had soared in value, and the locals now wanted it back. Thirty horses starved to death. The course at Wuhan was boarded up by 2013. Nanjing became a car park. Chengdu hosted an equestrian festival but no race meets. Even folk racing was suspended from the National Games after 2013, because new part-thoroughbred horses had transformed the sport into something that was beginning to look more Western than authentic. Everywhere local officials withdrew their support. What had happened?

In 2011, a new, nationwide Twelfth Guideline had been announced, with an emphasis on ending corruption and graft, and in 2012, the man who would enforce it, Xi Jinping – the 'most authoritarian leader since Chairman Mao' – came to power. Like Mao, he had his own version of the knight dream, admiring the lian of Song Jiang, the hero of *Water Margin*, who had tried to sweep away a decadent dynasty. Officials were the first to be targeted as their expense accounts shrank like dried corncobs and a series of corruption trials began. Chinese pride and communism was contrasted with Western decadence and corruption, and luxury goods sales fell by 20 to 30 per cent. Pedigree Tibetan mastiffs ended up on meat wagons. Golf, also condemned in the past by Mao, was now portrayed as a tool used to corrupt government officials and a source of illegal betting. Clubs were closed. Articles began to appear in the foreign press, suggesting that perhaps polo was not as popular as had been previously suggested. Xi Jinping duly announced a crackdown on overseas gambling, and the casinos in Macao, Singapore and the Philippines emptied of tuhao.

When the Tianjin Equine Culture City held its first meet in autumn 2013 it was not under the glittering phoenix's plate glass wings, but on a bare steppe in Inner Mongolia. White and scarlet embroidered yurts were pitched by the racecourse. The thoroughbred racehorses that took part were imported on a one-way ticket because of quarantine restrictions and sold at auction.

In Dubai, the patronage of the Maktoums could buoy horse racing for as long as there was oil. But east of the Taklamakan Desert, south of the grasslands, in the heart of China, which billionaire would pour yuan into a bottomless pit of yearlings with fragile legs, riding clubs with twenty-five matching black warmbloods, or half-empty polo clubs? How long would the Godolphins and Coolmores stay interested in a sport that never took off and how could the vulnerable horse industry survive with failing official support in a climate where, as the 'peasant champion' Li Zhengqiang worried in 2013, the media depicted it as an elite pursuit? That brief, glittering and reckless mianzi-dream of the new nobility was looking like a political liability.

IX

On the last day of the Sunny Times tournament, the wall of dirty grey cloud lifted like dry ice banished by colossal fans, leaving blue sky and benevolent white clouds over the new pitch. For the first time, I could see the mountains that framed the yellow grasslands: bright rock scooped and carved by long-gone runnels, covered here and there by a thin dark green baize of vegetation. Upright blue banners fluttered along the narrow birch-lined lanes leading to the playing field, and when my taxi dropped me off outside the gate with the SUVs and Land Rovers I found the local farmers waiting by parked scooters and tuk-tuks with red and white cherries in plastic bags.

I passed under the entrance arch, which had the names of luxury brands in Chinese on one side and English on the other, and was nearly run over by one of the student volunteers, who was zooming about hysterically on an untamed Segway. On this side of the green wire fence, the bucolic calm of the golden ranch house or the country lanes was scattered. A small drone buzzed overhead. Security guards paced. Golf carts chased more Segways. Martial music blared from tannoys, interrupted periodically by a reminder to check one's raffle tickets, as two first-class tickets to St Moritz and two weeks in a five-star hotel there were the top prize. Capering in and out of view were two people in panda costumes, for reasons never fully explained.

As elsewhere in China, the Sunny Times polo finals were somewhat overstaffed: three students pounced before I peeled back a glove to show my wristband. Smiling helpers handed me a gift bag with a reproduction of a Tang polo game on it and the slogan 'Enjoy quality life mingled with the elegance and romance'. The pitch was a dazzling dark green expanse – laid just three weeks before! – with yellow curls of birch leaves chasing across the furrows in the strong October breeze.

Along the length of the pitch were a series of white tents open like doll's houses, and I walked on past them, peering in to get a glimpse of the preparations. Someone on their hands and knees was polishing the metal platform on which a Range Rover stood. A British Polo Day manager was explaining to the elegant translator that the catering staff should not sit in guests' places in the food tent, while behind them the photographers got stuck into the champagne and lavender macarons and milled with the white expats up early from Beijing for the day. Tucked behind the first ranks of tents I found a smaller one frantic with ten chefs in toques who were all beating at flies on the ceiling with tea towels.

Xia Yang was being interviewed by a lady in brocade jodhpurs by the main pavilion, where a series of his bronzes, including Jalil, were set out. He looked boyishly delighted to be there with his horses, his club, his team, his tournament. His red and yellow shirt read 'Piaget China' in honour of his sponsor, the Franco-Swiss watchmaker. By the pavilion stood the VIP area, with golden chairs upholstered in mink-coloured velvet and pots full of palms. Although no VIP had arrived, a guard of honour of young men in pearly dark grey Mao jackets were already holding trays of red and white wine, orange juice and champagne.

One hundred and fifty metres away, across the expanse of the polo pitch, something curious was going on. A large crowd was arriving, some carrying what appeared to be posters. They were almost marching in formation. They came to a halt at their own row of white tents, and I squinted to watch them take their seats in an orderly fashion. They were so far from the action that it was as if they were in quarantine. Who were they? Locals? Mr Yang's assistant was not forthcoming, sending me instead into a black box-shaped construction covered with Piaget logos, where she said 'the presentation' would take place.

Inside, I found myself surrounded by the young Beijing fashion elite, whose style had a dressing-up box quality to it: purple velvet tracksuits, jeans and trilbies, pink pleated skirts, fur gilets, fascinators, hats with veils ... One exquisite young man wore a Kenzo baseball cap and a jacket with peaked shoulders. They wobbled on the lumpy carpet on heels, wedges and chopines, and pecked at smartphones with jade-coloured fingernails, eyes masked by gleaming black sunglasses. Piaget gift bags dangled from their arms, and in illuminated glass cases around us hung diamond-studded watches. 'I don't understand young people,' my taxi driver, Sun, had said to me that morning, 'they're all crazy.' They were exotic flowers for parched Yanqing – orchids perhaps more at home in this black box under the spotlights than out on the pitch.

The most beautiful of these flowers was a pale-skinned Snow White with scarlet lips and black hair, wandering among the glass cabinets as if in a fairytale forest. She wore a white dress cinched with a silver band, and red and white silk flowers tipped to one side of her head. Before her, bowing slightly, a cameraman walked deferentially backwards, with a camera dangling like the lamp of an angler fish over his shoulder, recording her every unselfconscious move. The crowd parted before her and closed behind her. She was the Hong Kong model, Angelababy, here to show us the new Piaget collection. A hostess in a floor-length gown on a small raised dais was waiting for her, beside a handsome man with a pencil moustache who was dressed like Clark Gable in *Mogambo*: this was China's most famous male model, Hu Bing.

Angelababy and Hu Bing both took turns talking to the hostess – about what, I cannot tell you, but it involved the display of wristwatches. They held their arms as though the watches were super heroes' weapons that would shoot lasers at the crowd. Xia Yang joined them for a photocall, and, as the VIPs lined up with the sombrely suited men from Piaget, the professional photographers elbowed the Beijing girls and boys and their whirring phones out of the way. The hostess shouted out commands: 'From left!' Yang, Angelababy, Hu Bing and the Piaget men swivelled left. 'From right!' They rotated right. 'From middle!' they turned into the full glare of the flashbulbs, Angelababy's beautiful face frozen for shot after perfect shot.

The match opened with a parade of the teams at the gallop, the leader of each carrying a huge flag representing their nation – the China-Argentina team added a banner from Piaget to the Five-Star Red Flag of the People's Republic. Behind them came a tapering parade of the young horses Yang was raising, junior members of the club wobbling on ponies secured by lead ropes, and the foals, who gambolled about loose, breaking ranks.

The vast plain of the digital screen at the end of the pitch was animated by a Land Rover advert in which a white man in a suave camel coat left a grand hotel to find his SUV waiting for him, faithful as a Labrador. I wondered how the marchers in the distant pavilion were faring – the parade was impossibly far from them, and the screen, too. The British Polo Day commentator, a retired lieutenant colonel of the Light Dragoons, called Xia Yang 'the precursor of Modern Chinese polo' and talked of polo as 'the spirit and essence of horse culture ... A living sport'. This, he stressed, 'was a significant Beijing sporting event. Here we have beautiful scenery and intelligent people, where the nomadic meets farming'. The steppe and the sown, all under the banner of wristwatches! 'The horse is a bridge between cultures. So what better way to combine the two than in such a beautiful location in the shadow of the Wall?' A montage of pouring champagne and of leather being manipulated by craftsmen played on the screen to the impassive mountains and the whirling birches.

But here at last, to my surprise, was my Heavenly Horse, and he was so obviously the tianxia ma that I wondered why I had ever even contemplated any of the other contenders. After the team parade came a tall, black Friesian and the bay warmblood I recognized from the horse fair, led docilely past the tents by their grooms. Behind them, his handler clutching his reins, was the magnificent Jalil himself – 'Impatient of all restraint/And of abounding energy', imperious and urgent, groomed to dark bay silk – the fifth most expensive yearling thoroughbred of all time, $9.7 million before he'd ever set foot on a racecourse. Jalil, the victor of two minor races and a few hundred thousand dollars thanks to a stint in Dubai, far from the Western courses where he was meant to win Guineas and Derbies. One of those exquisite duds thrown out by several centuries of blue-blooded breeding and monied faith in inheritance.

Sheikh Mohammed had shipped him and another stallion to stud to China in 2011 when the Tianjin Equine Culture City plan was at its height, a tribute horse from the West whose foals would fill racecourses yet to be built. He had just been sold to a local breeder, completing his one-way moon shot to a frontier where nothing was certain.

The Mongolian polo ponies pricked their ears, the Friesian subsided into dull black, the warmblood looked half carthorse, as Jalil pranced along the sidelines, past the all but indifferent Beijingers, or the expats to whom he was just another horse, and confronted his bronze likeness, his head high, nostrils flaring. He dismissed it as the cameras flashed and the people applauded. He called out to the other horses, the furry Shetlands, the polo ponies, any mares who might be nearby, and then, following the rest of the parade, he was led back across the pitch to his horsebox.

Piaget China-Argentina United won their final. The spectators chatted to one another and sipped champagne as the two teams had their distant skirmishes, puffs of dust exploding from the newly laid turf as the shadows lengthened in its furrows and the sun began to sink. At the award ceremony, a delighted Xia Yang sprayed Taittinger at his Argentines, confetti cannons exploded and bouquets of deep red roses were handed out. In the background, one of the Sunny Times grooms rode a tuk-tuk across the pitch, clearing up the apples of dung that were scattered about. The mountains were disappearing into smog once more. The strange far-off tents on the other side of the pitch were already deserted.

As I made my way out, the Segways were still zipping back and forth, though with a little less pep. I passed some delicate Beijing girls staggering on their platforms around a porta-potty; one of them opened the door and they all gazed in horror on the squat toilet. The sun was low and the SUVs had left the car park to blockade the birch avenue. I passed back through the gateway of luxury and found that the cherry sellers had gone. A man on the roadside pointed at my boots and said, 'You polo?' I shook my head. He pointed to the field behind him. Would I like a ride? Tied under a tree was a sere-looking bay, running to winter fur against the October wind, whose bridle was a strip of ragged tape running from his bit around his ears. His saddle had a metal rail on the front, and was worn right through on the skirts. It sat on an old green racing number cloth: lucky number 181.

WAR

Are Horses Warriors?

The kind of horse and of man is medlied.

Bartholomaeus Anglicus
(*De proprietatibus rerum*, Book 18)

I

South of the steppe but north of the Mesopotamian river valleys where the early, horseless farmers raised their primitive einkorn, emmer and barley crops, the people living in the foothills of the Taurus and Zagros mountain ranges discovered that copper could be smelted from brilliant green rocks in potters' kilns. They traded the bright, pinkish metal for Mesopotamian grains. When they began to alloy copper with arsenic and then tin to make durable, recastable bronze, they made tougher weapons and tools that earned them more grain and other goods. In the Bronze Age that followed, small raids and skirmishes gave way to the first recorded large-scale warfare among humans, to organized armies and mass armament production, and to a new creature, the warhorse.

The new metals flowed molten along old grooves of trade routes established by the first farmers, spreading north, south, east and west in pursuit of copper and tin. Like the iron rails laid down for the early railway horses to draw their loads along, metal accelerated the flow of goods. The proto-cities of the Mesopotamian valleys developed the earliest written languages – nicks made in clay with fingernails or styluses – to keep track of their trading. Land ploughed by animal draft produced surpluses of grain, and grew in value, supporting more workers, tenants and slaves. Hierarchies among men were enhanced, the difference between the wealthy and the poor bedded in, ownership and control reinforced across generations and territories. The Mesopotamians tried to use their new-found wealth to control the trade and extraction of metals, which inevitably led to the control of more people, whether their own citizens or slaves, or the peoples who lived around the mines. The circuit was reinforced: more grain, more power, more metal, more cities, more control, more metal, more war. With a further innovation, the control of horses joined this circuit, transforming battle and the course of history.

The chariots that rattled all the way across the grasslands to the Yellow River valley of the Shang began as crude wagons drawn by oxen in the Near East, rumbling up to the steppe at the end of the fourth millennium and returning – drawn by horses – to the Iranian plateau west of the Zagros around 2000 BC. Equine teeth, worn by bits, appear in the archaeological record, and soon in the Mesopotamian river valleys a new word, 'horse', or 'ass of the mountains', surfaces. These horses replaced the ass and mules that drew the light Near Eastern war chariots.

Metal was cast into pins that held the chariots together and made the axles run smooth, into bits to steer the horses as well as daggers, swords, arrowheads and spear tips. Thousands of years of smelting and casting bronze finally led to the use of iron, and the Hittites began extensive work with this metal around the time that Kikkuli's book on training chariot horses was compiled. The Han Chinese were already charioteers when they learned about iron from the steppe nomads, who arrived on their borders drawn by horses, having followed a thousand-mile long trading route that brought metals west across the grasslands from the Altai. The Chinese developed steel by the first century AD, sending it back across the steppe.

In the bellicose ages of Bronze and Iron, Babylon fell to Kassite chariots, and Minoan Crete to the Mycenaeans. The Zhou's 'sandalwood chariots were gleaming,/The teams of four were pounding', when they defeated the Shang at Muye. At Kadesh on the shore of Lake Homs, 5,000 Egyptian and Hittite chariots clashed. Across a tremendous spread of geography, time and culture – in the *Rig Veda* or *Metamorphoses* or *The Poetic Edda* – a sacred chariot or wagon was drawn by swift supernatural horses across the sky, carrying the disc of the sun on its heavenly course, or, elsewhere, the chariot bore the dead to the afterlife through an airlock of graves lined with pots, mirrors, jewellery and harness.

Warhorses were an expensive necessity for a king, and they were duly pampered: Kikkuli's chariot horses were massaged with expensive oil and stuffed with grain. The Babylonian warhorse whose flesh 'was not eaten' boasts of its superiority over the ox as 'I ... tread a pavement of kiln-fired bricks/ [Close to] king and counsellor my stall is located,/[The attendants make ready [for me] plants, the greenery of the earth/ ... they look after my magnificent drinking fountain ...'. In Ancient India, warhorses could only be owned by the aristocracy and royalty, were fed root vegetables stored in honey and given draughts of wine before battle.

Riders replaced chariots, and the invention of a leather and then a metal stirrup enabled riders to lean further from their seat to make shots or dodge sword blows, and eventually, to brace themselves against the impact of the lance they carried. Man and then horse were gradually coated over with bark, horn and leather: the Sarmatians on the steppe north of the Black Sea actually armoured themselves in scales chipped from the hoofs of their horses and sewn together with their sinews. Greek geographer Pausanias says it resembled 'the scales of a dragon' or 'a green fir-cone' and was proof against sword and arrow. Finally they made the scales from copper and bronze, and the horse itself was reskinned with metal.

Cavalry remained more expensive than infantry, creating a prestige that reinforced itself, and would, even when mounted soldiers were disadvantageous in battle, lead military authorities to go on promoting and funding their equestrian ranks. In *Politics* Aristotle talks of the city states of 'old times', saying that if their surrounding landscape was suitable for cavalry warfare, 'then a strong oligarchy is likely to be established. For the security of the inhabitants depends upon a force of this sort, and only rich men can afford to keep horses'.

This was status built by wealth but fortified by the performance of man – or woman, on the steppe – on the battlefield, and brightened by a mythology of heroes that inspired new heroism and myth in turn. Into this construction the warhorse also fitted, for, as the Duke of Newcastle put it in the seventeenth century, when in many ways little had changed, 'He is as much superior to all others creatures as man is to him, and therefore holds a sort of middle place between man and the rest of creation.'

As military technology and the conduct of warfare changed over the ages, warhorses have waxed and waned as new methods of fighting with horses evolved or as infantry became cheaper and more effective. From chariot to mounted archer to heavy armoured horse with lance and broadsword, and later cuirassiers and harquebusiers with their crude firearms, and finally the cavalryman's blood horse equipped with lance, sabre and rifle, at times the warhorse had a deadly efficiency, at other times disastrous inefficiency.

His battle work diversified: he became the indispensible sturdy packhorses, the mules, the donkeys and the teams that drew the cannon or the supply wagon as ably as they drew a farmer's cart or a London omnibus. He was the messenger horse running 40 miles between relay stations for Genghis Khan's army, the Takhi who became a trophy of war, and the Ferghana horses brought back across the Taklamakan Desert. He was Xenophon's carefully selected mount, prancing to his master's glory in a victory parade, Astley's Spanish horse from his old campaigning days, taking part in mock battles on the stage, and Chetak leaping at the elephant of Rajah Man Singh at Haldighati. He was the horse on the *Luttrell Psalter*, kicking out at a squire with a shield, and Totila's black steed dancing as his rider throws his javelin from hand to hand. He was even the poor carcass gutted at Eylau, or used by Napoleon's soldiers as a macabre shelter from the blast of the Russian

winter. Every element of the horse has been turned t
war and man, and he embodies the double face of b
clean heroism and the bloody, fearful reality.

II

The warhorse of the human imagination has always been a creature
that partakes in war itself, physically, psychologically, spiritually, as an
extension both of man's body, mind and courage. In the Book of Job,
God cites the creation of the warhorse as proof of His omnipotence,
demanding of the poor, benighted Job:

> Hast thou given the horse strength? Hast thou clothed
>   his neck with thunder?
> Canst thou make him afraid as a grasshopper?
>   The glory of his nostrils is terrible.
> He paweth in the valley, and rejoiceth in his strength:
>   he goeth on to meet the armed men.
> He mocketh at fear, and is not affrighted; neither
>   turneth he back from the sword.
> The quiver rattleth against him, the glittering spear
>   and the shield.
> He swalloweth the ground with fierceness and rage:
>   neither believeth he that it is the sound of the trumpet.
> He saith among the trumpets, Ha, ha; and he smelleth
>   the battle afar off, the thunder of the captains,
>   and the shouting.

The warhorse's relationship to his master is, as Grimm put it of
Teutonic heroes and their mounts, 'of necessary intimacy': 'You stake
... your valour and your fortune upon that of your horse,' wrote Michel
de Montaigne, 'his wounds or death bring your person into the same

danger; his fear or fury shall make you reputed rash or cowardly; if he have an ill mouth, or will not answer to the spur, your honour must answer for it.' In myth and in life, the warrior talks to his horse (which sometimes talks back). They march together, live off the same country, sleep perhaps in the same bivouac and – most intimate of all – face death together.

Pliny the Elder wrote that Bucephalus refused to let anyone other than Alexander ride him, and of a Scythian horse that trampled its rider's killer to death. Pliny's warhorses weep or pine away for their masters, or even hurl themselves down precipices when mounted by the enemy. 'These animals possess an intelligence which exceeds all description,' he wrote. 'Those who have to use the javelin are well aware how the horse, by its exertions and the supple movements of its body, aids the rider in any difficulty he may have in throwing his weapon.'

Later writers copied Pliny, and added further embellishments. The warhorse's motives and those of man were conflated, so the horse enhances man and the man the horse. In Ramon Llull's thirteenth-century The Book of the Order of Chivalry, the horse is the 'beast most suited and handsomest, most courageous and strongest to withstand troubles and most able to serve the man ... For after the horse, which is called cheval in French, is that man called chevalier, which in English is knight. Thus to the most noble man was given the most noble beast'. His contemporary Bartholomaeus Anglicus went further, saying, 'the kind of horse and of man is medlied'. Even the eighteenth-century French naturalist, the Comte de Buffon, believed that the horse loved war, for, 'equally intrepid as his master, the horse sees the danger, and encounters death with bravery; inspired at the clash of arms, he loves it, and pursues the enemy with the same ardour and resolution'.

This ennoblement of the warhorse doesn't just belong to European chivalric tradition. Somali clansmen wrote poems in praise of horses used in raids – horses for whom their owners would grovel under thorn bushes to find shrubs or grass – and who were described as 'sublime', 'beloved brother', 'an inheritance from blessed heaven'. 'If I don't see him for a brief season,' wrote Sayid Mahammed Abdille, 'I am smitten with an anxiety of longing,/and come close to dying from a nagging fear,/Is he not my very heart!' In Raage Ugaas's 'A Horse Beyond Compare', the horse saves his rider from 'spear that chops limbs,/half-blackened from its formidable iron shaft,/that, when it is shot at you, flails frightfully through the air'. 'I wonder:' asks the rider, 'is this horse of the kind of nobility as possessed by a holyman?'

In ancient Ghana, the Kayamagha royal family kept their horses in their own palace, where they slept on mats and were tended by three grooms each, who held copper pots to catch their urine. Mongolian warhorses were not used for meat when they had ceased to be useful, but retired to pasture or, if their master had died, buried alongside him. There are statues for the horses that died in the Mughal Wars, the Boer War, the Napoleonic Wars, the First World War, on the Eastern Front of the Second World War and in the Pacific theatre. Warhorses are stuffed or skeletonized and held in museums or formally interred, like Copenhagen, Wellington's mount at Waterloo, whose gravestone reads, 'God's humbler instrument though meaner clay/Should share the glory of that glorious day.' On the Western Front of the First World War, a Canadian officer lost one of his drivers and his team to the same shell burst, and 'knowing how a real man comes to love a real horse, I had him buried with a horse upon each side of him, and they now lie sleeping up in the Ypres salient. The horses were named Friend and Foe'.

In the 1950s, American marines awarded two Purple Hearts, a nineteen-gun salvo and a salute from nearly 2,000 servicemen to the first female staff sergeant of the United States Marine Corps: a pack mare called Reckless which had carried load after load unaccompanied to the front in the Korean War. Frederick George Scott, poet and chaplain in the 1st Canadian Division, thought back after the Great War to Dandy, a roach-backed, crock-kneed chestnut thoroughbred who carried him through the war: 'Sometimes I was on his back and sometimes he was on mine, but we always came home together,' he added of 'one of the best friends I had in the war'. When hostilities ended and he was ordered to leave Dandy for public auction in Belgium, he shot the horse, so he might 'enter horse's heaven as a soldier'.

And some warhorses *were* soldiers, doing their part in combat. The medieval Rochester Bestiary depicts two warhorses boxing like hares on the battlefield, and Usāmah ibn Munqidh, writing at the time of the First Crusade, saw two riders clash, kill one another, and their horses, having lost their riders, go on fighting one another. At the Battle of Fornovo near Parma in 1495, the horses fought 'with kicks, bites and blows no less than the men'. In Norse sagas and on Elizabethan streets, horses fought one another or other animals as proxies for men. According to the French war correspondent, Dick de Lonlay, the Cossacks' horses of 1877 went into combat 'with unrestrained rage: mane flowing, with bloodshot nostrils, he kicks and bites the enemies' horses with the greatest furore'.

The medieval theologian Albert the Great said that warhorses should not be castrated, for 'under battle conditions they show no hesitation in leaping over obstacles and attacking an array of the enemy, biting with their teeth and trampling with their shod hoofs'. Mares were used by Arabic cultures as they were quieter, but they were also praised as loyal. The Mamluks' horses not only picked up their masters' fallen weapons, but, according to the credulous Montaigne, 'by nature and custom ... were taught to know and distinguish the enemy, and to fall foul upon him with mouth and heels, according to a word or signal given'. In China, the third-century horseman Po Le mentions horses purchased because they were kickers. Claudio Corte, Elizabeth I's horse trainer, suggested giving a treat to a horse that attacks a soldier on cue.

But these hoofed war machines could be forgiven for having no concept of sides or the rules of combat. Montaigne said that some horses were trained to attack a man holding a drawn sword, 'but it often happens that they do more harm to their friends than to their enemies; and moreover, you cannot loose them from their hold, to reduce them again into order, when they are once engaged and grappled, by which means you remain at the mercy of their quarrel'.

The Mamluks' Chahri warhorses galloped on despite wounds, but were also notoriously savage, to a point where they were dangerous to their own side.

Warhorses kicked kings as well as stable boys, breaking ribs and skulls. The Duke of Wellington was nearly killed when he patted Copenhagen after seventeen hours in the saddle at Waterloo, and the exhausted stallion lashed out. Even Xenophon recommended muzzles for the stable. When it was not on the battlefield, the medieval warhorse's muzzle was often caged in an ornate *muselière* of bronze, on which writhed two-headed eagles, the leaves of a greenwood and the verses of the Bible.

But beyond the medals and the trappings and statues that a horse cannot possibly comprehend or appreciate, to Friend and Foe and Dandy and all the other weary horses and mules that struggle in the Flanders mud in photographs of the Great War, or the blue-grey Renaissance warhorses who sprawl in the foreground of Uccello's San Romano, surrounded by broken lance points, or the white-eyed, scared animals in the patriotic depictions of Balaclava, their mouths yanked open as their riders try to snatch them away from a whizzing sabre, war meant much what it does to any disillusioned human soldier: pain, fear and confusion.

Twenty thousand horses died at Waterloo in 1815, 10,000 at the Battle of Caravaggio in 1448, 7,000 in a single day at Verdun in 1916. In a mêlée horses struck out wildly at sources of pain, at sudden movements that came from behind or beside them, or before them. They met Chevaux de Frise or 'Friesian horses', frames bristling with spears, and stockades of halberds. They galloped into barbed wire. Spikes or caltrops were thrown under their hoofs, pikes slit their bellies, hidden pits opened beneath them. They felt the queasy unsteadiness of a body under their hoofs, and met arrows, spears, lances, pikes, maces, swords, bayonets, bullets, shells and rockets. In Flanders they inhaled mustard gas that blistered their skin and lungs.

There were long, spiked and high-wrought spurs that made knights' heels into weapons, and curb bits festooned with chains under the chin and long, curving shanks at the corners of the mouth. In the hands of a skilled, unhurried rider, the curb bits were precision instruments; pulled hard in the chaos of the battlefield they were levers that bruised the horses' poll, chin and mouth, cutting off the circulation in its tongue but stopping it fleeing. Some horses even had teeth removed to make space for bigger curbs.

At Waterloo the horses 'screamed at the smell of corruption', and at Crécy in 1346 the Burgundian chronicler Le Bel describes warhorses scattering 'like a litter of piglets'. At Satory in 1870 the French army made a demonstration of their new *mitrailleuse* machine guns by killing in just 180 seconds 300 horses fetched from nearby abattoirs, and the next day cut down 500 more in half the time.

Perhaps the strangest and most poignant story is that of the Household Cavalry horses which fought at Waterloo, and were subsequently cashiered out of service at auction. The king's surgeon, Sir Astley Cooper, purchased twelve of the worst wounded and had them taken to his home, Gadebridge House in Hemel Hempstead. With the help of his servants, he began to painstakingly remove bullets and grapeshot from their hides and muscle, treating them as carefully as he had his royal patients. Eventually they recovered enough to be turned loose to graze in the park.

One morning, looking out across his land, the surgeon saw the twelve horses form a line, shoulder to shoulder, then, without a cue, charge forward at a gallop. After a few strides they spun and retreated as formally as in a drill, and then broke from their line and careered about freely, in high spirits. After that day, Cooper watched every morning as his old cavalry horses, flecked white where their coats had grown back over their scars, enacted their enigmatic ritual and went to war together once more on the cool green parkland of the Home Counties, far from the smoke and horror of a shallow, Belgian valley.

I had travelled to Mongolia to see wild horses, to Versailles to see them dance. In Ohio and Massachusetts I'd seen working horses, in Beijing

Equus luxuriosus and in that dingy auction house the horse as pounds of flesh and hide. Where in the early twenty-first century could I safely see a horse go into battle, to try to understand how man persuaded the horse to go to war, to charge together, to fight, to stand steady in the face of fear, and what in the horse – that swift, large-eyed, prey animal that scattered across the grasslands – made it possible?

III

The grey Portuguese stallion stood motionless at the edge of the circular sand bullring, his dark hocks backed against the slender barrier of oxblood painted planks, his ears pricked, eyes looking straight ahead across the concentric circles ploughed into the sand, to a gate concealed opposite. Behind the gate was a dark entryway, sunk under the second raised tier of seating, and behind the entryway was the enemy that the horse was anticipating. Could it hear the other animal? Smell it?

The old Renaissance warhorse in little cream Uccello at Versailles was there in the grey's Roman nose, and the way he held himself – chin tucked slightly in, ears to the fore, legs a pillar at each corner – although his neck was high, tense and alert. His mane had been threaded with white and bright blue cloth that hung in small swags, like bunting on a tournament pavilion, and the skin under his white coat was slatey, showing through at the places where his limbs met his body, adding shadow to the tracery of his muscles.

He ignored the crowd, who beat little plastic fans at the stuffy summer night. There was a heatwave, and during the suffocatingly hot day, the air had sunk down into the cauldron of the arena, past the slender columns and Moorish arches that lined it, and now, even though the sky was darkening and the stars coming through in pinpricks, it was thick and unpleasant where I sat on a tipping, sweaty plastic seat just a few metres above and behind the oxblood barriers.

The bullfighter was a stout, steel-haired man in his fifties who wore a dimly eighteenth-century rig: glossy, charcoal-coloured coat embroidered with glitzy golden flowers on the chest and tails, a white shirt with thin lace cuffs and a tricorn hat whose crown was filled with flimsy ostrich feather. To a British eye, he looked rather like Baron Von Hard Up in a well-dressed pantomime. Any minute now he might summon Cinders into the arena and tell her she could not go to the ball. He was wedged into the deep-seated saddle – the sort of saddle that made it harder to knock a rider out at the point of a lance – filling the gap between the high cantle and pommel.

He, too, watched the barrier opposite, just at the point where its smooth planks were interrupted by a hinged door. His heels, in boxy metal stirrups, danced and dusted at the grey's sides, but otherwise he did not move, either. He seemed more animated than the horse, which cocked its tail and dropped a heap of firm dung on to the sand without shifting its hoofs, then twitched its tail, staring at the door. The rider held the reins in his left hand before his belly, his right extended at arm's length before him, holding poised and perpendicular a nasty spear whose steel shaft was wrapped in red and white frills, like a decorative cocktail stick, or the ruffles on a crown of beef. The arrow-shaped tip pointed at the arena sand, just to the right of the grey's shoulder. They waited. His heels danced. To their left stood a young man in a brilliant pink jacket and skintight breeches. He trailed a half circle of carefully arranged cape behind him on the sand, and had a small, round black hat, moulded at the sides into ears which sat above his own.

There was a cry and the gate flew open, clapping against the barrier, and out came the black bull, a surge of dark energy and muscle so thick that it guttered over his narrow rump. Its horns were broader than a man's wrists, sawn off at the tips and sheathed in leather, its eyes small, its bare muzzle wet. It swept around the circumference of the circle, its energy pulsing round the ring like a sound wave before it, sending the young man in pink tumbling over the barrier out of its way.

The bull reached the grey, which had been waiting for this and leapt forward to meet him, light where the bull was bulky, containing his body in a ball of collection that contrasted with the bull's low, relentless drive. The grey cantered almost on the spot, ears still straining forward, and when the bull's horns were feet away he feinted, jumping right so abruptly that his chest was still in the path of the bull, then, when the bull dipped its horns to thrust, darted his whole body left, and swerved past it on the other side, a foot away. His rider leant out of the saddle and jabbed the spear into the peak of muscle where the bull's thick, short neck met its sloping shoulders, and as the strike went home, the grey pinned its ears back, a trumpet played, the crowd applauded and the small javelin broke in two, leaving a long shaft slapping against the back of the bull, as the rider held the colourful handle, which now threw out a small white fluttering flag with the sponsor's logo on it.

As the horse cantered on, the bull spun to pursue it, head low, horns erect, and the rider leaned back in the saddle, jouncing against the horse's back, and trailing the white flag behind him to catch the bull's eyes. Now they cantered in tandem about the ring, the horse all tight discipline, its tail just inches from the bull's horns, the rider leaning back to taunt the bull, but judging by the millisecond just how close it came. Round they circled, the bull occasionally tossing at the horse's tail, flicking a few hairs up as the grey clamped it tight against its body. If the bull came too close, the rider pushed the horse a little faster, but never so fast as to escape. The horse's ears flickered loosely back and forth; it could no doubt see the bull behind it, and yet it obeyed its rider, letting him calibrate the distance. And just as the strange pursuit reached the second half of the circle, the rider spun the grey in a wild, ragged pirouette: 180 degrees, and for a second the horse's face was half a foot from that of the bull, eye to eye, ears to horns, and then, pressing from its fetlocks, its hoofs dug into the sand, the horse's body twisted against its hind legs, part leaping, part cantering, and it sprang away from the bull to complete the 360 degrees of the turn just as the bull rushed forward again.

This time the rider let the horse gallop on, and swung up his arm to punch the air with a cry, demanding the crowd's acknowledgement, which was lukewarm when it came. The bull stopped and gazed around itself, as if baffled by what the horse had just done – the flash and dazzle of it, tail, ribbons, great eyes, the lance point digging cruelly in as the shaft flopped against the bull's back. The first trickle of sticky blood was starting down its slabs of shoulders. The young *bandarilheiro* in pink leapt over the barrier with a teammate, and they shook their capes at the bull as the rider, on the other side of the ring, gestured to the crowd and the president, asking for more applause. Then these assistants withdrew, the rider circled the grey back into the bull's blurry field of vision, and they began again.

Even if the bullring is the closest you will come to the adrenalin and danger of the battlefield, it is more rite than mêlée, and while there is chivalry, it is not extended to the lancer's opponent, who is never expected to win. In Portugal, where I watched the mounted *tourada* or *rejoneo*, as it is called in Spanish, the bull is not killed in the arena. When enough darts have been planted in its back, the *forcados*, eight young men in cummerbunds and tight breeches, take to the ring, and provoke the weakened animal on foot, rugby-tackling its head and tugging on its tail. Then the bull is simply removed from the ring in the midst of a small herd of brown oxen driven by two picturesquely dressed countrymen with stocking caps and lances, pushed into a cattle crush where the spears are cut out of its shoulders, and perhaps slaughtered some days later, out of sight. It owes this doubtful reprieve to a nineteenth-century rule that made it a crime to kill the bull in the ring, although this ban did not stick until 1928, and even now some of the 100 Portuguese towns that organize bullfighting events are excused from it. Yet *cavaleiros*, as the Portuguese mounted bullfighters are known after the old knights, have still been jailed for killing a bull in Campo Pequeno, the Lisbon arena in which I sat in the stifling July of 2013.

The mounted bullfighter is older than the matador, but the origin of the *tourada* is misty. It first emerges in reliable records as a staged hunt and battle performed by the aristocrats and royalty we met at medieval tournaments and carrousels who charged head-on with lances on stocky horses, aiming for one fatal blow. In Spain, the bullfighter or *rejoneador*'s team is still called the *cuadrilla*, from the old quadrilles. Some suggest that wild cattle were originally hunted by spearmen in the Iberian peninsula – both Christian and Moorish. Others point to the beast fights of the Roman amphitheatre, or to more obscure pagan rites and bull cults linked to fertility, dating back to the pronghorned cattle that make their sooty way across cave walls – the aurochs the Hecks tried to recreate.

Despite many bans issued by both Iberian royalty and the Pope himself, the increasingly refined *tourada* was for centuries the Spanish or Portuguese nobleman's best display of his credentials: as a wealthy horseowner, as a warrior, as a knight, as a rider. One seventeenth-century nobleman of the Bragança family shod his eight bay bullfighting stallions in silver.

In Spain, the mounted bullfighter's valet, who had climbed into the arena as an earlier *bandarilheiro* to cape the bull away from his master, became the matador – the working-class hero who fought on foot – and the mounted bullfight declined in the eighteenth century. Meanwhile, in Portugal *tourada* continued to be the practice of the aristocrat, with the Marquis of Marialva formalizing the choreography of the various strikes and passes. The first *cavaleiro* I saw in Campo Pequeno on the grey was the veteran Joao Moura senior, and the embroidered tailcoat and feathered tricorn he wore dated from Marialva's day.

You still need to be fairly wealthy to be a mounted bullfighter, like the cavalry officers of Aristotle's oligarchy. Even if you can earn six-figure sums for a single *tourada*, you must support a full stable of horses both trained and in training – the most expensive of which can set you back many hundreds of thousands of euros – and you must pay your *bandarilheiros* and your grooms. It's a sport of dynasties: dynasties of bull breeders, of horse breeders like the famous Veigas, and of *cavaleiros* themselves. The bill on the night I attended Campo Pequeno included, alongside Moura, whose son is also a popular bullfighter, the baby-faced Manuel Ribeiro Telles Bastos, a fourth-generation *cavaleiro* whose grandfather lost all his land in the communist revolution of 1974, and who revived his family's fortunes in the arena, and the Spaniard Manuel Manzanares, who is the son of 'the maestro of the maestros', José María Manzanares. Manzanares wore the Spanish *traje corto* – a working cattleman's outfit from Andalusia of high-waisted trousers covered with leather chaps, a flat, grey broad-brimmed hat in place of the florid tricorn, and a cropped, plain blue jacket.

Country gentleman or courtly peacock, a bullfighter can be in and out of the saddles of up to six horses on a single evening. There's the horse used in the opening *paseillo* or parade, which prances and performs haute école movements for the crowd when the fighters are introduced to the president and the brass band in the gods blares. Then the horse for the first stage, who, like the grey, was swift enough to dodge the bull when it was still fresh and uninjured. With this horse, the *cavaleiro* strikes the blows that the picador makes in the matador's corrida, damaging the muscle until the bull can no longer hold up its head. Once this stage is completed, the rider swaps horses, while his assistant *bandarilheiros* keep the bull revving with their capes, and takes up a shorter set of darts – the horse need not be so fast or so lithe, but it must be braver, as it comes closer to the slower, bleeding, bewildered or enraged bull. Then comes the final stage, when the darts can be as short as six inches, and the horse must be courageous or obedient enough to pass within inches of the bull, so that his rider can lean out and drive home the spike.

The bravest horse is needed in the Spanish fight for the *tercio de muerte*, or death stage, when the *rejoneador* endeavours to kill the bull from horseback with a spear-like lance. If this does not work, he dismounts to stab the stumbling, staggering animal, now too weak to charge.

Moura's second horse was a golden bay, its mane twisted with green and red, which came bounding into the arena, surprising the bull, which jumped right round to face it despite its wounds. The bull was now wary, and slow to charge. According to the rules of the tourada, Moura could not strike until the bull initiated the attack, so, as the bull stood and twitched its head back and forth from the flickering fans of the audience to the horse, the bandarilheiros and the capes, Moura waved his spear and jerked his arms, crying, 'Hu! Hu! Hey!' to get its attention. At last the bull lunged at the horse, but when Moura jabbed in the dart to a fanfare and cantered away, the bull would not follow the horse, and the audience muttered behind their fans.

Moura's third horse must have been part Arab, a deep bronze chestnut with a blaze, its tail held kinked and high. Moura looked too heavy for it, and as it dodged and spun around the bull, with Moura leaning lumpenly out of the saddle, I feared for its balance. By now there were so many darts dangling from the bull's slick, red shoulders that it looked as though someone were making lace on its back. It pawed the ground and the crowd sighed in disappointment – a bull that paws is deemed a coward, trying to scare the horse without honestly attacking it. The chestnut finished its dance, Moura decided that he had poked the bull enough, and, as the bandarilheiros caped the bedazzled bull away, he made a quick parade around the ring, followed by a man pushing a wheelbarrow for dung. Then Moura and the chestnut departed, trailed by the workman and his barrow.

Each cavaleiro had a fresh bull, whose name, breeder and weight were given on a sign that was walked around the ring at the beginning of each bout. At over 600 kg, the bulls were both heavier and faster than the horses, but their front-loaded bulk and squat legs, set far under their torsos, made them blunt and clumsy as torpedoes in a small pond. The horses, taller, lighter, nimbler, trained, mocked the bull as they danced and pirouetted. Moura's grey knew the game in the way that the bull he faced did not, and he could learn across his career to expect the creature that came barrelling out from behind the barrier, whereas a bull that grows more wily in its twenty minutes in the arena is frowned upon and feared – and soon dead.

Bullfighting horses are praised like Pliny's warhorses for the way their 'exertions and the supple movements of [their] body, aids the rider' and for carrying them – and themselves, by no coincidence at all – out of reach of its horns. The celebrated *rejoneador* Pablo Hermosos de Mendoza says his mare Estella's vocation is the bullfight. At one *corrida* in Spain a horse ridden by Hermosos de Mendoza attacked the dying bull with its teeth, pushing it over and ripping off skin like a frenzied Cossack cavalry horse as its owner appealed to the crowd – see how my horse defends me! Normally, the horse is not encouraged to fight freestyle, although many think that when it pins its ears back as the spear goes home, it recognizes the spear and the arm that holds it as part of its own attack. Another *rejoneador*, Álvaro Domecq, said his mare Espléndida 'would have preferred that they kill her rather than that anything happen to me. She was a model of fidelity and loyalty, she understood me so completely'.

But the bull is, after all, trying to gore the horse, and the horse knows that. When the bull does not miss, and when his horns are neither sawn blunt nor covered, he can kill a horse. One website names the twelve horses injured or killed in a single year: Imperial, Fusilero (who was stabbed in the lung but lived), Romance, Zurbarin, Burrito, Salgueiro and Bálancin, who was gored at the Feria del Pilar, and died an hour later. YouTube yields you unwatchable 'bull beats horse' videos, of horns sinking deep into unprotected bellies after a horse is butted to the floor – Horongo, stabbed in the heart – or one horse, rider lost, galloping around an arena with its guts spilling from its side, pouchy as a pink silk parachute.

The anthropologist Kirrilly Thompson, who studied *rejoneo* in Andalusia, was told by one *rejoneador*, 'The horse knows by intuition that [the bull] has horns and that if it gets caught by those horns, it will be killed.' But he cast himself as a teacher, and conductor, and said that it was the rider's job to 'position the horse, then it [the horse] takes control'. The *cavaleiro* is, if he is lucky and skilled, the choreographer, with a repertoire of moves for engaging and attacking the bull. If the bull will not charge, horse and rider perform the elaborate haute école usually restricted to the opening parade, in a display of discipline in the face of savagery. They may piaffe before its eyes, or leap into the air in a courbette, a flamboyant gesture of *sprezzatura*. If the horse is struck, the rider is disgraced. Ultimately the *tourada* is about the power of man over both the *bravo* or wild bull and the *desbravando* – 'dewilded' – horse he has also mastered.

Ribeiro Telles's bull, Diablo, would not charge after a few spears had been stabbed into it. It stood, its belly snatching up like a fist as it panted, its tongue protruding from its awkwardly open mouth in a stiff, blue curl. It did not appear able to shut its mouth.

Telles waved his spear at it, then his hat. Then he turned his horse so that its quarters faced the bull and its forefeet rested on the low white rail at the base of the oxblood barrier. The horse looked out over the audience, ears sharp, and Telles removed his tricorn and set it on the horse's head. Now the bull moved, and Telles and horse leapt and turned in one smooth motion to face it, Telles slapping his hat on to his own head. The bull missed. Telles's dart missed, too.

It was clear from the audience's reaction that all this puncturing of confused bulls was an indifferent show, largely due to the understandable reluctance of the beasts to go on charging at the dancing figure that stabbed them. The three *cavaleiros* took on their second bulls after an intermission. The audience was beginning to drift away, but Diabrete, born in 2009 and 610 kg, was the bull that animated them. He was Moura's second bull, bounding into the ring as it grew dark, and the circle of lights crowning the roof's inner rim glowed brighter. Diabrete chased Moura's grey with the long, clever face right around the ring, its horns measuring the horse's quarters like a vice. When a *bandarilheiro* struck it with a cape, it leapt in the air. The president ordered music – an acknowledgement that Moura was now truly performing, and the reedy brass band trilled as Moura reappeared on a dark bay which cantered sideways, zigging across the sand, and then outran its opponent. It half reared before the bull, which did an answering do-si-do, and then, as the audience stilled their fans and drew forward, the bull charged, was struck, and then pursued horse and rider, who turned one, two, three, and – oh mercy – four pirouettes before its nose. The horse and bull crossed the ring in a serpentine, then a circle. When the bull stopped, Moura spun the bay in a double pirouette before it, his jowly cheeks glowing red, veins starting on his forehead and sweat trickling on the bay's side. They missed a pass, but one of the earlier darts appeared to have gone home false, perhaps striking a nerve, and now the bull's tongue was blue, and it would not chase the horse.

Moura had cheers, and some of the audience even stood to applaud; men Moura's age threw flowers or scarves for him to touch and throw back. But the bull took much persuasion to charge the *forcados* when Moura and his horse had left the ring.

The sixth bull, Dulala, was the last, and the cumulative limit of what I could stand on that hot night, already swallowing down a bellyful of spilling heartburn and disgust. Manzanares waited in the arena on a dark dappled grey stallion that called out, throatily, for the bull. The *cavaleiro* and the stallion galloped in rapid, sweeping circles to warm up, large and then small, and then cut across the centre to change leg, like gymnasts stretching and chalking their hands. Then they retired to the spot opposite the gate to wait.

Dulala was the fiercest bull of the night. He flew from the gate like a killer whale, almost marine in his smoothness, the slip of muscle under flat coat, barrelling up the edge of the arena like a furious wave, scattering the *bandarilheiros* and drawing Manzanares's horse up on its toes. The crowd cooed in anticipation, but then the moment shattered: Dulala stumbled and crashed into the sand. Other bulls had fallen that night, but Dulala could not get up. Something in him was broken. He thrashed furiously on his side, trotters punching at air, he rolled up on to his knees and butted at nothing, then, hind legs pumping, ploughed himself forward, nose shovelling sand. The crowd began to boo and hiss. Manzanares and the dark grey left the ring. Two nervous men ran out with a rope, reluctant to get near to the bull as it struggled. The audience shouted for it to be killed. Like a dying gladiator, it staggered up, and the gate was opened to let in the herd of brown oxen. They gathered up their dark brother, and took him hobbling from the ring.

IV

'I think they don't like the bull being there,' said the ethologist and horse trainer Lucy Rees when I asked her why the *cavaleiros'* horses challenged the bull. 'I once took a colt from a good bullfighting line up from Spain to Britain and I stayed in a place on the way and they said, "Put him in the paddock – there's a donkey in the field next door." And the colt looked at the donkey, and he was frightened, so he jumped in with the donkey.' She grinned, pushing her long grey hair back out of her eyes. 'That for any other horse is not really logical, it should be cowering in the furthest corner, but no, he jumped in in order to go for the donkey. He just didn't want it on the other side of a fence.'

We sat in Lucy's kitchen-bedroom-sitting room in her small house on a hillside in Extremadura in Spain, among sacks of lemons, a pile of firewood for the stove and her dozing pack of four dogs, Lukas, Gato, Sumo and Tess, all of whom had been retrieved by Lucy after local hunters abandoned them. The walls were lined with books – books on animal behaviour, tomes on horse history, the *Mabinogion* (half-eaten by a mouse) and novels – and the bleached bones of animals Lucy had found in the dry, rocky foothills of the Sierra de Gredos. Swallows flew in and out of the room's three open windows, carrying beakfuls of insects to their young in the mud nests that lined the roof beams.

Lucy, now in her early seventies, has worked with Spanish and Portuguese horses for decades, teaching her own *doma natural* or 'natural horsemanship'. Her method is not that of the marquee-name horse whisperers with DVDs, 'buck stopper' harnesses and 'carrot stick' batons, but is informed by her lifelong study of horse behaviour as an ethologist.

Fifteen minutes from her house, above sloping cherry orchards, Lucy kept a herd of piebald Pottokas – a fine-limbed, semi-feral pony from the Basque Pyrenees – on 3,000 acres of open heath thick with wild lavender and bracken. Some believe the Pottokas represent a continuum from the wild horses that once inhabited Europe alongside the Takhi and Tarpan. At Lascaux, one chamber of the great caves is decorated with a fat little Pottoka-like piebald running with bay and dun mates. Lucy has pointed out that black and white colouring makes the horses hard to spot against a white background, suggesting it was an Ice Age adaptation: dazzle camouflage for endless snow. The herd are the subjects of her ongoing Pottoka Piornal project for students, horse people and scientists to observe horse behaviour.

Campo Pequeno had not given me a deeper understanding of anything other than the horse's fear and man's force, so I had come to Lucy to understand the mystery of what carried a horse obediently into the ring and on to the battlefield, and why, indeed, any horse did what we asked of it.

That morning I had been woken by the sound of hoof on stone as Lucy's twenty-five-year-old Lusitano stallion, Iberico, poked his head through the open front door and, yawning, watched Lucy as she slept on her raised bed in the kitchen, waiting for her to wake up. I got up and pulled on some jeans to take him a treat of some dried fruit I'd dug from the bottom of my rucksack, and he had gently taken it, then stood dozing as I stroked the silky underside of his greying muzzle. When turned out with cattle, Iberico preferred to chase and organize them rather than attack them. He spent his days grazing in the field around Lucy's house, coming to the kitchen for company on occasion. 'Someone once said to me, "He only lacks words,"' said Lucy, 'but he doesn't need words. Why would he?'

'Most of the bullfighting horses are pretty highly selected before training,' she explained later as the kettle boiled on the stove. 'The really great bullfighters end up working with just about any horse, but when they select which colts to train they put them in a big school, and charge at them with a *tourinha*' – a set of horns mounted on what looks like the front half of a bicycle – 'and of course the colt runs away at first, then two or three more times. But then he starts going, "Don't do that to me, I don't like it." So they do it again and he says, "I told you not to." And the good horses, after a couple more times, will actually come at you. Those are the ones they train. They use tame bulls for the next stage.' She once looked after a Lusitano which would charge her car when she drove it into his field, and rip off the windscreen wiper.

'I've been to different bullfighting stables and the training for the horses is hugely different and attitudes are hugely different,' Lucy went on as she poured steaming water into mugs. 'Pablo Hermosos's attitude is that the horse has to be really well-schooled. Other people just put really strong tack on the horses to make sure "he's bloody well going to do what I ask when I ask him". You see faces ripped off with serretas,' she grimaced. A serreta is a noseband with a toothed lining that digs into the horse's face when pressure is applied.

I asked if she had ever been asked to rehabilitate a bullfighting horse that had broken down and lost its nerve. 'Yes, I had one with a hole in his bum where the bull got him. He was really, really frightened of being ridden. I put him in my riding classes with someone to lead him and with no saddle or bridle. He just completely calmed down.

'You can habituate horses to almost anything,' Lucy shrugged. 'The first rule of being a horse is don't trust anything unless it's proved safe. So you start off as a one-day-old foal by being startled by grasshoppers. But you can't spend your whole life running away from grasshoppers and butterflies and rabbits, so you are quite quick to distinguish between things that can help in times of need, things to stick with, things that you just don't pay any attention to, and baddies. It means you've got a huge capacity for habituation.'

This process of gradual habituation has been exploited over centuries by humans to prepare warhorses, and helped to explain some of their 'courage' in the battlefield. In his *Militarie Instructions for the Cavallrie* of 1632, John Cruso advises putting oats for horse on the skin of a drum, setting up a suit of armour on a pole and riding into it 'that he may overthrow it, and trample it under his feet: that so (and by such means) your horse (finding that he receiveth no hurt) may become bold to approach any object'. Guérinière also fed horses while accustoming them to the firing of a pistol, though, 'Some horses are so fearful that their ears stand straight forward, they roll their eyes, tremble and sweat with fright, hold a mouthful of feed between their jaws without chewing and throw themselves against the manger and across the bars.' Alternatively, he tied a horse between the two pillars of the manège, let it examine the pistol, then cocked the hammer, and let it understand that. First he fired away from horse, so it could smell the smoke, then increased the charge and came closer until he was able to shoot as he sat on the horse's back. Even in Flanders in the Great War, the horses soon became accustomed to the shattering boom of shellfire and continued to pull their wagons as houses, roads and people disappeared into blasted mudscapes.

In the 1960s, Lucy Rees studied zoology in London before moving to Sussex as a postgraduate, specializing in neurophysiology, neuro-anatomy and ethology, but after four years she left academia for the mountains of North Wales, because, as she told me, 'at that time you could not admit that animals had emotions. Animals didn't feel fear, they "exhibited aversive reactions". That gave you a licence to do an awful lot of stuff to them. It was Descartes who started all that off, animals as mechanical'.

In Wales, a friend asked her to look after some horses, and gradually, because local children pestered her to let them ride, she became a trainer of both horses and riders. She broke in the half-feral ponies that were raised on the mountains, and gave riding lessons in which the students were led up the hillsides, riding bareback, with no reins. People gave her difficult horses to rehabilitate, like a Lipizzaner stallion called Maestoso Sitnica III, who had fetched up in England after a chequered career that began in an elite Yugoslavian stud and continued through various continental dressage establishments.

Maestoso's head was scarred and he both bit and bolted – refusing to be ridden. Lucy gave him a barn to share with a donkey and some pigs for company, and later turned him out with a herd of mares. Instead of confining him to the arena, she began to ride him across country – though at first she had to mount him in the barn and sit out a series of wild courbettes as he surged out into the open – asking friends with mares to wait at various distances from the barn to entice the stallion along. From a miserable and neurotic indoor-schooled horse he became a gentle and outgoing trekker who could cover a hundred miles in two days, and still be fresh and ready to go out again on the third day.

On the strength of these experiences and her studies, she wrote *The Horse's Mind*, a handbook that combined ethology, horse care and quotes from Shakespeare and Boswell, and which sold well around the world. Although it was published in 1984, before the cavalcade of 'natural horsemanship' trainers began, it has barely aged, and it is still on reading lists for equestrian studies at universities. Like her work with Maestoso, it leads from the horse itself, by thinking what a horse might really want and need, and how best to gain its trust and interest.

Rees's is a free-flown, loose-footed equitation, which still moves the horse through the formal stages of dressage, skilfully turning his instincts to the service of a cooperative partnership. 'People say to me about their horses, "He doesn't want to work." Bloody hell, they're not born with a Protestant work ethic. What is work to them? It's not "work", it's just moving, so you should make it interesting for them. Otherwise you're having to put all this bloody pressure on them all the time, and they're thinking,' she dropped her voice to a whisper, '"shut up". I think it should all be much more playful.'

Like Xenophon, she seeks 'exactly the airs that [the horse] puts on before other horses'. She'd explained to me earlier how she waited for a horse she was training to freely offer the movements of haute école: a young stallion passing a field of mares will begin to perform the passage, and, by cueing it a little earlier each time and maintaining the passage when the mares are passed, the action could later be produced at a signal, independently of the female audience. Pirouettes began in fooling around with cattle, and dodging to herd them. Iberico had recently, despite his greying muzzle, insisted on passaging for much of a riding holiday they took in the hills, and if he and Lucy disagreed on which route they took, 'The first I'm aware is the ground disappearing, and then I feel the kick out back' as the old stallion exploded – untaught – in a capriole. She laughed. 'And all the peasants are cowering in the hedgerows!'

After a divorce, Lucy left Wales to travel, mainly to America, working as she went. 'I was handling horses in all sorts of situations, racehorses, polo ponies, trekking horses, mad horses.' She learned different schools of thought and riding, but eventually grew frustrated with these codes. 'It was all about being correct, but then I started thinking well, actually, in the end a horse is just a horse. Whether it's big or small, it's still *Equus caballus*, so I got more interested in wild horses.'

On the Orinoco's floodplains in Los Llanos in northern Venezuela, she joined a study of the herds of feral horses who survived in extreme heat for six months of the year, and then another six months in a metre of water when it rained and the river burst its banks. 'They get eaten alive by mosquitos, and millions of all sorts of biting things, and life is hard there,' she said. The horses were also under frequent attack from pumas, and were often disturbed by the *vaqueros* ('a sort of honorary puma') who herded cattle on Los Llanos. Many populations of feral and wild horses have been studied worldwide, but few have given ethologists the chance to regularly see their reactions to predators, even though the horse is, as Lucy put it to me, 'this famous prey animal'.

'I saw a lot of stampedes and startles. I found it really, really, really difficult to describe what was going on. Because you could have ten horses or twenty horses or 100 horses or all 150 and the cows, and the stampedes still worked the same way. So it became perfectly obvious that nobody was giving any directions, which made me wonder how a stampede worked in the first place.'

The work of the computer animator Craig Reynolds gave her her paradigm shift. In the 1980s, Reynolds began to write programs that simulated the movements of groups of birds by considering each virtual 'boid' in his programmed flocks as an independent actor. The boids were governed by three basic rules, as Lucy explained: 'One is when there's a fright come together, then synchronize – do what everybody else is doing – but also don't collide.' Tap in a series of numbers to delineate their behaviour, let the program run and the boids on screen created lifelike flocks and murmurations.

'I started looking at the horses like that, and it absolutely worked,' went on Lucy. 'I could see more and more things, how one stallion gets startled, then he snaps into alarm posture, all the mares take notice and they all just come together behind him, and then if he turns round and starts running, they're already turning round and starting running. If he trots, they trot. If he relaxes, they relax. But it's not actually that the stallion is directing things – domestic horses mostly don't have stallions, but they still manage to move together in the same way – it's just that he's rather more alert to signs of danger, which is characteristic of having testosterone.'

Like the boids, horses had a sense of individual space that governed their movements and the way they accepted – or rejected – the approaches of others. 'If you try to get up to a frightened horse, the air goes thick about a metre away from them,' said Lucy, circling her hands around her own body to mark the space. Even in the fearful rush of a stampede, that space was maintained.

Lucy showed me some footage of the horses in Venezuela trotting high-kneed through the floodwater. 'From the side, it looks like they're all higgledy-piggledy, but when they turn you can see the spaces beautifully.' The horses on her laptop screen moved from right to left in a frieze of bay, chestnut and grey, and then, without any indication of where the movement began, a few horses had suddenly turned chest-on to the camera, and then all of them were trotting towards the viewer, each surrounded by its equally visible and invisible force field. It reminded me of some of the moves in Bartabas's productions – regular but irregular, flowing but formal.

From this stampede behaviour, Lucy began to expand her understanding of that instinctive response and its cohesion, synchrony and space, to the rest of horse behaviour.

'You see that cohesion and synchrony all the time. You see it so often that you don't even realize you are seeing it. They're usually all eating at the same time apart from the foals, who might be sleeping, as they need to sleep more, not graze, and then there's somebody watching out. When they go off to their favourite resting place they will rest together. When they go off on a march they all march together. They're always synchronizing.' Like the Household Cavalry horses that Sir Astley Cooper watched performing their old drill together without a spur or a cannon in sight, horses simply liked to move in harmony with their peers.

A strong strand of equine ethology and popular horsemanship holds that this coherence is the result of a series of hierarchical relationships between horses, in which some horses are dominant over others, like the chickens in the Norwegian scientist Thorleif Schjelderup-Ebbe's 'pecking order' experiments in the 1920s. Its theorists have counted the times horses in a group pull threatening faces at one another, or make another hostile movement like lifting a heel or even kicking. Like the nineteenth-century naturalists who saw the Takhi stallions in Uzbekistan as 'sultans', they assumed that the stallion was the leader, and then, when the results didn't fit, a 'lead mare' was added to head the troops while the stallion brought up the rearguard. But even this combined patriarchal/matriarchal vision did not quite work with the data drawn from mustangs and New Forest ponies and *rosso* of the Camargue. Exceptions emerged constantly.

When Lucy tried to work out which horse led each band in Los Llanos, she 'started realizing that nobody was fixed, nothing. Who provokes the change in group activities? Whoever wants to most, whoever notices the puma. The default setting is do what everybody else is doing, but if you've got a particular need, like a lactating mare on a hot day, then you move in a very definite way to water'.

It was not that one horse was the brain of the operation and bullied or inspired everyone else to follow, it was, more generally, a peaceable system of group consensus. The horses knew that they were safer together, and generally, even when scattered in a landscape, they came together to varying degrees – like the Takhi that Laétitia, Marco and I had startled at Hustai, which grouped together and vanished as a flock down a fold in the hill. And the horses maintained their space: foals, who are allowed into their mother's personal space, learn to respect the space of others when they approach other mares in the herd for milk and are rebuffed. The 'threatening faces' usually resulted in the unwanted, would-be buddy getting out of the space of the threatener, and not a rigid pecking order. A horse might graze with its head all but entwined with that of a special mate, or lip at the base of their necks in an embrace, but these affectionate gestures did not really add up to a grand hierarchy, just individual preference. After all, as Lucy pointed out, humans don't love every other human willy-nilly, and there's no reason why horses should either: 'We're sociable too, but we don't have to like everybody, do we?

'Those three rules – cohesion, space and synchrony – are operating all the time but it's rather like a light with a dimmer switch, it goes up and down. When you see foals and youngsters play, it's usually all about synchrony. Colts do play-fight, but an awful lot of what they do is, for example, walking along side by side with a sort of twinkle in their eye, and then all of a sudden they're both galloping. And they turn and they stop and they stand on their back legs or they do something else, but it's like they're trying to be synchronized swimmers all the time.'

Theories of pecking orders and hierarchies and lead mares and stallions were, according to Lucy, driven by human projections on to animals, and by ill-thought-out experiments in less than natural conditions. The theory behind social dominance began with other animals, and was then transplanted on to horses. 'As far as I can see it was the primatologist Solly Zuckerman who really did it in the 1930s, looking at baboons and then chimps in the London Zoo. Again, they'd been collected from anywhere, shoved in a huge cage together, food all in a pile, only a certain number of resting places, and you start getting this very powerful alpha macho figure who just gets all the resources that he wants. He's actually attractive to everybody too, and they can switch off his aggression with submissive gestures, which don't actually occur so much in the wild as they do in caged animals. When they can't get away they do these gestures.'

No clear picture emerged in the literature of dominance in horses when one compared different populations of horses – the exceptions blurred the hierarchies carefully sketched by scientists. In one American herd, stallions drink first when there is less water; in a herd in Namibia, they do not. A New Forest pony stallion appeared to get the pick of food but was dismissed by a mare in season that he wanted to mate with. Being an 'alpha macho' figure was complicated in horse-kind. Lucy's working partner, Victor Ros, suggests that horses develop different 'cultures' depending on where they live, which might account for the variation. Equine society is richer and more complex than we have arrogantly supposed.

'In the OED for "dominant" it says supreme authority, governor, ruler. Authority is the one with the power or right to enforce obedience, so the dominant is the one who gives the orders and that's the way *we* think, isn't it? We're attracted by dominants, because they have the most resources. You don't want to make an enemy of them otherwise you're never going to eat. I'll pick your fleas whenever you want, darling!' Lucy grinned. 'But horses of course don't behave like that. At the first sign of aggression or dominance they go away. If dominance theory worked, then you'd put a wild horse in a stable, beat hell out of it and then it would do what you wanted for the rest of its life. And it doesn't work like that. The more pressure you put on them, the more they explode.'

———

If the arrival of domestic horses along with all that bronze and grain and iron helped disrupt human societies, then we have also caused serious disruptions in the societies of horses, and the way they deal with their own strife and stresses. The aggression rates of domestic horses are calculated by some to be up to twenty times higher than those of feral or wild horses. Beyond our habit of transforming them into little elephants, solar engines, national symbols, dancers and war machines, the very conditions we make horses live in create much of the tension and hierarchy observed by researchers.

Like the baboons' and chimps' keepers, we create shortage and restrict movement. We wean foals early and separate them from the mares that would teach them correct social behaviour. Some of these foals, rushed off milk and on to processed cubes of grain and supplements, learn to hook their teeth on fences or stable doors and gulp down air – a compulsion known as cribbing. We isolate stallions like Maestoso, making them miserable and aggressive – even to the point of mutilating themselves by biting their own chests. We give man-made herds the territory of a box or paddock to defend instead of rolling, open country, throwing together horses who might not have chosen one another in the wild, then stopping them fleeing from conflicts with fences, so we see them becoming conciliatory like the chimpanzees, or being bullied, or fighting.

Whereas in the wild a horse spends 60–70 per cent of its day cropping grass, and covers many kilometres as it does so, stabled horses develop ulcers and chew their boxes, or weave on their feet as though mimicking the rocking movement of a grazer. 'Imagine,' Lucy went on, 'spending twelve hours with nothing in your tiny stomach.' Aggression is learned in the stable and at the manger – in the wild, nothing could be more democratic than a slope of grass, but, 'Say we're domestic horses and we're both going for this bucket or pile of hay, and I say to you, get out of my space, and you do. What do I get? A bloody great enormous reward. You can train a horse with little bits of carrot, what does half a bucket of oats do? A lot. So the next time we go up to it I'm going to say to you earlier on, "Go away." And then actually, because this is the way training works, I'm going to say, "Get away" at any other moment, too.'

Many of the studies on outright aggression among feral or wild horses are also complicated by their conditions, partly because few populations are genuinely untouched by man. On the marshes of the Camargue, on the Atlantic coast of America and in the New Forest, feral horse populations are frequently disrupted by round-ups, or by having extra food provided in harsh seasons, or stallions culled or added to 'improve' the type. The most macabre result is the occasional infanticide of foals by stallions, which has been observed in populations interrupted by the addition or removal of males, like a Takhi group kept in a semi-reserve in China, where more than four fifths of foal deaths were caused by attacks from stallions.

Lutz Heck may have fantasized about the 'grim schelch' as 'fiercer' and 'more dangerous' than any other animal, but most of the fights between adult horses involve bluster and display and very little contact. They roar to show their lung power and size, arch their necks to make themselves intimidating, strike the air with their forelegs, rear and tango around one another, and then usually one decides it is outfaced and runs away, perhaps after a short boxing match, or an attempt to bite under the elbows of the rival and bring them down.

Lucy said she had only seen real fights when a full-grown eight-year-old bachelor Pottoka took on a band's stallion – usually the old king was deposed with little bloodshed after being run ragged by a tag team of younger males, who then divided up his mares between them. At Hustai, nearly 20 per cent of the stallions – those broken-eared 'soldiers' of Dr Usukhjargal Dorj – that died over twenty years were killed by injuries from fights, but not because they were hunted down or trampled to death by their rivals, but due to infection from smaller wounds.

In any case, winning a band of mares was no guarantee of foals: mares sometimes left to be with the old stallion, went elsewhere or simply rejected the advances of the new self-appointed sultan. Mares may even do without a stallion altogether for periods of time. And stallions didn't need to fight to get hold of mates – simply by hanging out at the edges of bands they could entice away a filly that the stallion was soon going to throw out of the band anyway, or even a mare which simply fancied the bachelor more. Lucy described stallions as bodyguards whose job perk was to sire foals. Ekain, one of the most successful stallions in her Pottoka herd, was neither mature nor especially large, but he was an excellent, solicitous father to his foals, and so the mares chose him. The lone stallion silhouetted on the hillside was a myth: some form lifelong partnerships with other stallions, and manage their bands together. Adult stallions with harems have even been known to pay social visits to bachelor herds, and to play with them.

Above all, horses sought society. A lonely horse is an unhappy one, and one at risk. At Hustai, one stallion, Ares, had lost his harem then tried to seduce away the park's five riding geldings and sequester them several kilometres away. Humans made a poor substitute for horses. 'What are we compared to a herd?' asked Lucy. 'We show up and ride them for an hour a day. You have to spend twenty-four hours with them. That's a herd, sleep, eat and live with them, but in all the places where I've worked, the first way of breaking in a horse is conflict.'

In the name of mastery, horses are hauled about with ropes, have legs tied up, are tethered short to rails, have their bucks sat out and are punished. Benevolent trainers of the past like Xenophon, Pluvinel and Guérinière mention others who had horses whipped, clubbed, curbed, spurred, menaced at the testicles with cats tied to sticks or rods covered in hedgehog skins. Riders may have forsaken angry cats today, but they still ride with the spurs and the curbs and the whips, the serratas and the crank nosebands fastened so tight that a horse cannot open its mouth to relieve the pressure of the bit. 'Draw reins' are used as pulleys to haul down a horse's chin to its chest, the rider bracing backwards with their whole weight until the horse's neck is so constricted where it meets the skull that the animal can barely swallow. Poles set with spikes are raised to rap the knees of horses as they leap fences, so they will jump carefully, and gaited horses thresh their legs high after their pasterns have been wrapped in poultices soaked with blistering chemicals and hung with heavy chains. Even the most benevolent trainers ask a horse to do what should seem alien to it – to take metal in its mouth and allow a rider on to its back. Why did horses even cooperate with us, I asked Lucy, even if our methods were sometimes brutal?

'The horse'll go on fighting until there's a moment of release, he stops to catch his breath and if the rope slackens off at that moment he goes, "Ah, it's worse when I fight,"' she explained. 'They're not stupid. They want to avoid pressure and fighting at all costs, so the moment they have a way of avoiding it, they choose it.' The release just happened to coincide with what we often want from them, or what we call 'obedience'. Their flight from pressure brought about their domestication and usefulness, but it was their love of harmony that continued it. This, Lucy said, was what was fundamental to understanding the horse.

'They want things to be smooth between them and what's going on around them. They don't want even the most minimal sign of conflict. They don't want pressure, they want flow. They want that lovely feeling of all being in synchrony, in harmony, that's when it gets good for them and they start getting really pleased about things. They are happy when they feel safe in your hands. He finds it rewarding when he feels you're pleased on top of him. They really like that. Your body's got a little bit of tension in it as you ask for something, then they do it and good boy, there's a release of tension. Even unbroken or wild horses see that in your body.'

Horses' survival had depended on their quicksilver reading of the non-verbal signs from the beings around them, horse or man, as we took them from what Lucy, quoting E. E. Somerville, called their 'monumental idleness' in the wild and offered them instead that incomprehensible Protestant work ethic. While we ascribed human motives of dominance or obedience to them, they were in fact only trying to find that smoothness, and applying what we taught them – however we intended it.

The Russian political theorist Peter Kropotkin began his writing career as an evolutionary naturalist who, following Darwin and the Russian Karl Kessler, believed that it was 'Mutual Aid and Mutual Support' that drove evolution, not a Hobbesian battle against all-comers for survival. 'Sociability is as much a law of nature as mutual struggle,' he wrote, drawing on his observations of wildlife in Siberia and Manchuria. Although the landscapes and climates of those regions were inexpressibly harsh, he did not see animals, birds or insects squandering their energy on competing against rivals over the scant resources. Domination was not the goal.

By coming together, animals thrived and evolved: 'Those animals which acquire habits of mutual aid are undoubtedly the fittest. They have more chances to survive, and they attain, in their respective classes, the highest development of intelligence and bodily organization ... mutual aid ... favours the development of such habits and characters as insure the maintenance and further development of the species, together with the greatest amount of welfare and enjoyment of life for the individual, with the least waste of energy.' Intra-species struggle, which occurred when resources were scarce, simply weakened a species as a whole.

In *Mutual Aid: A Factor of Evolution*, Kropotkin described the observations made of wild horses on the Steppe by another naturalist, Kohl: 'Neither the wolf nor the bear, not even the lion, can capture a horse or even a zebra as long as they are not detached from the herd. When a drought is burning the grass in the prairies, they gather in herds of sometimes 10,000 individuals strong, and migrate. When a snow storm rages in the Steppe, each stud keeps close together, and repairs to a protected ravine. But if confidence disappears, or the group has been seized by panic, and disperses, the horses perish and the survivors are found after the storm half dying from fatigue. Union is their chief arm in the struggle for life, and man is their chief enemy.'

This understanding of animals, though it sits ill with the Cartesian coldness that drove Lucy away from the university laboratory to Wales, is one that pre-Darwinian writers used to feel less embarrassed about expressing. Aristotle believed that, 'The whole race of horses appears to have warm natural affections', and even the Comte de Buffon, who thought horses loved human battles, admired the horse as 'they never make war with [other animals], nor with themselves. They never quarrel about their food ... They live in peace because their appetites are simple and moderate, and having enough there is no object for envy'.

And what have we given horses in this deal of domestication in return for combat, wealth, industrial complexity and our own hierarchies? The survival and spread across five continents of the world – more than they ever achieved alone – of a beleaguered species that fetched up on the Steppe and the Iberian peninsula at the end of the Holocene, an endurance achieved at the cost of uncountable horse lives, and as a conscript in the battles and economy of man, as work-horse, warhorse and food. The taller, faster, finer, heavier domestic horse, protected, fed grain, wine and vegetables soaked in honey, standing on a brick floor by its magnificent water fountain, massaged with oil, fitted with the bronze muselière and the gas mask.

By domesticating horses we have brought them stratification, scarce resources and limited Lebensraum, as we did for ourselves when we swapped the leisurely existence of the hunter-gatherer for that of the farmer and eventually the Bronze Age city dweller. We are afflicted by shortages of our own creation, and rules we have imposed upon ourselves.

'Horses are self-organizing anarchists,' said Lucy, as Gato the greyhound stretched and shifted on her bed, and the swallow fledglings began their liquid squabble in the nest over our head. An adult bird had just arrived with a beak crammed with insects, and perched at the mud lip of the nest, undecided which mouth to stuff it into. 'Horses don't want to hoard things, they don't want to own bits of territory, there's enough for everybody, they don't fight over air and they don't fight over grass, but then they all come together in this collective defence, which just seems to me a beautiful model. Of society. Of how to live.'

V

In this Age of the Horse, equids are still caught up in warfare beyond the bullring, even as combat evolves into forms unimaginable in the earliest civilizations. Horses and donkeys have been used to carry concealed bombs or mines in Afghanistan, Gaza, Iraq, Colombia and even on Wall Street, where in 1920, a horse was blown to smithereens after delivering an anarchist's bomb to J. P. Morgan's front stoop, killing thirty-eight and sending a 100-foot-high mushroom cloud into the sky (its hoofs flew for several blocks in all directions; its head landed by the limestone stairs). The Royal Army Veterinary Corps transported supply mules behind Japanese lines in Burma by roping them into inflatable dinghies and dropping them out of planes, and over 1,000 Tennessee mules were airlifted to carry packs for the mujahideen in Afghanistan in 1987. In Iraq in 2003, asses brought rocket launchers within range of the oil ministry for an attack, and Libyan donkeys ferried weapons along familiar paths between villages as the uprising that removed Gaddafi from power played out. In Tahrir Square in 2011, supporters of President Mubarak galloped the horses and camels that usually carried tourists to the Pyramids into the crowds of protestors, waving whips and sticks as they went. In August 2015, Boko Haram launched a mounted attack on three villages in northern Nigeria, gunning down worshippers in a mosque, scattering people into the bush and then disappearing like bandits.

The US Armed Forces and their allies have also fought on horseback in this century. In the Darya Suf and Balkh Valleys in Afghanistan in October 2001, 1,500 cavalrymen of the Afghan Northern Alliance galloped across a mile of rising and falling hollows towards the village of Bishqab, into Taliban bullets, and towards old Soviet tanks, armoured personnel carriers and anti-aircraft cannon that could fire 4,000 rounds a minute. They came in two waves, carrying walkie-talkies, rocket-propelled grenades and machine guns, and backed by an equal infantry force. A hullock away from the Taliban lines, they dismounted, stood on their reins and opened fire with machine guns and grenades. The second wave of horsemen passed through, reins in their teeth to free their hands for combat, taking the battle line forward. As the Taliban began to run before them in fear, the Northern Alliance men beat them with gun butts, slashed with knives or shot them in the back, while American planes dropped bombs on the tanks. In the hills behind the plain, twelve Americans from the 5th Special Forces Group directed the dropping of the bombs, travelling between locations on tough, crabby Lokai pony stallions, their equipment loaded on mule trains.

'What about the horses?' one of the Americans had asked as the men prepared for the Battle of Bishqab. 'How will they react when the bombs start dropping?'

'They will not be nervous,' the Northern Alliance warlord replied.

'Why?'

'Because they will know that these are American bombs.'

When Americans honoured the commandos of this first plunge into Afghanistan, they erected a bronze statue of a split-hoofed Afghan pony, jaw fighting at its bit, caught with its weight back on its hocks and its mane and tail blown forward, like the emperor's mount in Jacques-Louis David's famous 'Napoleon Crossing the Alps'. On its small wooden saddle is a lanky American Green Beret in a sun hat, M4 assault rifle hanging from his shoulder and binoculars in his right hand. It stands at the foot of the One World Trade Center skyscraper, not far from the sombre, sunken fountains that mark the footprints of the old Twin Towers. The Afghan war is represented at Ground Zero not by the Hellfire missile or the M1 Abrams tank, but the energy of a pony, and an appeal to old, horseback warriors.

VI

Arlington was overwhelming. The cemetery sprawled over 624 acres, each plot seamed with long rows of white headstones marking the graves of American servicemen and women from every war since the Revolution. In the legend of Cadmus, the first king of Thebes, the hero strews dragons' teeth and when they strike the Boeotian soil, they are transformed into warriors. At Arlington, Virginia, warriors went into the ground and in their place rose row after row of blunt, pale tooth stubs. The grass around the graves was thick, bristle-cut perfection, even in a sweltering southern summer.

I left the visitor centre behind and set off on foot, draining my small bottle of water as I went, following a macadamed track as it rose and fell over the contours of the cemetery grounds. At each fork in the path, visitors were shucked off, following the pointing arms of discreet signposts indicating the Kennedy graves, the memorial to female services veterans, the tomb of the unknown soldier or the Iwo Jima monument. Golf carts ferried those who could not walk. It was a strange tourist attraction, each row an anthology of life stories where private grief is alloyed with patriotism and national loss.

History telescoped in the cemetery and in the neighbouring capital of Washington DC. From the slopes of the cemetery I looked out across the cars buzzing on the potholed George Washington Memorial Parkway, over the broad Potomac to the giant's throne where a colossal Abraham Lincoln sat in a Doric temple, and down the National Mall, set with weighty markers of American history: Washington's obelisk, a stern Martin Luther King, the black granite walls etched with the names of the 58,000 killed in Vietnam, the memorials to the dead of Korea and both World Wars.

Before the domed Capitol at the farthest end of the Mall was the nation's largest equestrian monument, which I'd visited the day before: General Ulysses S. Grant sits on a fine, alert thoroughbred on a Vermont marble cliff of a pedestal, guarded by four lions couchants.

He is flanked on one side by cavalry and the other by artillery – who both charge chaotically on marmoreal bases that put a viewer at ground level, able to take in the knots and straps of the harness and the faces of the bronze men and horses as they surge forward. The artillery horses, like the horses that draw the sun chariots of Indo-European religions, are leaving the earth behind them, leaping upwards and forwards into the unknown. One cavalry horse has fallen shoulder first to the ground, its head buckling against its neck, and its rider, falling alongside it, clings to it in an embrace before he is trampled underfoot.

As I walked on through Arlington, the rest of the visitors dropped away. At certain turns of the road over a hillock I would find myself alone in a section and, it seemed for a moment, in the entire cemetery. I was beginning to feel as though I was the only living person in Virginia when in a far corner of the grave field I found a gardener asleep on the grass in front of his ride-on mower, and picked my way quietly past him and through a gate in the low stone wall that marked the boundary of the cemetery and Joint Base Myer-Henderson Hall, a settlement of military chapels, concession stores and neat, clapperboard All-American houses where the top brass lived with their Stars and Stripes hanging over the porch. I found the low red barn I was looking for next to another with a sign reading 'Army Substance Abuse Program'.

The barn was home to the men and horses of the Caisson Platoon of The Old Guard, or 3rd US Infantry Regiment, who took part in parades and, more importantly, provided the consolation of ritual to the families of some 5,000 veterans a year, at a rate of thirty funerals a day, six days a week. The caskets or urns at all Full Honors funerals at Arlington are borne on a 1918 caisson or artillery wagon, drawn by a team of six grey or black horses, three of them ridden by Caisson Platoon men, and led by another mounted rider.

These horses continued the ancient equine task of mediating between this world and the next that harked back to those Bronze Age graves in the Steppe and the Shang tombs with their cowrie shells and jade figurines, where horses were lain alongside the dead, ready to carry them by chariot to the afterlife. When medieval Europe tried to erase its pagan past, the sacrificial warhorse was instead draped in fine trappings and harness for the funerary procession of its master, before being presented alive to the church. Now the consecrated warhorse has become the 'cap' or caparisoned horse of Arlington, who follows the cortèges of those who have reached the rank of colonels or higher – and of presidents – saddled and bridled and with a set of empty boots reversed in his stirrups. Consecrated to battle like the chariot horse the Romans offered to Mars on the ides of October, the horse still lends a non-denominational sanctity to the military funeral.

'I can't imagine being the officer who says, "Your husband can't have the horses," not when his father, his grandfather, his great grand-father did,' said Sergeant Ford, the Platoon's affable head of Caisson Operations, sitting in a little wooden cabin of an office at one corner of a barn. 'It's probably the most rewarding thing I've ever done in my life. To see the difference that we make in the lives of family members, to be able to send the army's message that you are not alone in your grief, your loved one mattered to us too, we grieve as well.

'If we had no money, no power, no gas for the trucks, nothing else, Arlington cemetery would continue to do Full Honors funerals every day and we would continue to put our horses into the cemetery.'

Ford had been an infantryman – a machine gunner and squad leader in Iraq and Afghanistan, and had applied to join the Caisson Platoon on his return. He had never worked with any large animals before, let alone ridden, but had fallen in love. When I asked him why he liked to work at Arlington, he beamed, 'The horses. The horses.'

Each old-fashioned stall in the 106-year-old barn had a fan at work overhead, and the yellow brick floor was cool underfoot. The horses dozed or snuffled at hay, waiting for lunch, which they knew would come – routine is the lodestone of both the stabled horse and the soldier. Patton, Roosevelt, Minnie, Mickey, Jerry, Sure Fire and their companions were friendly, contented creatures who regarded with open ears and bright eyes anyone – and in particular Sergeant Ford – who approached their box.

I dropped into the rhythm of the barn for a few hours, scritching horses' necks, talking to the farrier hammering shoes for a caisson horse and to Sergeant Reuben Troyer, who trained the new horses. He came from an Amish background in Holmes County, near the Horse Progress Days gathering, and told me that the caissons used by the platoon were serviced by Amish wheelwrights. I met Sergeant York, the Standardbred that had carried Ronald Reagan's tan cavalry boots at his state funeral. Later I followed a tour conducted by one of the soldiers, learning about the intricacies of the US Army's last mounted artillery instruction manual, printed in 1942, and admiring the gleaming brass-faced parade tack used for the inauguration of President Barack Obama.

Both Ford and the farrier wore plain metal bands around their wrists. The bands indicate a soldier who has lost 'a brother' in combat. Sergeant Ford explained why he'd applied to join the platoon: 'In 2005–6 in Iraq my best friend was killed in combat and he was buried in Arlington Cemetery. When I came home from my deployment I came to visit him with my wife and our kids – it was my first experience of Arlington. You read the numbers … ' he spread his hands, trailing off, thinking of the stretch of white grave-teeth. 'At the time it was 300,000 but now we're over 400,000 service members buried in Arlington. You've read those numbers, you know in your head, you've seen the pictures, but then, oh my God. Until you're standing at the bottom of Section 60 looking up at the hill and all you can see in every direction is our honoured dead. That really had a profound impact on me.'

Section 60 was where the casualties of Iraq and Afghanistan lay under headstones that, unlike the plain white stubs of the other sections, had been decorated with seashells, empty bullet cases, seasonal ornaments and painted pebbles, draped with dog tags and rosaries and guarded by teddy bears and whisky bottles. Laminated photographs of the deceased and their families were taped by the chiselled names and the words 'Operation Enduring Freedom', with children's paintings and sonograms of new babies. In the autumn of 2013, the cemetery authorities cleared away everything, and left the stubs behind.

'Our troops' were everywhere I went in America in the summer of 2014, fresh from the Horse Progress Days and the green organic health of Natural Roots. They paraded in blinking red LED lights on a display board suspended over the corridors of LaGuardia airport that urged 'Support our troops' as I hurried to catch a flight. They were gummed to the inside of supermarket doors I pushed my way through with armsful of groceries – 'Safeway supports our troops' – and to car bumpers on the freeways, 'Enjoy your freedom? Thank our troops', 'Pray for our Troops'. At Reagan National, veterans and servicemen could jump the security queues; in DC itself they got discounts on museum admissions, sports events, restaurant meals, rental cars and hotel stays. I felt as I never had before that I was travelling in a military nation, but these flags were the reminders required for a population almost untouched by the wars of the twenty-first century – fewer than 1 per cent of Americans had actually served in Iraq or Afghanistan.

Troops were being slowly withdrawn from Afghanistan after their removal from Iraq in 2011, but a month earlier they had returned to Iraq as ISIS emerged from the ruins of Operation Iraqi Freedom. A war that was meant to be over was revived, but there was no appetite to put American troops on the ground once more. The fear aroused by ISIS was balanced by a spreading scepticism about any new intervention. That scepticism was a product of the same national hyper-awareness of 'our troops' – troops who seemed not so much a brave wave of steely-jawed heroes as a vulnerable population cradled by bumper-sticker messages, families and Congressional questions.

There were approximately 2.2 million active and reservist military personnel in the country, and almost exactly ten times as many veterans from all US conflicts: 22 million – a nation in itself. About 4 million of them were incapacitated physically or mentally to varying extents by their time in service and received financial compensation.

In 2013, the number wounded in Iraq and Afghanistan alone was believed to have passed 1 million, but the authorities had ceased to publish statistics. Beyond the headstones of Arlington and the laminated photos and whisky miniatures of Section 60 stood millions who had survived at a cost. The burned, the shot, the broken, could be pieced together with plastic skin, blood transfusions, titanium, electrical circuitry and the most advanced medical knowledge in human history, but the aftermath of that destruction and rebuilding was lifelong.

The signature of the Iraq and Afghanistan conflicts was traumatic brain injury or TBI, a brain that has been rattled in the owner's very skull – perhaps, some believed, as a consequence of armoured vehicles and helmets that were too tough for the less solid humans inside them. The symptoms come from inside the head, where they could not be amputated or pushed away: the sinister irritation of tinnitus which reminds you that you are sliced away from the real world; headaches that cripple; a short-term memory that leaves you stranded and mentally woolly; a breakdown of control over a body that had once been tightened by discipline, by actions done again and again in drills or gyms and that were now done, if at all, only with immense effort and concentration.

And then there were those whose damage could not yet be assessed by pulsing their skulls in an MRI machine. Post-traumatic stress disorder or PTSD was estimated to occur in up to 18 per cent of veterans of Operation Enduring Freedom and Operation Iraqi Freedom – a low estimate given the reluctance of many military men and women to report their concerns. Half of the wars' veterans knew someone who had been killed in action. Those with severe PTSD lived – survived, usually – in a loop of memory created in their own neurones, in which the stored, cumulative force of months or even years on tour, as others were killed or maimed around them, replayed daily. These were wars in which the enemy was often concealed, out of uniform. The man who killed you could be a policeman you had worked with all week, or the stranger who planted a mine under the surface of a dirt road.

And then there was the arbitrariness of any long, drawn-out combat: friendly fire deaths were high; swapping seats in a Humvee meant that one man survived and another died; and a day in the sick bay for something minor left a lifetime of survivor's guilt.

After the servicemen and women returned home, the war came with them. PTSD could mean for veterans the hot, appalling compression of sensual memories relived as hallucinations, the sound of the explosion that burst their ear drums, the bullet that cooled their skin and killed the man next to them, the smell of skin on fire, the sight of dead civilians, the taste of sweat and blood, memories whose grain was magnified by horror, and that returned in nightmares so vivid that men and women woke screaming, or with their hands round the throat of their partner.

Enemies long left behind in Iraq appeared in Nebraska living rooms, threatening families many thousands of miles from Helmand. Veterans slept with guns by their beds, or saw an insurgent in the SUV that carved them up on the freeway. They wrestled with deep depression, thick guilt and anger that was of little use in peacetime. Some took to alcohol or narcotics, legal or illicit. The prescription drugs from doctors were dual-edged blades: sleeping pills and painkillers can become as rooted in the body as the pain they are intended to numb. The damage rippled out in waves through spouses, partners, children, families.

Under the Department of Veterans Affairs, the branch of government dedicated to managing the pensions and medical benefits of ex-servicemen and women, this diffuse and deep trauma had had to be bureaucratized, with panel interviews and scores for eligibility which totted up frequency of suicidal thought, wives punched and substances swallowed. Soldiers were dealt with by doctors, surgeons, psychiatrists, psychologists, social workers and miscellaneous therapists, sometimes bounced from one specialist to another, dragging behind them ever weightier files of reports and assessments.

They were rebranded as warriors: there was a 'Warrior Transition Unit', a 'Wounded Warrior Program' and a charity called the Wounded

Warrior Project, entreating men and women shaped by the stoic culture of the forces that to acknowledge weakness and pain was to be strong. Even if they could no longer walk, or were reduced to tears by being unable to control their body and do what was expected of them as men and fighters, they were now asked to understand that they were still brave, still warriors and that, as was so often repeated, the hardest battle was at home. Veterans complained of being made to crawl through narrow or shifting bureaucratic hoops; some were excluded altogether; many crumpled and took their own lives.

The summer I went to Arlington, questions were being raised in the press about the Department of Veterans Affairs (VA). Delays meant veterans often went unsupported, waited years for psychiatric evaluations or had their applications rejected. That July there was a widely quoted statistic that every day twenty-two veterans killed themselves, and although the study behind this tally was limited, the headlines that talked of an 'epidemic' had some justification: male suicide for the Iraq War military survivors was one third higher than the army's historical average, and half of the wars' veterans knew a comrade who had tried – successfully or not – to kill themselves. In the youngest group of veterans, aged between eighteen and twenty-four, the suicide rate had lurched up suddenly.

The $1.6 trillion wars had ended, but the damage done would tail on decades into the future, when the million wounded would continue to require government support. The survivors of Vietnam were still enduring PTSD, and there was no reason to believe that this would not also be the case for the fighters in the War on Terror. Billions had been invested in their care since 2002, and some estimates suggested that the true costs of the wars would reach $6 trillion dollars. At each budget, billions more were added to the purse of the VA. The week I visited Arlington, the parents of veterans who had killed themselves appeared before a House Committee on Veterans Affairs, and a Democrat representative had quoted the extra money that the VA was receiving for mental health services, but in sorrow added, 'It's not working. We have to figure out why.'

VII

'I can feel the top of my right leg, down to about here, and then below that there's enough to move my leg but I can't feel anything.' Greg was a big man, with sandy hair and freckled forearms. He wore a Manchester United baseball cap and a T-shirt which read 'Freedom. Wounded Warrior Project'. He had served in Kosovo. 'I got into a fight and I hurt discs in my lower back and ruptured three in my neck. They tried to cut out the herniations in my lower back but it left me paralyzed from the waist down for seven days. I finally got the feeling back in my left leg, but only partially in the right.'

The 'fight' Greg breezed over had taken place when his security team had been guarding Serbians as they came into an ethnic Albanian town under US protection, to shop and to go to the Orthodox church. A former insurgent leader who had been released on bail had thrown a piece of rubble at an elderly Serbian woman, hitting her in the head. 'Real nice guy,' Greg commented. His team had given chase, only to find themselves trapped in an alleyway. They got their man, but a riot broke out. Greg was trying to hold both his gun and the prisoner secure, and someone in the crowd picked up a plank and brought it down on his head with all their force.

'When they did the brain scan, they diagnosed me with multiple traumatic brain injuries – probably other things I'd done in the military – and found that my hearing and my sight was messed up.' He had a spinal cord stimulator that was meant to galvanize nerve endings so he could move his right leg. It also eased the shooting pains that had plagued him, along with headaches, since that market day in Kosovo, but recent storms had affected the stimulator and stopped him sleeping for days.

A few miles beyond the chaotic freeways around DC, deep in the Virginia countryside at Fort Belvoir, the Caisson Platoon keeps another troop of ten horses pressed into service from auctions in Texas and Oklahoma. The Caisson Platoon Equine Assisted Program or CPEAP provides physical and psychological recovery programmes for veterans, and the horses were purchased by the army not to lead parades or draw caissons, but instead to carry the burnt, the broken and the lost to relief of a sort. The Wednesday before I visited the long red barns at Fort Myer in Arlington, I'd driven out to see them and their riders, and here I sat with Greg after his medical check in a hut lined with veterinary charts of horses.

'I was a little nervous at first,' he went on, 'especially when they showed me my horse, Duke.' Duke was a steady grey Percheron. 'I was like, holy cow, this thing's huge, but I just went on working with him. He tested me the first time in the round pen, he didn't want to do what I wanted him to do, and I kept working at him, and now he goes along with the programme and we had our click. He's really, really easy for me. And I'm surprised. They told me he's actually hard to ride but he's pretty gentle with me. I can't wait for every Wednesday to come out here.'

When the CPEAP was founded in 2006 it was the first of its kind, and by 2014 there were over 200 similar programmes across the country, although unlike the CPEAP, they ran from civilian stables. To prove to the VA that there would be few extra expenses in such a programme, the founders, a Vietnam veteran called Larry Pence and a retired naval commander called Mary Jo Beckman, worked at Fort Myer, and used the greys and the blacks that hauled the caissons. Now, eight years on, and showing results, they had their own horses and were still running on volunteer power.

Like most of the servicemen and women, Greg bounced between multiple specialists and therapists. 'I do vision therapy, I have brand new hearing aids, I did hearing therapy, I did occupational therapy over at the TBI clinic, physical therapy, and they call it speech therapy, but for me it's more learning how to do cognitive processing, so I do memorization, drills to see how much I can process, if I can do multiple things at a time.' Duke's broad back and slow gait gave him gentle movement, loosening up his back, hips and pelvis, which were normally braced against the pain. 'I was a five or six pain level last time I came out here, and when I left I was a three. It really helps that much.' He tried to schedule his speech therapy for the day after a riding session, because he did so much better. His chronic dizziness had dropped from a six to a three.

After talking to Greg, I went back out into the sunny July day, to squint with a hand shielding my eyes at the grounds of the fort: the large, airy covered school, a wooden wheelchair ramp, a round pen and the post and rail-fenced paddocks, where I could see groups of the resting caisson horses moving methodically across the grass, grazing as they went. Soldiers from the platoon, dressed down in T-shirts and fatigues, sped around on small tractors, keeping up the steady work of maintaining fences and clearing manure, or sat at benches, chatting to the physiotherapists and behavioural therapists from Walter Reed National Military Medical Center. Larry Pence, in a maroon polo shirt, jeans and US Army baseball cap, was an unfocused centre to the activity. He'd served thirty years, his wife twenty, and his son had been deployed in Iraq and Afghanistan.

'It's an American problem,' he said, of the veterans' care. 'America sent these people to war so America has to figure out how to repair these people, because every one of them, whether they stay in service or not, wants to be a contributing member of society again. That's just the kind of people they are.

'Our overriding mandate here is to maintain a calm, peaceful and serene environment. The horse does all of the work. It's a natural relief of anxiety, the horse has a calming effect. We're not bringing anyone out here to have a pony ride, and we're not trying to make them Olympic equestrians either. We're trying to help them recover, rehabilitate and the horse is the vehicle we use to do that.'

—

The Greek physician Galen included horse riding and chariot driving with massage, walking, sleep and gymnastics as ways to maintain health in his 'science of hygiene', and there are enough testimonies to the psychological boost of riding along the lines of the famous anonymous saying that 'the outside of a horse is good for the inside of a man'. Many's the old warrior who was sent, either by physician or by inclination, out on horseback for air and exercise. Horses were taken to Oxford Hospital in England to provide riding for the wounded veterans of the Great War, and men who had lost limbs often reached the hunting field once more by adopting a side-saddle.

For America, it was Larry's war, Vietnam, that first saw the army systematically use horses in a rehabilitation programme. At the Fitzsimmons Army Hospital in Colorado in the late 1960s, Colonel Paul W. Brown presided over a programme endeavouring to rehabilitate over 1,000 servicemen who had returned home from Vietnam and Korea missing arms, legs or eyes. The hospital provided modified physical activities like waterskiing, skiing, golf, dancing, swimming and scuba diving. One of the social workers on site was Mary Woolverton, who kept Morgan horses – a quintessentially American breed used by Confederate and Union cavalry in the Civil War – and when the ski season ended in spring 1968, she brought some of them to the Fitzsimmons parade ground for the men to ride. It was a success.

To those robbed of physical ability and forced into passivity, horses gave four legs more powerful than any prosthetic. Those who had lost only their lower legs were admitted to the fledgling programme first and had a relatively easy time of it. Later those with above-knee amputations rode with adapted saddles. There was a ripple effect, as patients enthused about it to friends, and more wanted to join.

Colonel Brown had reservations about letting two men who had lost both legs above the knee ride, but, once they'd been lifted from their wheelchairs and on to the horses' backs, and secured with straps, they were cantering within minutes. 'After a few sessions they were riding at a gallop, whooping in triumph at their accomplishment,' Brown later wrote. 'They were very proud of this – one told me,

"Colonel, when I'm in this saddle, I'm taller than you are."' Black and white photographs show men and horses crossing the parade ground at the double, manes flying, eyes lit up.

Jim Brunotte was a military policeman who had lost one arm, both legs above a major joint and one eye when his jeep hit a landmine near Long Binh in Vietnam eight weeks into his tour of duty. He could drag himself on to his horse's back from his wheelchair unassisted, and had a saddle with holsters for his leg stumps. He, too, progressed from walking, trotting, cantering and galloping to trick riding. Brunotte joined other men in rodeo displays and roping exhibitions. The men competed in trail riding events, winning against the able-bodied. When Brunotte left the army, he bought a 367-acre ranch in California where he organized riding retreats for the disabled. Another man who was blind and had lost both legs told Brown that riding was the 'only part of his rehabilitation program that had given him any real encouragement to face the future', even though he needed to be led by a sighted rider and was insecure in his seat, never getting above a trot.

Mary Jo Beckman was in the audience when Brown spoke at a 1997 therapeutic riding conference. Originally she wanted to organize a programme for the disabled children of services members, but when this did not work out, she joined forces with Larry Pence's wife, and together they secured $50,000 from the Department of Veterans Affairs, and got the programme underway with the help of local non-military therapy groups. Instructors were provided by the Professional Association of Therapeutic Horsemanship International, the organization that developed out of the North American Riding for the Handicapped Association and which Mary Woolverton had gone on to run.

Some veterans needed only one ten-week course, others could return for over a year. Most had never worked with horses before. Before each session a military medic in desert combat fatigues took their blood pressure, pulse and vitals. They answered a short questionnaire on their pain levels, their emotions, dizziness grade and their anxiety about working with horses. They answered the same questions at the end of the session. 'It's usually dropped on all counts,' said Pence. 'Unless it's a hot day.'

Work began on the ground. Dizziness was tackled by leaning slowly about the horse to reach each spot while grooming it. Next they moved on to trying to control a loose horse in the round pen. 'That's balance, mobility and sequencing short-term memory – hand and eye coordination,' explained Larry. 'There's a certain number of steps you go through, and you have to do them the same way each time, give the same commands to the horse so you are not confusing it. It helps the student to get confident in their ability to be in charge of a 1,200-lb animal, pure body language. They're alone in the pen with them.'

We leaned on the fence of the covered school while the team from the hospital chatted and checked clipboards with lesson plans, watching their patients as they took part in another groundwork class conducted by a volunteer soldier from the Caisson Platoon. He held Skeeter, a chestnut gelding with three socks and a blaze, who watched him closely as the soldier sent him away to the end of his lead rope, then made him circle back in. 'As soon as he gives, I release. Horses learn from release of pressure. When I started with him I had to yank hard, but now he just releases right away.' He stood at Skeeter's flank, asking him to flex his head back to look at him.

Four of the veterans attempted to go through the same moves with their horses as the soldier called out advice. Other men from the Caisson Platoon in orange mesh tabards stood on hand, waiting to help out. 'You can imagine some days some of these soldiers are in the cemetery burying their fallen comrades, and in here, they get to help them recover,' Larry added. 'There's some serendipity there, I think, but again there's that natural relief and emotional benefit from talking to your fellow servicemen. Soldiers tell soldiers everything.'

The next step was mounted work to build core strength, balance and posture. CPEAP consulted the hospital specialists each week to personalize the programme. Larry had left my side to head to a sand school where a series of cones and poles were set out, and a woman on a little golden bay was moving slowly around them, accompanied by an instructor and two men from the Caisson Platoon who flanked the horse. Larry had told me the servicewoman, Michelle, who was in her sixties, had been 'a full-bird colonel in the air force, an aero-space medicine doctor – and she came here because she couldn't read a printed page any more. Double vision. Can't read a computer screen. It's a very humbling thing for the TBI veterans – yesterday I knew the hell what I was doing, today I forgot to brush my teeth this morning.'

They had spent five straight weeks working on her visual acuity. Numbers and letters were pasted to the poles in the school, and she worked slow figures from one pole to another, pausing sometimes to raise one leg, or to lean a certain number of degrees to one side, challenging her balance. Eventually, she'd been able to work with her eyes closed. When her session was finished, she leant forward on to the bay's neck, patting it over and over, and then slowly dismounted without help, as the men stood around ready to step in, Larry with a hand ready to catch her. She stood for a while by the bay's head, rubbing its neck, and then led it across the school to the gate.

Some of the physical activities at CPEAP could, it was obvious, have been done with gym equipment, or by physiotherapists manip-ulating the limbs and bodies of the men and women. It was the response of the horse that transformed the treatment – Greg's feeling of satisfaction at getting hefty Duke to do as he was told, but also the neutrality which Larry summed up: 'Horses don't know if you have an amputated left leg or your brain is screwed up, black, red, green they don't care, all they care about is how you treat them.' Beneath the games of pushing the horse to step this way and that, walk around poles or halt so you could pick a beanbag up off a drum, there was a deeper, truer sympathy.

At a DNA level we have more in common with horses than with even our 'best friend' and longest domesticated companion, the dog. We have similar facial musculature, and even though horses have fewer distinct facial expressions than us, they have more than chimpanzees, our closest relative. We also share with horses, as we do with many other creatures, chemicals like oxytocin, cortisol, testosterone and oestrogen that shape our behaviour and responses.

For the wounded warriors, this connection was reinforced. Humans afflicted with PTSD tend to have raised levels of dopamine, a neurotransmitter that drives a range of emotional behaviour from motivation and learning to addiction, paranoia and hyper-vigilance. It's also dopamine that fires when the horse feels the release of the rope that Lucy and the Caisson soldier described, and begins to associate it with the signal to move or stand. When horses come under stress, they produce more dopamine in their brains, leading to the development of the strange, unsettling vices like weaving or cribbing that are as addictive and soothing to them as opiates or alcohol. Earlier I'd seen Skeeter – from habit begun long before the army found him in a Texas auction lot – sink his teeth into the fence around the school, lean back and inhale, a gesture that had long since ceased to have any association with stress for the gelding. He did it for the endorphins. One of the soldiers holding him gently chided, 'Skeeter, that's like huffing in a paper bag, quit it.'

Like the wounded warriors, horses see things that no one else sees. They act before they ascertain what a plastic bag in a hedge actually is, or whether a raised hand will hurt them. They appear to have an enhanced sense for the heart rates of creatures around them, even through the leather and wood of a saddle: in research carried out in Sweden, the heart rates of horses increased in parallel with those of their human handlers or riders when the humans passed a point in an obstacle course where they'd been told an umbrella would be flashed open. The humans physically anticipated a shock which never occurred, and the horses reacted to the human's fear.

But this didn't really explain how the CPEAP programme at Fort Belvoir worked – that sensitivity and response should have meant that the horse and the veteran's hyper-vigilance would just trigger an escalation of fear in one another, the horse ready to flee after being alerted to an enemy that neither it nor the soldier could locate. Where did the 'natural relief of anxiety' and the 'calming effect' that Larry and Greg mentioned come from?

A month earlier, I'd read about another study conducted at the University of Guelph in Ontario, where ten draught geldings were turned loose one at a time into a round pen in which stood a blindfolded person. Some of the human subjects were experienced with horses, and comfortable with them. Others were afraid of them. Observers watched the reactions of the horses and measured their heart rates and those of the humans during the encounters. To test whether the horses simply reacted to a raised human heartbeat rather than to emotional distress, the researchers got the two human participants who were comfortable with horses to run vigorously before entering the pen, ensuring that their pulses were raised. With the confident humans, the horses began to raise their own heart rate, but when they moved into the round pen with the most nervous humans something strange happened – they lowered their heads, moved less around the blinded humans and reached a state of calmness. Their pulses fell.

This was the sympathy of the herd that brings all its members to one level, to quieten or to grow alert together, in cohesion, synchrony and space, as Lucy had explained in her swallow-filled kitchen. And humans responded in turn – there it was in Greg's falling blood pressure, the lifting of chronic dizziness and the anxiety scores that fell away. Animal fear was overcome by what was also, it turns out, an animal instinct. Survival lay in a sympathy between humans and horses that ran like moving cogs – separate but locked together. The horses watched us and sensed our bodies change as our emotions responded to their own, and they waited for our fists to unclench.

'This is my second twelve-week iteration,' said John, who must have been in only his mid-twenties. He stood braced as if to attention in his black bait-shop T-shirt and his mirrored sunglasses in the middle of the empty riding school, leaning perceptibly away from me and my tape recorder. 'If I could spend three or four years doing this I would, but you have to leave so other people come in and do the same treatment as you. It's awesome, and the volunteers are awesome, the instructors are awesome.'

He answered each question with the quaint formality of the Marine Corps, 'Ma'am', 'Yes, ma'am'. 'My physical therapist Miss Annette she recommended it because of being out in the wildlife and close to the horses, to relieve anxiety. I got PTSD, I had a lot of anxiety, a lot of stress, and being with the horse is totally unlike I ever imagined doing. When I first came out here I was kind of hesitant, not because I was scared of the horse, it was more that I didn't know how the horse would react. Once I got familiar with Skeeter, I started to enjoy coming here, a lot. The mobility of the horse, how he moves helps the rotation of your pelvic. I have back problems.'

'Where did you serve?' I asked, beginning to realize that I had made the mistake of standing between him and the safe group of the hospital therapists and Caisson Platoon soldiers, and wondering if the small interview felt like too much pressure, but John went on, in words that I guessed he'd had to trot out on many occasions for the uncomprehending: 'I was in Afghanistan in oh-nine. I did Operation Strike of the Sword, and then I went to Afghanistan in 2011. My first deployment was the first major HELO offensive since Vietnam. And my second was an IED-heavy area. So I got blown up a couple of times, got shot at a couple of times. A little bit of TBI. The symptoms from that is like memory, more cognitive memory, the psy aspect of it, I get headaches, sometimes I'll get dizziness, but I think it kind of factors into the PTSD a lot.'

I said something about the way that dizziness could cause panic, and panic lead to breathing shallowly, and more dizziness – it was a loop of each interaction that became impossible to untangle. Mind or body? Both?

'Ma'am.'

I thought I could feel the panic coming off him by now, like the atmosphere around a scared horse as it feels pressure tightening on it, and ended my questions. Later, listening back to the recording, his voice seemed calm enough, which didn't fully explain my response:

without thinking, I'd turned sideways so that I was not confronting or blocking him, as I'd once read about a horse whisperer doing as he tried to approach a frightened animal. Before he left me with a final, polite 'Ma'am' to join the others at the far side of the school, John told me he only felt safe in his home and on a fishing boat, and here, at the stables.

'This is a safe haven.'

Later in the school, a new riding session began. The two warriors sat loose-legged on thick, Stars-and-Stripes saddlecloths that were cinched into place on their horses' backs. Young women held the horses' heads, and on either side, in a knight's guard of honour, were broad-shouldered Caisson Platoon volunteers with ornately tattooed arms and shaved necks under their forage caps. The destriers dozed through the muted, gentle parade around the school, and the soldiers surreptitiously stroked them behind the girth as they plodded on. Above their heads, the warriors bent and stretched tentatively at the instructor's request, and then each in turn leaned gingerly over to lift a bandana off a pole as their mounts walked steadily past it. The rider on the palomino pulled his fists up to his chest to halt his mount, unsure of the gesture and the response, and his horse flicked its ears back and forth as if questioning, came to a standstill and then relaxed once more.

A few years before I stood in the manège in Arlington, before I set out on the six bridleways across Mongolia, France, America, Portugal, Spain and China, I'd come across two lines in a book by the philosopher John Gray. At the time they had struck me, and I knew they would be part of the book ahead of me, but I hadn't known how they would fit into what was then an unmapped territory of branching *carriles* that intersected in impossible complexity. I would trace my own paths along them, I'd thought, and somewhere on the way the *carriles* would fall into the outline of a figure I could recognize. Perhaps at Versailles, where the *écuyères* communed in monkish contemplation with the cream-coloured dancing horses, or in a corner of New England where a small group of people and work-horses were trying to live according to the rhythms of animal and planet. In the end it was from Lucy in Extremadura and the men and horses of the Caisson Platoon that I found the connection. 'If you seek the origins of ethics,' Gray had written, 'look to the lives of other animals. The roots of ethics are in the animal virtues. Humans cannot live well without virtues they share with their animal kin.'

Acknowledgements

Heartfelt thanks to The Society of Authors for their help with visas and for giving me a grant from the Author's Foundation that made travel to Mongolia possible.

Thank you to everyone who welcomed me into their projects, workplaces or homes and talked to me about their lives with horses.

Thank you also to Christian Mögwitz for helping me fact-check Bernhard Grzimek, to the very patient Waltraut Zimmerman, to Caroline Humphrey, Jean Khalfa, Lee Boyd, Richard Reading, Kai Artinger and to Nataliya Yasynetskaya at Askania-Nova. The Deutsche Dienstelle was also invaluable.

Thank you to Rick Walker, Kean Wong, Tony Sebok, Karl de Meyer and Eric Dickson for providing spare bedrooms.

Thank you to Desmond Tumulty and Dima Mixalicyn for Russian translation.

Thank you to Ingeborg Formann at Österreichische National-bibliothek, Joelle Hayden at APHIS and Wayne Olson at the US National Agricultural Library, and all the other librarians and archivists who helped me.

Thank you to Jen Porto, Paul Festa and Ed Ward for musical counsel.

Thank you to Justin E H Smith for answering some stupid questions.

Thank you to Aimee Male, Sarah Everts, John Borland and Michael Scott Moore for War.

Thank you to Yuhang Yuan for help with Chinese visas, knights, celebrities and confirmation that the Chinese for BMW does indeed translate as 'treasure horse'.

Thank you to Melissa Perez, Gwyneth Talley and Kathryn Renton.

Thank you to Mark in China, Feng Ke and interpreters 'Dora' and 'Alana'.

Thank you to Angus MacKinnon, to James Nightingale, Karen Duffy, Margaret Stead and the team at Atlantic Books.

Thank you to my agent, Judith Murray at Greene and Heaton.

Thank you to my family.

Bibliography

EVOLUTION AND DOMESTICATION

Achilli, A, A Olivieri, P Soares, H Lancioni, B Hooshiar Kashani, U A Perego, and others, 'Mitochondrial Genomes from Modern Horses Reveal the Major Haplogroups That Underwent Domestication', *Proceedings of the National Academy of Sciences of the United States of America*, 109/7 (2012), 2449–54

Bendrey, R, 'From Wild Horses to Domestic Horses: A European Perspective', *World Archaeology* (2012) <http://centaur.reading.ac.uk/33691/>

Bennett, D, and R S Hoffman, 'Equus Caballus', *Mammalian Species*, 628 (1999), 1–14

Brubaker, T M, E C Fidler, and others, 'Strontium Isotopic Investigation of Horse Pastoralism at Eneolithic Botai Settlements in Northern Kazakhstan', 2006 *Philadelphia Annual Meeting*, 2006

Cieslak, M, M Pruvost, N Benecke, M Hofreiter, A Morales, M Reissmann, and others, 'Origin and History of Mitochondrial DNA Lineages in Domestic Horses', *PLoS One*, 5/12 (2010), e15311

Curry, A, 'Archaeology: The Milk Revolution', *Nature*, 500/7460 (2013), 20–22

Dobos, A, 'The Lower Palaeolithic Colonisation of Europe. Antiquity, Magnitude, Permanency and Cognition', *PaleoAnthropology*, 2012

'Early Stone Age Hafted Spear Points from South Africa' <http://johnhawks.net/weblog/reviews/archaeology/lower/wilkins-2012-kathu-pan-spear-points.html>

'Endangered Horse Has Ancient Origins and High Genetic Diversity, New Study Finds' <http://www.sciencedaily.com/releases/2011/09/110907163921.htm>

'Fossil Evidence of Laminitis in Ancient Horses, TheHorse.com' <http://www.thehorse.com/articles/33506/fossil-evidence-of-laminitis-in-ancient-horses>

'From Ancient DNA, a Clearer Picture of Europeans Today', *New York Times*, 30 October 2014

Gardiner, J B, R C Capo, and others, 'Soil Trace Element Evidence for Horse Corralling during the Copper Age in Northern Kazakhstan', 2008 *Joint Meeting of The Geological Society of America …*, 2008

Harding, D, S Olsen, and K Jones Bley, 'Reviving Their Fragile Technologies: Reconstructing Perishables from Pottery Impressions from Botai, Kazakhstan', *Annu. Meet. Soc. Am. Arch.*, 65th, Philadelphia, 2000

'Harnessing Horsepower – Anthony and Brown' <http://users.hartwick.edu/anthonyd/harnessing horsepower.html>

'Horse Evolution' <http://www.talkorigins.org/faqs/horses/horse_evol.html>

'Horse Evolution Over 55 Million Years' <http://chem.tufts.edu/science/evolution/horseevolution.htm>

Howe, T, 'Domestication and Breeding of Livestock: Horses, Mules, Asses, Cattle, Sheep, Goats and Swine', in *The Oxford Handbook of Animals in Classical Thought and Life*, ed. by G L Campbell (Oxford, 2014)

Jansen, T, P Forster, M A Levine, H Oelke, MHurles, C Renfrew, and others, 'Mitochondrial DNA and the Origins of the Domestic Horse', *Proceedings of the National Academy of Sciences of the United States of America*, 99/16 (2002), 10905–10

Kelekna, P, *The Horse in Human History* (Cambridge, 2009)

Levine, M A, 'Domestication, Breed Diversification and Early History of the Horse', ... *Workshop: Horse Behavior and Welfare*, June, 2002

——, 'Eating Horses: The Evolutionary Significance of Hippophagy', *Antiquity*, 1998

Levine, M A, C Renfrew, and K V Boyle, *Prehistoric Steppe Adaptation and the Horse*, 2003

Lindgren, G, N Backström, J Swinburne, L Hellborg, A Einarsson, K Sandberg, and others, 'Limited Number of Patrilines in Horse Domestication', *Nature Genetics*, 36/4 (2004), 335–36

Ludwig, A, M Pruvost, M Reissmann, N Benecke, G A Brockmann, P Castaños, and others, 'Coat Color Variation at the Beginning of Horse Domestication', *Science* (New York, N.Y.), 324/5926 (2009), 485

Ludwig, A, M Reissmann, N Benecke, R Bellone, E Sandoval-Castellanos, M Cieslak, and others, 'Twenty-Five Thousand Years of Fluctuating Selection on Leopard Complex Spotting and Congenital Night Blindness in Horses', *Philosophical Transactions of the Royal Society of London B: Biological Sciences*, 370/1660 (2014)

MacFadden, B J, *Fossil Horses: Systematics, Paleobiology, and Evolution of the Family Equidae* (Cambridge, 1992)

'Olsen – Botai – Horses and Humans: Carnegie Museum of Natural History' <http://www.carnegiemnh.org/science/default.aspx?id=16610>

Olsen, S, B Bradley, D Maki, and A Outram, 'Community Organization among Copper Age Sedentary Horse Pastoralists of Kazakhstan', *Beyond the Steppe and the Sown, Colloquia Pontifica*, 2006

Olsen, S, A Brickman, and Y Cai, 'Discovery by Reconstruction: Exploring Digital Archeology', *SIGCHI Workshop*, 2004

Olsen, S L, 'Expressions of Ritual Behavior at Botai, Kazakhstan', in *Proceedings of the Eleventh Annual UCLA Indo-European Conference*, 2000, 183–207

Olsen, S L, 'This Old Thing? Copper Age Fashion Comes to Life', *Archeology*, 61/1 (2008), 46–47

Outram, A K, and A Kasparov, 'Patterns of Pastoralism in Later Bronze Age Kazakhstan: New Evidence from Faunal and Lipid Residue Analyses', *Journal of Archaeological Science*, 39 (2012), 2424–35

Outram, A K, N A Stear, and A Kasparov, 'Horses for the Dead: Funerary Foodways in Bronze Age Kazakhstan', *Antiquity*, 85/327, January 2011, 116–128

Outram, A K, N A Stear, R Bendrey, S Olsen, A Kasparov, V Zaibert, and others, 'The Earliest Horse Harnessing and Milking', *Science* (New York, N.Y.), 323/5919 (2009), 1332–35

Petersen, J L, J R Mickelson, A K Rendahl, S J Valberg, L S Andersson, J Axelsson, and others, 'Genome-Wide Analysis Reveals Selection for Important Traits in Domestic Horse Breeds', ed. by J M Akey, *PLoS Genetics*, 9/1 (2013)

'Punctured Horse Shoulder Blade' <http://humanorigins.si.edu/evidence/behavior/food/punctured-horse-shoulder-blade>

Schubert, M, H Jónsson, D Chang, C Der Sarkissian, L Ermini, A Ginolhac, and others, 'Prehistoric Genomes Reveal the Genetic Foundation and Cost of Horse Domestication', *Proceedings of the National Academy of Sciences of the United States of America*, 111/52 (2014)

Secord, R, J I Bloch, S G B Chester, and D M Boyer, 'Evolution of the Earliest Horses Driven by Climate Change in the Paleocene-Eocene Thermal Maximum', *Science*, 2012

Simpson, G G, *Horses: The Story of the Horse Family in the Modern World and through Sixty Million Years of History* (New York, 1951)

Sommer, R S, N Benecke, L Lõugas, O Nelle, and U Schmölcke, 'Holocene Survival of the Wild Horse in Europe: A Matter of Open Landscape?' *Journal of Quaternary Science*, 26/8 (2011), 805–12

'Spots, Stripes and Spreading Hooves in the Horses of the Ice Age – Tetrapod Zoology', *Scientific American Blog Network* <http://blogs.scientificamerican.com/tetrapod-zoology/spots-stripes-and-spreading-hooves-in-the-horses-of-the-ice-age/>

Vilà, C, J A Leonard, A Gotherstrom, S Marklund, K Sandberg, K Liden, and others, 'Widespread Origins of Domestic Horse Lineages', *Science* (New York, N.Y.), 291/5503 (2001), 474–77

Warmuth, V, A Eriksson, M A Bower, J Cañon, G Cothran, O Distl, and others, 'European Domestic Horses Originated in Two Holocene Refugia', *PLoS One*, 6/3 (2011)

Warmuth, V, A Eriksson, M A Bower, Graeme Barker, E Barrett, B K Hanks, and others, 'Reconstructing the Origin and Spread of Horse Domestication in the Eurasian Steppe', *Proceedings of the National Academy of Sciences of the United States of America*, 109/21 (2012)

'Why Did Horses Die out in North America?' *HorseTalk*, 29 November 2012 <http://horsetalk.co.nz/2012/11/29/why-did-horses-die-out-in-north-america/#axzz3pfUzULtx>

Wilkins, A S, R W Wrangham, and W Tecumseh Fitch, 'The "Domestication Syndrome" in Mammals: A Unified Explanation Based on Neural Crest Cell Behavior and Genetics', *Genetics*, 197/3 (2014), 795–808

WILDNESS

Africanus, L, *The History and Description of Africa Vol III* (London, 1896)

Alliance of Religions and Conservation, 'Mongolian Buddhist Environment Handbook' <http://www.arcworld.org/downloads/Mongolian Buddhist Environment Handbook.pdf>

'An SS Booklet on Racial Policy', *German Propaganda Archive* <http://research.calvin.edu/german-propaganda-archive/rassenpo.htm>

Atwood Lawrence, E, *Hoofbeats and Society: Studies of Human–Horse Interactions* (Bloomington, 1985)

Bandi, N O, *Takhi: Back to the Wild* (Ulaanbaatar, 2012)

Bell, J, *Travels from St Petersburgh in Russia to Various Parts of Asia* (Edinburgh, 1806)

'Berliner Zoo: Urmacher Unerwünscht', *Der Spiegel*, 1954 <http://www.spiegel.de/spiegel/print/d-28956824.html>

Bold, B O, *Eques Mongolica* (Reykjavik/Toronto, 2012)

Bouman, J, I Bouman, and A Groeneveld, *Breeding Przewalski Horses in Captivity for Release into the Wild* (Rotterdam, 1982)

Boyd, L, and K A Houpt, *Przewalski's Horse: The History and Biology of an Endangered Species* (Albany, 1994)

Branigan, T, 'Mongolia: How the Winter of "White Death" Devastated Nomads' Way of Life', *The Guardian*, 20 July 2010

——, 'Mongolia: "The Gobi Desert Is a Horrible Place to Work"', *The Guardian*, 20 April 2014

Byron, G G, 'Mazeppa, A Poem', *The Internet Archive*, 1819 <https://archive.org/details/mazeppaapoem02byrogoog>

Capitolinus, J, 'Historia Augusta', (Cambridge, MA, 1924)

Chimedsengee, U, A Cripps, V Finlay, G Verboom, V Munkhbaatar Batchuluun, and V Da Lama Byambajav Khunkhur, *Mongolian Buddhists Protecting Nature: A Handbook on Faiths, Environment and Development* (Ulaanbaatar, 2009)

Cornay, J E, 'De La Reconstruction Du Cheval Sauvage Primitif', in *Libraire de La Faculté de Médecine*, ed. by P Asselin (Paris, 1861)

Daszkiewicz, P, and J Aikhenbaum, 'Aurochs, Le Retour D'une Supercherie Nazie', *Courrier de l'Environnement de l'INRA*, 33 (1998)

'Desertification in Mongolia, 2013 UNEP Report' <http://www.unep.org/wed/2013/docs/Desertification-in-Mongolia-30.5.13.pdf>

Die Geschichte Berlins, 'Heck, Lutz' <http://www.diegeschichteberlins.de/geschichteberlins/persoenlichkeiten/persoenlichkeitenhn/491-heck.html>

DiNardo, R L, *Mechanized Juggernaut or Military Anachronism? Horses and the German Army of World War Two* (Westport, 1991)

Fauvelle, C, 'Le Cheval Sauvage de la Dzoungarie', *Bulletins de La Société d'Anthropologie de Paris*, 10/3 (1887), 188–206

Fijn, N, 'The Domestic and the Wild in the Mongolian Horse and the Takhi', in *Taxonomic Tapestries The Threads of Evolutionary, Behavioural and Conservation Research*, ed. by A M Behie and M F Oxenham (Canberra, 2015)

Fox, F, 'Endangered Species: Jews and Buffaloes, Victims of Nazi Pseudo-Science', *East European Jewish Affairs*, 31/2 (2001), 82–93

'Genealogy: Families Prieb, Stark, Mehlmann, Fein, Falz, Falz-Fein' <http://www.stammbaum-familie-prieb.de/StBFFAnlage_engl.htm>

Gill, V, 'Chernobyl's Przewalski's Horses Are Poached for Meat', *BBC Nature* <http://www.bbc.co.uk/nature/14277058>

Gongorin, U, 'Sacred Groves in Mongolia: Country Report', in *Conserving the Sacred for Biodiversity Management*, ed. by P S Ramakrishan (New Delhi, 1998)

Gordon, W J, *The Horse-World of London* (London, 1893)

Grzimek, B, *Wild Animal, White Man: Some Wildlife in Europe, Soviet Russia and North America* (London, 1966)

Hagenbeck, C, *Beasts and Men: Being Carl Hagenbeck's Experiences for Half a Century Among Wild Animals* (London, 2012)

Hamilton Smith, C, *The Nauralist's Library: The Natural History of Horses* (Edinburgh, 1841)

Harmon Snow, K, 'The Bare Naked Face of Capitalism: Foreign Mining, State Corruption, and Genocide in Mongolia', *I C Magazine*, 2014 <https://intercontinentalcry.org/goldman-prizewinner-gets-21-years-resistance-genocide/>

Heck, H, 'The Breeding-Back of the Tarpan', *Oryx*, 1/7 (1952)

Heck, L, *Animals: My Adventure* (Norwich, 1954)

Heiss, L, *Askania-Nova: Animal Paradise in Russia* (London, 1970)

Heissig, W, *The Religions of Mongolia* (Berkeley, 1980)

Hepter, V G, *Mammals of the Soviet Union* (Washington DC, 1988)

Herodotus, *Herodotus: The Histories* (London, 2003)

Humber, Y, 'Mongolia $1.25/day Labor Amid $4K Purses Stirs Discontent', *Bloomberg*, 2013 <http://www.bloomberg.com/news/articles/2013-02-04/mongolia-1-25-day-labor-amid-4k-purses-stirs-discontent>

Humphrey, C, 'Horse Brands of the Mongolians: A System of Signs in a Nomadic Culture', *American Ethnologist*, 1/3 (1974), 471–88

——, *The End of Nomadism? Society, State and the Environment in Inner Asia* (Cambridge, 1999)

'International Wild Equid Conference: Book of Abstracts' (Vienna, 2012) <https://www.vetmeduni.ac.at/fileadmin/v/fiwi/Konferenzen/Wild_Equid_Conference/IWEC_book_of_abstracts_final.pdf>

Irvine, R, 'Thinking with Horses: Troubles with Subjects, Objects, and Diverse Entities in Eastern Mongolia', *Humanimalia*, 6/1 (2014), 62–94

Jennings, J J, *Theatrical and Circus Life: Secrets of the Stage, Green-Room and Sawdust Arena* (Chicago, 1886)

Jezierski, T, and Z Jaworski, *Das Polnische Konik* (Hohenwarsleben, 2008)

Kaiman, J, 'Mongolia's New Wealth and Rising Corruption Is Tearing the Nation Apart', *The Guardian*, 27 June 2012

Kavar, T, and P Dovč, 'Domestication of the Horse: Genetic Relationships between Domestic and Wild Horses', *Livestock Science*, 116/1–3 (2008), 1–14

von Manstein, E, *Lost Victories: The War Memoirs of Hitler's Most Brilliant General* (Minneapolis, 2004)

'Nabokov Family Web #187' <http://dezimmer.net/NabokovFamilyWeb/nfw01/nfw01_187.htm>

Noble Wilford, J, 'In Mongolia, an "Extinction Crisis" Looms', *New York Times*, 6 December 2005

Orlando, L, A Ginolhac, G Zhang, D Froese, A Albrechtsen, M Stiller, and others, 'Recalibrating Equus Evolution Using the Genome Sequence of an Early Middle Pleistocene Horse', *Nature*, 499/7456 (2013), 74–78

'Review of *Mazeppa*', *Morning Chronicle*, 5 April 1831

'Review of *The Wild Horse and the Savage*', *North Wales Chronicle*, 17 February 1835

Ridgeway, W, *Origin and Influence of the Thoroughbred Horse* (Cambridge, 1905)

le Roux, H A, *Acrobats and Mountebanks* (London, 1890)

Russell, J, 'The Compelling Imagery of Hans Baldung Grien', *New York Times*, 22 February 1981

Samojlik, T, 'The Bison: Rich Treasure of the Forest', in *Conservation and Hunting: Bialowieza Forest in the Time of Kings*, ed. by T Samojlik (2005)

Sax, B, *Animals in the Third Reich: Pets, Scapegoats, and the Holocaust* (London, 2000)

Saxon, A H, *Enter Foot and Horse: A History of Hippodrama in England and France* (New Haven and London, 1968)

Sayed, A, 'Dzud: A Slow Natural Disaster Kills Livestock – and Livelihoods – in Mongolia', *World Bank Blogs*, 2010 <http://blogs.worldbank.org/eastasiapacific/dzud-a-slow-natural-disaster-kills-livestock-and-livelihoods-in-mongolia>

Schama, S, *Landscape and Memory* (London, 1996)

'The First Circus', *Victoria and Albert Museum*, 2011 <http://www.vam.ac.uk/content/articles/t/the-first-circus/>

UNFPA Mongolia, 'Mongolia Has Launched the Main Findings of Its 2010 Population and Housing Census', 17 July 2011

Varro, *De Re Rustica Vol II* (Cambridge, MA, 1934)

Veit, V, '"In Autumn Our Horses Are Well-Fed and Ready for Action" – the Ch'ing Empire and Its Mongolian Cavalry', in *SOAS War Horse Conference* (London, 2014)

Wagner, M A, *Le Cheval Dans Les Croyances Germaniques: Paganisme, Christianisme et Tradition* (Paris, 2005)

Wiener, L, *Anthology of Russian Literature from the Earliest Period to the Present Time* (New York, 1902)

Wit, P, and I Bouman, *The Tale of the Przewalski's Horse: Coming Home to Mongolia* (Utrecht, 2006)

World Bank, 'Mongolia: Improving Public Investments to Meet the Challenge of Scaling up Infrastructure' <http://www.worldbank.org/en/news/feature/2013/02/27/mongolia-improving-public-investments-to-meet-the-challenge-of-scaling-up-infrastructure>

Zimmerman, W, *International Przewalski's Horse Studbook*, (Cologne)

Борейко, В Е, *Аскания-Нова: тяжкие версты истории 1826–1997* (Kiev, 2001)

История городов и сел Украинской ССР. Херсонская область (Kiev, 1983)

Письма из MaisoRusse. Сестры Анна Фальц-Фейн и Екатерина Достоевская в эмиграции (St Petersburg, 1999)

Салганский, А А, И С Слесь, В Д Треус, and Г А Успенский, *«Аскания-Нова» (опыт акклиматизации диких копытных и страусов)* (Kiev, 1963)

Треус, В Д, *Акклиматизация и гибридизация животных в Аскании-Нова: 80-летний опыт культурного освоения диких животных и птиц* (Kiev, 1968)

CULTURE

Alt, A, and S Nauleau, *La Voie de l'Ecuyer* (Arles, 2008)

Athenaeus, and C D Yonge, *The Deipnosophists, Or Banquet Of The Learned Of Athenaeus* (London, 1854)

Bartabas, *Manifeste Pour La Vie D'Artiste* (Paris, 2012)

———, *Mazeppa*, 1993

'Bartabas: Instinct Cavalier', *Cheval Magazine*, November 2011

Battista Tomassini, G, *The Italian Tradition of Equestrian Art* (Franktown, 2014)

Bhakari, S K, *Indian Warfare – An Appraisal of the Strategy and Tactics of War in Early Medieval Period* (Delhi, 1983)

Bondeson, J, 'The Dancing Horse', in *The Feejee Mermaid and Other Essays in Natural and Unnatural History* (Ithaca, 1999)

Bregman, M R, J R Iversen, D Lichman, M Reinhart, and A D Patel, 'A Method for Testing Synchronization to a Musical Beat in Domestic Horses (Equus Ferus Caballus)', *Empirical Musicology Review*, 7/3–4 (2012)

Castiglione, B, and T Hoby, 'Full Text of Sir Thomas Hoby's Translation of The Book of the Courtier by Baldassare Castiglione', *Sheffield Hallam University Website*, 1561 <http://extra.shu.ac.uk/emls/iemls/resour/mirrors/rbear/courtier/courtier.html>

Cavendish, W, *A New Method and Extraordinary Invention to Dress Horses, And Work Them according to Nature; As Also, to Perfect NATURE by the Subtilty of ART; Which Was Never Found Out, but by the Thrice Noble, High and Puissant PRINCE William Cavendishe* (1657)

'Chasseur', 'Third Letter from Paris', *The Sporting Magazine*, 2 (1830), 155

Cowart, G J, *The Triumph of Pleasure: Louis XIV and the Politics of Spectacle* (Chicago, 2008)

DeCastro, J, *The Memoirs of J. Decastro, Comedian* (London, 1824)

Dixon, K R, and P Southern, *The Roman Cavalry: From the First to the Third Century* AD (London, 1997)

Falk, A, *The Psychoanalytic History of the Jews* (Madison, NJ, 1996)

Garcin, J, *Bartabas, Roman* (Paris, 2004)

Greening, L, and C Carter, 'Auditory Stimulation of the Stabled Equine; the Effect of Different Music Genres on Behaviour', in *International Society for Equitation Science* (Edinburgh, 2012)

de la Guérinière, F R, *School of Horsemanship* (London, 1994)

Hamilton, J, *Marengo, the Myth of Napleon's Horse* (London, 2000)

'Hippika Gymnasia', *Wikipedia* <https://en.wikipedia.org/wiki/Hippika_gymnasia>

Hippisley Coxe, A, *A Seat at the Circus* (London, 1951)

Homer, and S Butler, *Illiad* (London, 1898)

Huth, F, *Works on Horses and Equitation: A Bibliographical Record of Hippology* (London, 1887)

Hyland, A, *The War Horse: 1250–1600* (Stroud, 1998)

Jando, D, 'Philip Astley', *Circopedia* <http://www.circopedia.org/Philip_Astley>

Kane, K, 'Royal Hanoverian Creams', *The Regency Redingcote*, 2010 <https://regencyredingote.wordpress.com/2010/04/30/royal-hanoverian-creams/>

Kautilya, and R Shamasastry, *Arthasastra* (Mysore, 1967)

Kellock, E M, *The Story of Riding* (Newton Abbot, 1974)

Landry, D, *Noble Brutes: How Eastern Horses Transformed English Culture* (Baltimore, 2009)

Le Mercure François Ou La Suitte de L'histoire de La Paix Commençant L'an 1605 Pour Suite Du Septénaire Du D. Cayer, et Finissant Au Sacre Du Très Grand Roy de France et de Navarre Louis XIII (Paris, 1611)

'Living with Animals 2: Interconnections' (Richmond, 2015)

Loch, S, *Dressage: The Art of Classical Riding* (London, 1990)

———, *The Royal Horse of Europe* (London, 1986)

Lopes, M S, D Mendonça, T Cymbron, M Valera, J Da Costa-Ferreira, A Da Câ, and others, 'The Lusitano Horse Maternal Lineage Based on Mitochondrial D-Loop Sequence Variation', *Animal Genetics*, 36/3 (2005) 196–202

Loring Payne, F, *The Story of Versailles* (New York, 1919)

Luís, C, C Bastos-Silveira, J Costa-Ferreira, E G Cothran, and M M Oom, 'A Lost Sorraia Maternal Lineage Found in the Lusitano Horse Breed', *Zeitschrift Für Tierzüchtung Und Züchtungsbiologie*, 123/6 (2006), 399–402

'Lusus Troiae', *Wikipedia* <https://en.wikipedia.org/wiki/Lusus_Troiae>

Mackay-Smith, A, J R Druesedow, and T Ryder, *Man and the Horse: An Illustrated History of Equestrian Apparel* (New York, 1984)

MacNeille Nelson, N, 'Courses de Testes et de Bague and the Cultural Legitimization of Louis XIV's Personal Rule, 1661–1671' (Haverford College, 2008)

Marshall, J, *Taxila: An Illustrated Account of Archaeological Excavations Carried out at Taxila, Etc.* (Cambridge, 1951)

Mayor, A, *The Amazons: Lives and Legends of Warrior Women across the Ancient World* (Princeton, 2014)

Nelson, H, *François Baucher: The Man and His Method* (London, 1992)

——, *The Écuyère of the Nineteenth Century in the Circus* (Cleveland Heights, 2001)

Nyland, A, *The Kikkuli Method of Horse Training* (Mermaid Beach, 2009)

van Orden, K, *Music, Discipline, and Arms in Early Modern France* (Chicago, 2005)

Patel, A D, 'The Evolutionary Biology of Musical Rhythm: Was Darwin Wrong?' *PLoS Biol*, 12/3 (2014)

de Pluvinel, A, *L'Instruction Du Roy, En l'Exercice de Monter a Cheval Par Messire Antoine de Pluvinel, Son Sous-Gouverneur, Conseiller En Son Conseil d'Etat, Chambellan Ordinaire, et Son Ecuyer Principal* (Amsterdam, 1668)

Poppiti, K, 'Galloping Horses: Treadmills and Other Theatre Appliances in Hippodramas', *Theatre Design and Technology*, 41/4 (2005), 45

Poppiti, K D, 'Pure Air and Fire: Horses and Dramatic Representations of the Horse on the American Theatrical Stage' (New York University, 2003)

Powell, J S, 'Music and the Scenic Portrayal of Gods, Men, and Monsters in Corneille's Andromède' (Oxford, 2001)

Raber, K L, '"Reasonable Creatures": William Cavendish and the Art of Dressage', in *Renaissance Culture and the Everyday*, ed. by P Fumerton and S Hunt (Philadelphia, 1999)

Raber, K, and T J Tucker, *The Culture of the Horse: Status, Discipline and Identity in the Early Modern World* (London, 2005)

Roche, D, and D Reytier, *A Cheval: Ecuyers, Amazones & Cavaliers* (Versailles, 2007)

Rowlands, S, 'Maroccus Extaticus: Or Bankes' Bay Horse in a Trance', in *The Four Knaves, a Series of Satirical Tracts by Samuel Rowlands*, ed. by E P Rimbault (1597)

Royo, L J, I Alvarez, A Beja-Pereira, A Molina, I Fernández, J Jordana, and others, 'The Origins of Iberian Horses Assessed via Mitochondrial DNA', *The Journal of Heredity*, 96/6 (2005), 663–69

Saxon, A H, *Enter Foot and Horse: A History of Hippodrama in England and France* (New Haven and London, 1968)

——, 'The Circus as Theatre: Astley's and Its Actors in the Age of Romanticism', *Educational Theatre Journal*, 27/3 (1975)

Shehada, H A, *Mamluks and Animals: Veterinary Medicine in Medieval Islam* (Leiden, 2013)

Sidney, P, 'The Defense of Poesy', in *English Essays* (New York, 1909)

Spawforth, A, *Versailles: A Biography of a Palace* (New York, 2008)

'Sprezzatura', Wikipedia <https://en.wikipedia.org/wiki/Sprezzatura>

Talley, G, 'Fantasia: Performing Traditional Equestrianism as Heritage Tourism in Morocco (in Preparation)' (University of Los Angeles)

Thirsk, J, Horses in Early Modern England: For Service, for Pleasure, for Power (Reading, 1977)

Thomas, K, Man and the Natural World: Changing Attitudes in England 1500–1800 (London, 1983)

Tobey, E M, 'The Legacy of Federico Grisone', in The Horse as Cultural Icon: The Real and the Symbolic Horse in the Early Modern World, ed. by P Edwards, K Enenkel, and E Graham (Leiden, 2011)

Tomassini, G B, and S Chiverton, The Works of Chivalry <http://worksofchivalry. com/en>

Tucker, T J, 'From Destrier to Danseur: The Role of the Horse in Early Modern French Noble Identity' (University of Southern California, 2007)

Velten, H, Beastly London: A History of Animals in the City (London, 2013)

Watanabe-O'Kelly, H, 'The Equestrian Ballet in 17th Century Europe – Origin, Description, Development', German Life and Letters, 36 (1983)

———, 'Tournaments and Their Relevance for Warfare in the Early Modern Period', European History Quarterly, 20 (1990)

———, 'Tournaments in Europe', in Spectaculum Europaeum (1580–1750) Theatre and Spectacle in Europe (Wolfenbüttel, 1999)

Xenophon, The Art of Horsemanship (London, 2007)

POWER

Alexander, A, Horse Secrets (Philadelphia, 1909)

'Amish Farmers' Success Goes Against The Grain', New York Times, 4 September 1986

Barclay, H, The Role of the Horse in Man's Culture (London, 1980)

Bewick, T, A General History of Quadrupeds (Newcastle Upon Tyne, 1800)

Burnes, A, Travels into Bokhara. Being an Account of a Journey from India to Cabool, Tartary and Persia. Also, Narrative of a Voyage on the Indus from the Sea to Lahore (London, 1834)

Caldwell, E, 'Estimate: A New Amish Community Is Founded Every 3.5 Weeks in U.S.', Ohio State University Research News, 2012

Carpenter, F, 'Horse Meat for Food', The National Tribune, 19 January 1893

'Chevaux de Trait Français. Menacés D'extinction Faute de Consommateurs', Le Télégramme, 2010 <http://www.letelegramme.fr/ig/generales/france-monde/france/chevaux-de-trait-francais-menaces-d-extinction-faute-de-consommateurs-17-12-2010-1152036.php>

Chivers, K, History with a Future: Harnessing the Heavy Horse for the Twenty-First Century (Peterborough, 1988)

———, The Shire Horse (London, 1976)

———, 'The Supply of Horses in Great Britain in the Nineteenth Century', in Horses in European Economic History: A Preliminary Canter, ed. by K Thompson (1983)

Collins, E, 'The Farm Horse Economy of the Early Tractor Age 1900–1940', in *Horses in European Economic History: A Preliminary Canter*, ed. by F Thompson (1983)

Courteau, D, 'Horse Power: A Practical Suggestion That Would Transform the Way We Live', *Orion Magazine* (September 2007)

Davis, R, 'The Medieval Warhorse', in *Horses in European Economic History: A Preliminary Canter*, ed. by F Thompson (1983)

Dealers Tricks, *Sporting Magazine*, September 1839

De Decker, K, 'Bring Back the Horses', *Low-Tech Magazine* <http://www. lowtechmagazine.com/2008/04/horses-agricult.html>

Derry, M, *Horses in Society: A Story of Animal Breeding and Marketing Culture, 1800-1920* (Toronto, 2006)

Downes, A, and A Childs, *My Life with Horses: The Story of Jack Juby MBE, Master of the Heavy Horse* (Tiverton, 2006)

'Dray Horse Falls into Pit near Treasury, *Morning Chronicle*, 14 October 1834

Edwards, P, K Enenkel, and E Graham, *The Horse as Cultural Icon: The Real and the Symbolic Horse in the Early Modern World* (Leiden, 2011)

Enrle, L, *English Farming Past and Present* (London, 1936)

Ewart Evans, G, *Horse Power and Magic* (London, 2008)

Fiennes, C, *Through England On a Side Saddle in the Time of William and Mary* (London, 1888)

Frey, C, and M Osborne, *The Future of Employment: How Susceptible Are Jobs to Computerisation?* (Oxford, 2013) <http://www.oxfordmartin.ox.ac.uk/ publications/view/1314>

George, C, 'City P.C. George H. Hutt, Police Poet, and the Issue of Horse Cruelty', *Reflections of a Ripperologist*, 2011 <http://blog.casebook.org/ chrisgeorge/2011/09/08/city-pc-george-h-hutt-police-poet-and-the-issue-of-horse-cruelty/>

Główna, S, 'Renaissance of Working Horses. Benefits of Using Horses in Farming and Forestry', *International Coalition to Protect the Polish Countryside*, 2007 <http:// icppc.pl/index.php/pl/icppc/pl/home/14-english/projects/230-working-horses.html>

Gordon, W J, *The Horse-World of London* (London, 1893)

Hamilton Smith, C, *The Nauralist's Library: The Natural History of Horses* (Edinburgh, 1841)

Hart, E, *Heavy Horses: An Anthology* (Stroud, 1994)

Herold, P, P Schlechter, and R Scharnhölz, 'Modern Use of Horses in Organic Farming', *Fédération Européenne Du Cheval de Trait Pour La Promotion de Son Utilisation* <http://www.fectu.org/Englisch/Horses in organic farming.pdf>

Hodak, C, 'Les Animaux Dans La Cité: Pour Une Histoire Urbaine de La Nature', *Genèses*, 37 (1999), 156–59

Holden, B, *The Long Haul: The Life and Times of the Railway Horse* (London, 1985)

Hollows, D, *Voices in the Dark: Pony Talk and Mining Tales* (2011)

Hornsey, I, 'Industrial Revolution', in *The Oxford Companion to Beer*, ed. by G Oliver (Oxford, 2011)

'Horse Fed Turkish Delight', *Illustrated Police News*, August 1900

'Horseless Carriages' *York Herald*, 29 October 1895

'Horses and Stables', *Camden Railway Heritage Trust* <http://www.crht1837.org/history/horsesstables>

Jackson, Lee, 'Victorian London' <http://www.victorianlondon.org>

James, R, 'Horse and Human Labor Estimates for Amish Farms', *Journal of Extension*, 45/1 (2007)

Kean, H, *Animal Rights: Political and Social Change in Britain since 1800* (London, 1998)

Kendell, C, *Horse Powered Traction and Tillage: Some Options and Costs for Sustainable Agriculture, with International Applications*, 2011 <http://www.fao.org/ag/againfo/themes/animal-welfare/aw-awhome/detail/en/item/51858/icode/>

Kidd, J, *The New Observer's Book of Horses and Ponies* (London, 1984)

Lawrence, D H, *Apocalypse* (London, 1995)

Leslie, S, *The New Horse-Powered Farm: Tools and Systems for the Small-Scale Sustainable Market Grower* (White River Junction, 2013)

Lesté-Lasserre, C, 'Does Horses' Waste Help or Hinder the Environment?' *TheHorse.com*, 2013 <http://www.thehorse.com/articles/32259/does-horses-waste-help-or-hinder-the-environment>

Lizet, B, *La Bête Noire* (Paris, 1989)

M'Fadyean, J, 'The Prophylaxis of Glanders', *Journal of Comparative Pathology and Therapeutics*, 18 (1905), 23–30

Major, J, 'The Pre-Industrial Sources of Power: Muscle Power', *History Today*, 30/3 (1980)

McGilvray, C, 'The Transmission of Glanders from Horse to Man', *Canadian Journal of Public Health*, 35/7 (July, 1944) 268–275

McKenna, C, *Bearing a Heavy Burden* (London, 2008)

Miele, K, 'Horse-Sense: Understanding the Working Horse in Victorian London', *Victorian Literature and Culture*, 37/01 (2009), 129

Moore-Colyer, R, 'Aspects of Horse Breeding and the Supply of Horses in Victorian Britain', *The Agricultural History Review*, 43/1 (1995)

——, 'Aspects of the Trade in British Pedigree Draught Horses with the United States and Canada', *The Agricultural History Review*, 48/1 (2000), 42–59

——, 'Horses and Equine Improvement in the Economy of Modern Wales', *Agricultural History Review*, 39/2 (1991), 126–42

Morris, E, 'From Horse Power to Horsepower', *Access*, 30, 2007

Norton Greene, A, *Horses at Work: Harnessing Power in Industrial America* (Cambridge, MA, 2008)

Oswald, F, 'Health Hints for Old and Young', *The National Tribune*, 11 April 1889

Pinney, C, 'The Case for Returning to Real Live Horse Power', in *Before the Wells*

Run Dry – Ireland's Transition to Renewable Energy, ed. by R Douthwaite (Dublin, 2003)

Power O'Donoghue, N, *Ladies on Horseback: Learning, Park-Riding, and Hunting, with Hints upon Costume, and Numerous Anecdotes* (London, 1881)

Prothero, R, 'The Stock-Breeder's Art and Robert Bakewell (1725-1795)', in *English Farming Past and Present* (Cambridge, 2013)

Reguzzoni, A, 'Small Farmers Crave Horse Power', *Grist*, December 2011 <http://grist.org/sustainable-farming/2011-12-06-small-farmers-crave-horsepower/>

Richardson, C, *British Horse and Pony Breeds and Their Future* (London, 2008)

Ritvo, H, 'Processing Mother Nature: Genetic Capital in Eighteenth-Century Britain', ed. by J Brewer and S Staves (London, 1995)

Rydberg, T, and J Jansén, 'Comparison of Horse and Tractor Traction Using Emergy Analysis', *Ecological Engineering*, 19/1 (2002), 13–28

'Scrapiana', *The Essex Standard, and Colchester, Chelmsford, Maldon, Harwich and General County Advertiser*, 11 December 1835

'Second Horse Age Is Here, and It Will Stay, Says Sir W. Gilbey', *Gloucestershire Echo*, 13 October 1939

Sewell, A, *Black Beauty* (Norwich, 1879)

Stewart, J, and B Allen, *The Stable Book: Being a Treatise on the Management of Horses, in Relation to Stabling, Grooming, Feeding, Watering and Working. Construction of Stables, Ventilation, Stable Appendages, Management of the Feet. Management of Diseases and Defective Horses* (New York, 1858)

Sutter, J, 'Despite Horses and Buggies, Amish Aren't Necessarily "Low-Tech"', CNN, 2011 <http://edition.cnn.com/2011/TECH/innovation/06/22/amish.tech.brende/>

Tann, J, 'Horse Power 1780–1880', in *Horses in European Economic History: A Preliminary Canter* (1983)

Tarbell, I, 'The Meat of Paris', *Pittsburgh Dispatch*, 10 April 1892

'The Tricks of Horse Dealers', *New Zealand Herald*, 16 May 1868

The Victorian Web <http://www.victorianweb.org/>

Thirsk, J, *Horses in Early Modern England: For Service, for Pleasure, for Power* (Reading, 1977)

Thomas, K, *Man and the Natural World: Changing Attitudes in England 1500–1800* (London, 1983)

Thompson, F, 'Horses and Hay in Britain 1830–1918', in *Horses in European Economic History: A Preliminary Canter* (1983)

——, 'Nineteenth-Century Horse Sense', *The Economic History Review*, 29/1 (1976)

Tollefson, Jeff, 'Intensive Farming May Ease Climate Change', *Nature*, 465/7300 (2010), 853

Various, 'Proceedings of the 6th International Colloquium on Working Equids' (New Delhi, 2010)

————, 'Proceedings of the 7th International Colloquium on Working Equids' (London, 2014)

Velten, H, Beastly London: A History of Animals in the City (London, 2013)

'Watchlist: Equine', Rare Breeds Survival Trust <http://www.rbst.org.uk/Rare-and-Native-Breeds/Equine>

'Whimsical Horse', The Essex Standard, and Colchester, Chelmsford, Maldon, Harwich and General County Advertiser, 31 May 1834

'With Gas Costs Rising, Farmers Take to Mules', NPR, 2008 <http://www.npr.org/templates/story/story.php?storyId=90840231>

Wykes, D, 'Robert Bakewell (1725–1795) of Dishley: Farmer and Livestock Improver', The Agricultural History Review, 52/1 (2004), 38–55

Youatt, W, The Horse: With a Treatise on Draught and a Copious Index (London, 1831)

MEAT

'A Horse Dinner', The Times, 7 February 1868

'A Horseflesh Food Plant at Rockford', True Republican, 7 February 1923

'Abandoned Horses Are Slaughter Rejects, Say Advocates', HorseTalk, 20 November 2011 <http://horsetalk.co.nz/2011/12/20/abandoned-horses-slaughter-rejects-advocates/#axzz3rrMXG2Jx>

'Alleged Dynamiter Made Two Dashes from Court Room', True Republican, 19 December 1925

'Alleged Sledgehammer Attack Bid to Euthanize Animal', HorseTalk, 31 May 2012 <http://horsetalk.co.nz/2012/05/31/alleged-sledgehammer-attack-bid-to-euthanize/#axzz3rrMXG2Jx>

'An Enthusiast on Horse-Flesh', Opelousa Courier, 22 February 1890

'Animal Control Investigate Severely Emaciated Horse Abandoned on Los Angeles Street', The Examiner, 2010 <http://www.examiner.com/article/animal-control-investigate-severely-emaciated-horse-abandoned-on-los-angeles-street>

Animal Plant Health Inspection Service, USDA, Part 88: Commercial Transportation of Equines for Slaughter (Washington DC, 2014) <http://www.gpo.gov/fdsys/pkg/CFR-2014-title9-vol1/pdf/CFR-2014-title9-vol1-part88.pdf>

Animals' Angels, Horsemeat Imports into the EU and Switzerland (Zurich, 2014)

'Arrested for Selling Horse Meat', Pike County Press, 17 January 1896

Associated Press, 'Horse Found with Brand Cut out of Hide', MSNBC.com, 2009 <http://www.nbcnews.com/id/31856204/ns/us_news-life/t/horse-found-brand-cut-out-hide/>

————, 'More Horses Being Let Go, Officials Say', Las Vegas Review Journal, 2 December 2008

Barclay, H, 'Commentary on Marvin Harris' Good to Eat', Anthropos, 84/1 (1989)

————, The Role of the Horse in Man's Culture (London, 1980)

Bicknell, A, 'Hippophagy: The Horse as Food for Man', *Journal of the Society of Arts*, XVI/801 (1868), 349–59

'Big Howl from the Papers over the Canning of Horsemeat in Oregon', *San Francisco Call*, 23 April 1895

Bogdanich, W, J Drape, D Miles, and G Palmer, 'Death and Disarray at America's Racetracks', *New York Times*, 24 March 2012 <http://www.nytimes.com/2012/03/25/us/death-and-disarray-at-americas-racetracks.html?pagewanted=all&_r=1>

Booth Thomas, C, 'T. Boone Pickens to the Rescue', *Time*, July 2006

Bouchet, G, *Le Cheval à Paris de 1850 à 1914* (Paris, 1993)

'Butter Made from Horse Bones', *The Tiffin Tribune*, 8 August 1872

'Cab Horse Cutlets with Whine Sauce', *Perrysburg Journal*, 6 June 1918

Cain Oakford, G, 'Canadian Slaughterhouses Resume Deliveries from U.S.', *Daily Racing Form*, 2012 <http://www.drf.com/news/canadian-slaughterhouses-resume-deliveries-us>

Carpenter, F, 'Horse Meat for Food', *The National Tribune*, 19 January 1893

CBC News, *Truth About Canadian Horse Slaughter – Dr. Temple Grandin*, 2011 <https://www.youtube.com/watch?v=_orfuh2mtos>

Chai, C, and B Leffler, 'Tainted Meat: Banned Veterinary Drugs Found in Horse Meat', *Global News*, 2014 <http://globalnews.ca/news/1193995/tainted-meat-banned-veterinary-drugs-found-in-horse-meat/>

'Chain Pulls Horsemeat after New Cruelty Charges', *The Local*, 2014 <http://www.thelocal.ch/20140312/discount-chain-pulls-horsemeat-after-cruelty-charges>

Chap. 555 An Act to Provide for the Inspection of Live Cattle, Hogs, and the Carcasses and Products Thereof Which Are the Subjects of Interstate Commerce, and for Other Purposes (1891)

'Chappel Brothers', *RockfordReminisce.com* <http://www.rockfordreminisce.com/Chappel_Brothers.html>

'Charge Horses Died on Way to Packing Plant at Rockford', *True Republican*, 28 December 1927

'CHEER UP! Nobody's Said Anything about Horse-Meatless Days', *The Evening World* (19 March 1918)

Colberg, S, and C Casteel, 'U.S. Horse Slaughter Plants in the Very Early Stages of Planning, Proponent Says', *NewsOK*, 2011 <http://newsok.com/article/3626718>

'Condition of the South', *Ottawa Free Trader*, 6 June 1868

Couturier, L, 'Dark Horse', *Orion Magazine*, July 2010

Cowan, T, *Horse Slaughter Prevention Bills and Issues* (Washington DC, 2013) <http://nationalaglawcenter.org/wp-content/uploads/assets/crs/RS21842.pdf>

Cullom, S, *Agricultural Appropriation Bill: Report to Accompany HR6351*, 1898

——, 'Letter of the Secretary of Agriculture, Agricultural Appropriation Bill' (Washington DC, 1898)

Dent, A, and D Machin Goodall, *The Foals of Epona* (London, 1962)

Dierkens, A, 'Réflexions Sur L'hippophagie Au Haut Moyen Age', in *Viandes et Sociétés: Les Consommations Ordinaires et Extra-Ordinaires*, ed. by HASRI (Paris, 2008)

Dobos, A, 'The Lower Palaeolithic Colonisation of Europe. Antiquity, Magnitude, Permanency and Cognition', *PaleoAnthropology*, 2012

Dodman, N, N Blondeau, and A M Marini, 'Association of Phenylbutazone Usage with Horses Bought for Slaughter: A Public Health Risk', *Food and Chemical Toxicology*, 48/5 (2010), 1270–74

Donovan, R, *Conflict and Crisis: The Presidency of Harry S. Truman, 1945–1948* (Columbia and London, 1977)

Doyle, C, *Observations of a Horse Slaughterer Killer, Part I* <https://www.youtube.com/watch?v=GfzX4Fx5xuE>

——, *Observations of a Horse Slaughter Killer, Part II* <https://www.youtube.com/watch?v=m8ZNiRV5-Mw>

——, *Observations of a Horse Slaughter Killer, Part III* <https://www.youtube.com/watch?v=jmzsiUoP_58>

'Dozens of Horse Carcasses Found in Field near Highway 65', *KGET.com*, 2012

Drape, J, 'Doping at U.S. Tracks Affects Europe's Taste for Horse Meat', *New York Times*, 8 December 2012

Drouard, A, 'Horsemeat in France: A Food Item That Appeared during the War of 1870 and Disappeared after the Second World War', in *Food and War in Twentieth Century Europe*, ed. by I Zweiniger-Bargielowska (Farnham, 2011)

'Envelope Flaps', *St Paul Daily Globe*, 17 September 1895

Eurogroup for Animals, 'Press Release: Eurogroup Welcomes Importation Ban of Horsemeat from Mexico' (Brussels, 2014)

'Europe Christmas to Eat Our Horses', *True Republican*, 12 December 1925

European Commission, 'Commission Publishes Encouraging Second Round of EU-Wide Test Results for Horse Meat DNA in Beef Products: Measures to Fight Food Fraud Are Working', press release (2014)

——, DG(SANCO) 2010-8522 – MR FINAL, *Final Report of an Audit Carried out in Canada from 23 November to 06 December 2010* (Brussels, 2010)

——, DG(SANCO) 2010-8524 – MR FINAL, *Final Report of a Mission Carried out in Mexico from 22 November to 03 December 2010* (Brussels, 2010)

——, DG(SANCO) 2011-8913 – MR FINAL, *Final Report of an Audit Carried out in Canada from 13 to 23 September 2011* (Brussels, 2011)

——, DG(SANCO) 2012-6340 – MR FINAL, *Final Report of an Audit Carried out in Mexico from 29 May to 08 June 2012* (Brussels, 2012)

——, DG(SANCO) 2014-7223 – MR FINAL, *Final Report of an Audit Carried out in Mexico from 24 June to 04 July 2014* (Brussels, 2014)

——, DG(SANTE) 2014-7216 – MR FINAL, *Final Report of an Audit Carried out in Canada from 02 to 15 May 2014* (Brussels, 2014)

Eurostat, *EU Horse Production Annual Data* (Brussels, 2012)

FAQ on Phenylbutazone in Horsemeat, 2013 <http://www.efsa.europa.eu/en/faqs/phenylbutazone>

'FBI Most Wanted' <https://www.fbi.gov/wanted>

Ferrières, M, *Sacred Cow, Mad Cow: A History of Food Fears* (New York, 2006)

Fetter, H, 'No, Horse Racing Can't Be Saved—Even by a Triple Crown Winner ', *The Atlantic*, 2015 <http://www.theatlantic.com/entertainment/archive/2014/05/no-horse-racing-cant-be-savedeven-by-a-triple-crown-winner/371255/>

Fitzgerald, A, 'A Social History of the Slaughterhouse: From Inception to Contemporary Implications', *Research in Human Ecology*, 17/1 (2010), 58–69

Food Safety and Inspection Service, and USDA, 'FSIS Directive: Ante-Mortem, Postmortem Inspection of Equines and Documentation of Inspection Tasks' (Washington DC, 2013)

'Foodsavers Eat Horse Meat', *Tacoma Times*, 20 December 1917

Forsyth, J, 'Texas Drought Leaves Heartbreaking Toll of Abandoned Horses', *Reuters*, 2011 <http://www.reuters.com/article/2011/12/03/us-horses-abandoned-idUSTRE7B2oLF20111203>

Freidburg, S, *Fresh: A Perishable History* (Cambridge, MA, 2009)

'Frog Eaters and Hippophagists: Paris Letter to the New York Times', *Nashville Union and America*, 28 April 1875

Gabaccia, D, *We Are What We Eat: Ethnic Food and the Making of Americans* (Cambridge, MA, 1998)

Gavin, A, *Dark Horse: A Life of Anna Sewell* (Stroud, 2004)

Géraud, M, *Essai Sur La Suppression Des Fosses D'aisances, et de Toute Espèce de Voiries, Sur La Manière de Converter En Combustibles Les Substances Qu'on Y Renferme, Etc.* (Amsterdam, 1786)

'Getting Rid of Horses', *Fayetteville Observer*, 17 November 1870

Gordon, G, 'Investors Buy Redmond Slaughterhouse', *The Oregonian*, 22 January 1998

Gordon, W J, *The Horse-World of London* (London, 1893)

Grandin, T, K McGee, and J Lanier, *Survey of Trucking Practices and Injury to Slaughter Horses*, 1999 <http://www.grandin.com/references/horse.transport.html>

'Hadith – Book of Foods (Kitab Al-At'imah)', *Sunnah.com* <http://sunnah.com/abudawud/28/54>

Harris, M, *Good to Eat: Riddles of Food and Culture* (London, 1986)

'He Sold Horse Steaks', *Omaha Daily Bee* (18 December 1890)

Hill, D, 'Can American Pharoah Save Horse Racing? ', *New Yorker*, 2015 <http://www.newyorker.com/news/sporting-scene/american-pharoah-won-the-triple-crown-now-what>

Hoehing, C, *Ueber Die Verwendung Der Thierischen Ueberreste Unserer Hausthiere, Das Pferdefleisch-Essen Und Die Aufhebung Der Kleemeistereien* (Stuttgart, 1848)

Holland, J, and L Allen, 'Analysis of Factors Responsible for the Decline of the U.S. Horse Industry: Why Horse Slaughter Is Not the Solution', *Kentucky Journal of Equine, Agriculture, & Natural Resources Law*, 5 (2012)

'Horse Meat Feast', *Kansas City Journal*, 6 March 1898

'Horse Meat Sale Rises as Meat Shortage Grows', *Chicago Daily Tribune*, 17 November 1942

'Horse Meat to Be Put on Sale Here Thursday for First Time', *Washington Post*, 2 February 1943

'Horse Meat Won't Hurt', *The Science News-Letter*, 43/1 (1943), 5–6

'Horse Passports: Up to 7,000 Unauthorised Documents Issued', BBC News, 2013 <http://www.bbc.com/news/science-environment-21430330>

'Horse Racing Reform' <http://horseracingreform.org/>

Horse Welfare: Action Needed to Address Unintended Consequences from Cessation of Domestic Slaughter (Washington DC, 2011)

'Horseflesh for Food', *The Anaconda Standard*, 1 January 1895

'Horsemeat as Food', *The Islander*, 11 April 1895

'Horsemeat Export', *The Sun*, 15 December 1897

'Horsemeat Is the Latest Delicacy in New York', *Chicago Eagle*, 24 February 1917

'Horsemeat Steak Sir? Yessir, Immejut, Sir', *Washington Times*, 17 January 1916

'Icelandic Horsemeat' <www.skagafjordur.is/displayer.asp?cat_id=4927>

'In Our Great Recession, Unwanted Horses Are Taken to a Wild Herd Started in the Great Depression', *The Rural Blog*, 2012 <http://irjci.blogspot.de/2012/02/in-great-recession-domestic-horses.html>

'Incidents of Abandoned Horses at Sand Wash Basin Concern BLM', *Craig Daily Press*, 2011 <http://www.craigdailypress.com/news/2011/jan/27/incidents-abandoned-horses-sand-wash-basin-concern/>

International Equine Business Association, *The Promise of Cheval*, 2012

'Kalona's Creek Will Not Be Named Horse Meat Forbidden in Japanese', *Kalona News*, 2011 <http://www.kalonanews.com/articles/2011/12/26/news/doc4ef8751c28b27843772826.txt>

'Kaufman Zoning: Horse Slaughter Information Page' <http://www.kaufmanzoning.net/>

Kenning, C, 'Free-Roaming Horses a Growing Problem in E. Kentucky', USA Today, 2015 <http://www.usatoday.com/story/news/nation/2015/02/10/stray-horses-a-growing-problem-in-kentucky/23199035/>

'La Viande de Cheval: De Qualités Indiscutables et Pourtant Méconnues', *Cahiers de Nutrition et de Dietologie*, 23/1 (1988), 25–40

Lambert, W, *Babylonian Wisdom Literature* (Indiana, 1996)

Landrieu, M, S541 *A Bill to Prevent Human Health Threats Posed by the Consumption of Equines Raised in the United States* (2013) <https://www.govtrack.us/congress/bills/113/s541>

Lárusson, H, 'About the Consumption of Horse Meat', *Skaga Fjordur* <www.
skagafjordur.is/displayer.asp?cat_id=4927>

Lees, P, and P-L Toutain, 'Pharmacokinetics, Pharmacodynamics, Metabolism,
Toxicology and Residues of Phenylbutazone in Humans and Horses',
Veterinary Journal (London, England: 1997), 196/3 (2013), 294–303

Leffler, B, M Rowney, and S O'Shea, '16×9 Investigation: Canada's Horse Slaughter
Industry under Fire', *Global News*, 2014 <http://globalnews.ca/news/1186346/
tainted-meat-canadas-horse-slaughter-industry-under-fire/>

Leteux, S, 'Is Hippophagy a Taboo in Constant Evolution?' *Menu: Journal of Food
and Hospitality*, 2012, 1–13

'Letter to the Editor', *New York Daily Tribune*, 15 September 1856

Leuchtenburg, W, 'New Faces of 1946', *Smithsonian Magazine*, November 2006

Levenstein, H, *Revolution at the Table: The Transformation of the American Diet* (Oxford,
1993)

Levine, M A, 'Eating Horses: The Evolutionary Significance of Hippophagy',
Antiquity, 1998

'Like a Kimberley Diet', *Washington Weekly Post*, 27 February 1900

Ling, V, *The Lower Paleolithic Colonisation of Europe* (Oxford, 2011)

'London's Disabled Horses', *The Columbus Journal*, 29 May 1889

'Long Links of Horse', *The Evening World*, 13 December 1889

Marler, B, 'Could Horsemeat (and Kangaroo Meat) Get into the US Food Supply
– Too Late, It Already Did', *Marler Blog*, 2013 <http://www.marlerblog.com/
case-news/could-horsemeat-and-kangaroo-meat-get-into-the-us-food-
supply-too-late-already-did/#.VkSNV6S4ITk>

Mayhew, H, *London Labour and the London Poor* (1968)

'Meat Fraud Climax Nearing in Houston', *Sweetwater Reporter*, 9 July 1948

Meiklejohn, HR 1454 *A Bill to Prevent Discrimination in the Shipment and Transportation
of Live Stock on Vessels to Foreign Countries, and to Provide Penalties for Its Violation*
(1895)

Mendoza, M, 'Trail's End for Horses: Slaughter', *LA Times*, 1997 <http://articles.
latimes.com/1997-01-05/news/mn-15653_1_wild-horses>

Merritt, R, 'Arson Declared at Packing Plant', *Bulletin*, 22 July 1997

Moran, J, 'Officials: Informant Violated Protection', *The Register Guard*, 2011

'National Agricultural Statistics Service', USDA <http://www.nass.usda.gov/
About_NASS/>

'National Can Save $36,000,000 a Year by Eating Cab-Horse Steak', *The Day Book*,
4 January 1916

'New Advocate Against Horse Slaughter? Start Here...', *Canadian Horse Defence
Coalition*, 2015 <https://canadianhorsedefencecoalition.wordpress.
com/2015/03/19/new-advocate-against-horse-slaughter-start-here/>

'New Yorkers Now Eat Flesh of Noble Horse', *Chicago Eagle*, 15 January 1916

'No Demand for Horses', *San Francisco Call*, 5 December 1896

Oakford, G C, 'Canadian slaughterhouses resume deliveries from U.S.,' *Daily Racing Form*, 16 October 2012

Official Journal of the European Communities, *Council Directive 96/23/EC* (Brussels, 1996)

Ogle, M, *In Meat We Trust: An Unexpected History of Carnivore America* (Boston, New York, 2013)

'Old Dobbin Served Up on a Platter! Imagine!', *University Missourian*, 20 January 1916

Ormsby, M, 'Ottawa refuses to say whether drug-tainted horse meat entered food chain', *Toronto Star*, 29 March 2013

Oswald, F, 'Health Hints for Old and Young', *The National Tribune*, 11 April 1889

Otter, Chris, 'Hippophagy in the UK: A Failed Dietary Revolution', *Endeavour*, 35/2–3 (2011), 80–90

Outram, A, 'Logging a Dead Horse: Meat, Marrow and the Economic Anatomy of Equus', *University of Durham and University of Newcastle Upon Tyne Archaeological Reports*, 20 (1996)

'Packing Business', *The Pullman Herald*, 6 April 1895

'Packing Plants in Which Horses Are Prepared for Food', *True Republican*, 6 August 1927

'Permit to Put up Horse Flesh by-Products', *True Republican*, 26 July 1924

Philipps, D, 'Fate of Wild Horses in Hands of BLM, Colorado Buyer', *The Denver Post*, 2012 <http://www.denverpost.com/ci_21654948/fate-wild-horses-hands-blm-colorado-buyer>

Pierre, E, 'L'hippophagie Au Secours Des Classes Laborieuses', *Communciations*, 74 (2003), 177–200

'Questions remain over short-lived slaughter halt,' *HorseTalk*, 17 October 2012 <http://horsetalk.co.nz/2012/10/17/questions-remains-temporary-slaughter-halt/#axzz3vj44sDgp>

Raia, P, 'House Committee Passes Slaughter Ban Amendment', *TheHorse.com*, 2014 <http://www.thehorse.com/articles/33963/house-committee-passes-slaughter-ban-amendment>

Rosebraugh, C, *Burning Rage of a Dying Planet: Speaking for the Earth Liberation Front* (Herndon, VA, 2004)

Saint-Hilaire, G, *Lettres Sur Les Substances Alimentaires et Particulièrement Sur La Viande de Cheval* (Paris, 1856)

'Sale of Horse Meat for Humans Licensed', *Los Angeles Times*, 3 March 1943

'Sausages Made from Horse Meat', *The Daily Dispatch*, 5 December 1857

'Secretary Rusk Put a Stop to the Export of Horse Meat', *The Morning Call*, 22 December 1891

Shenk, P, 'To Valhalla by Horseback? Horse Burial in Scandinavia during the Viking Age' (The Center for Viking and Medieval Studies at the University of Oslo, 2002)

'Shipping Diseased Horse Meat as Family Beef', *Salt Lake Herald*, 22 December 1891

Simoons, F, *Eat Not This Flesh: Food Avoidances from Prehistory to the Present* (Madison, WI, 1994)

Sinclair, U, *The Jungle* (New York, 1906)

'Sirloins and Pork Chops Though Heavens Fall', *Bisbee Daily Review*, 7 June 1919

Sonner, S, 'Four Charged in Corral Firebombing', *Columbian*, 10 April 2006

'St Louis Likes Horse Meat', *Evening Public Ledger*, 22 March 1917

'State Asylum Is Quiet Again after Battle', *Daily Illini*, 19 May 1931

Stillman, D, *Mustang: The Saga of the Wild Horse in the American West* (Boston, New York, 2008)

Strom, S, 'USDA May Approve Horse Slaughtering', *New York Times*, 2013 <http://www.nytimes.com/2013/03/01/business/usda-may-approve-horse-slaughter-plant.html?_r=0>

'Stuttgart, Vom 24 April', *Libausches Wochenblatt*, 29 April 1842

'Tacomans in Arms', *San Francisco Call*, 26 May 1895

Tarbell, I, 'The Meat of Paris', *Pittsburgh Dispatch*, 10 April 1892

Tessier, A, 'Mon Royaume Pour Un Cheval! Une Histoire de L'hippophagie Française', *Les Cahiers de Gastronomie*, 2010 <http://lescahiersdelagastronomie.fr/2010/03/mon-royaume-pour-un-cheval%E2%80%89-une-histoire-de-lhippophagie-francaise>

'The Animal Liberation Front Apparently Is Taking Blame for the Arson That Destroyed a Redmond Horse-Slaughtering Plant', *Bulletin*, 27 July 1997

'The Horseless Age Is Surely Creeping up on Us', *Perrysburg Journal*, 28 March 1918

'The Last Year 40,000 Horses Slaughtered in Rockford Plant', *True Republican*, 15 February 1928

'The Rumor Is Said to Be Abroad in Germany', *Washington Evening Star*, 5 February 1895

The Truth Behind United Horsemen's Pro Slaughter Campaign, 2011 <http://www.youtube.com/watch?v=_ZRvhgoJsQs>

Thomas, K, *Man and the Natural World: Changing Attitudes in England 1500–1800* (London, 1983)

Tim Sappington's Message for Animal Activists <https://www.youtube.com/watch?v=_l5pXviVBig>

'To Buy the Horsemeat Cannery', *Daily Capital Journal*, 31 October 1896

'United Horsemen', *United Horsemen*, <https://www.facebook.com/United-Horsemen-390246801093132/>.

United Horsemen. Advocates of Horse Welfare, 2011 <https://www.youtube.com/watch?v=oK5WHr_XZ-8>

United Press, 'Capture of Con Thwarts Plan to Dynamite Plant', *Urbana Daily Chronicle*, 7 November 1927

Unwanted Horse Coalition, *2009 Unwanted Horses Survey* (Washington DC, 2009)

Veterinarians for Equine Welfare, *Horse Slaughter – Its Ethical Impact and Subsequent Response of the Veterinary Profession*, 2008

'Violations Documented at the Cavel Horse Slaughter Plant in Illinois', *Animals' Angels, North America*, 2014 <http://www.animalsangels.org/issues/horse-slaughter/foia-requests/violations-documented-cavel-horse-slaughter-plant-illinois>

Wagner, S, 'Presentation by Susan Wagner, President and Founder of Equine Advocates', in *International Equine Conference* (Alexandria, VA, 2011)

Waller, D, 'Horse Slaughtering: The New Terrorism?' *Time*, September 2006

'Was Horse Meat Plant Fired by Incendiary?' *True Republican*, 4 November 1925

Weil, K, 'They Eat Horses, Don't They? Hippophagy and Frenchness', *Gastronomica*, 7/2 (2007), 44–51

'What We Are Talking About', *The Sun* (9 December 1889)

Wilson, B, *Swindled: From Poison Sweets to Counterfeit Coffee – the Dark History of the Food Cheat* (London, 2008)

'Woman Who Crawled Inside Gutted Horse Carcass Wanted to "Feel One" With Nature' (2011) <http://www.foxnews.com/us/2011/10/28/woman-who-crawled-inside-gutted-horse-carcass-wanted-to-feel-one-with-nature/>

YouGov, *Would You Eat Horse Meat?*, 2013 <https://today.yougov.com/news/2013/02/28/would-you-eat-horse-meat/>

WEALTH

'2013 International Symposium: Common Development of Sports and Modern Society' (Beijing, 2013)

'A Gallop Through China's Horse Culture', *Chinese Ministry of Culture*, 2003 <http://www.chinaculture.org/gb/en_curiosity/2004-01/07/content_45457.htm>

'All Bets Off', *The Economist*, September 2012 <http://www.economist.com/node/21563708>

Atsmon, Y, 'Tapping China's Luxury-Goods Market', *McKinsey Quarterly*, April (2011)

Balfour, F, 'Billionaire Developer Lures Rich Chinese to Gated Polo Community', *Bloomberg Business*, 24 March 2014 <http://www.bloomberg.com/news/articles/2014-03-24/billionaire-developer-lures-rich-chinese-to-gated-polo-community>

Beckwith, C, 'The Impact of the Horse and Silk Trade on the Economies of T'ang China and the Uighur Empire: On the Importance of International Commerce in the Early Middle Ages', *Journal of the Economic and Social History of the Orient*, 34/3 (1991), 183–98

Berry, M, *The Chinese Classic Novels* (London, 1988)

Besio, K, and C Tung, *Three Kingdoms and Chinese Culture* (Albany, 2007)

Bretherton, M, 'China Rides into Its First Equestrian Olympics', *Horsebytes*, 2008 <http://blog.seattlepi.com/horsebytes/2008/08/06/china-rides-into-its-first-equestrian-olympics/>

Browne, A, 'China's World: Kicking the Luxury Habit', *Wall Street Journal*, 5 November 2013 <http://www.wsj.com/articles/SB1000142405270230348250457917728248577522>

Burkitt, L, 'China: The Next Big Horse Racing Center?' *Wall Street Journal*, 5 June 2014

Chehabi, H, and A Guttman, 'From Iran to All of Asia: The Origin and Diffusion of Polo', *Journal of the History of Sport*, 19 (2002)

Chin Hu, Hsien, 'The Chinese Concepts of Face', *American Anthropologist*, 46/1 (1944) 45-64

Chinese Horse Industry White Paper 2012 (Beijing, 2012)

'Classical Reinterpretation: Rolls-Royce Majestic Horse Collection', *Luxury Insider*, 2013 <http://www.luxury-insider.com/luxury-news/2013/10/classical-reinterpretation-rolls-royce-majestic-horse-collection CNT>

Cooke, B, *Imperial China: The Art of the Horse in Chinese History Exhibition Catalog*, ed. by Kentucky Horse Park (Seoul, 2000)

Creel, H, 'The Role of the Horse in Chinese History', *The American Historical Review*, 70/3 (1965), 647–72

Crump, J, *Chan-Kuo Ts'e* (1979)

Daley, B, 'China's Era of the Horse', *Horse Canada*, 2011

Delacour, C, 'The Role of the Horse and the Camel in Chinese Expansion Along Western Trade Routes', *Orientations*, 32/1 (2001)

Dickinson, S, and A Kipfer, 'China Online Gambling. Illegal But Everywhere', *China Law Blog*, 2014 <http://www.chinalawblog.com/2014/10/china-online-gambling-illegal-but-everywhere.html>

'Dongguan "Peasant Champion" Li Zhenqiang Continue to Be the Son Inherited His Father's Olympic Dream', *Jinyang–Yangcheng Evening News*, 22 August 2014

Duncan, M, '"Sport of Kings" Looks to China's Elite', *Reuters*, 16 February 2011 <http://www.reuters.com/article/us-china-polo-idUSTRE71F0YD20110216>

Eimer, D, 'Chinese Tycoon Xia Yang Inspired by Prince Charles to Restore Polo to Communist China', *Daily Telegraph*, 25 October 2008

Ellis, J, *Cavalry, the History of Mounted Warfare* (Barnsley, 2004)

Elverskog, J, *Our Great Qing: The Mongols, Buddhism, and the States in Late Imperial China* (Honolulu, 2006)

'Equestrian Team Creates New Guinness Record', *CNTV*, 2012 <http://english.cntv.cn/program/cultureexpress/20120112/111526.shtml>

Frank, R, 'China Has a Word for Its Crass New Rich', *CNBC*, 2013 <http://www.cnbc.com/2013/11/15/china-has-a-word-for-its-crass-new-rich.html>

———, 'How Many Chinese Billionaires? Take a Guess', *CNBC*, 2014 <http://www.cnbc.com/2014/03/04/how-many-chinese-billionaires-take-a-guess.html>

Friedman, D, 'The Bling Dynasty', *GQ*, January 2015 <http://www.gq.com/story/chinas-richest>

Gao, Y, 'The Retreat of the Horse: The Manchus, Land Reclamation, and Local Ecology in the Jianghain Plain (ca 1700s to 1850s)', in *Environmental History in East Asia: Interdisciplinary Perspectives*, ed. by T Liu (London, 2014)

Godfrey, M, 'Horses' <http://horses.markgodfrey.eu/#home>

———, 'Racing in Asia', in *The Cambridge Companion to Horse Racing*, ed. by R Cassidy (Cambridge, 2013)

Goodrich, C, 'Riding Astride and the Saddle in Ancient China', *Harvard Journal of Asiatic Studies*, 44/2 (1984), 279–305

Gulik, R, *Hayagriva: Tha Mantrayanic Aspect of the Horse-Cult in China and Japan* (Leiden, 1935)

Harper, E, 'The Origin of Polo – the Game in Ancient China', *Badminton Magazine*, May 1898

Harrist, R, 'The Legacy of Bole: Physiognomy and Horses in Chinese Painting', *Artibus Asiae*, 57/1/2 (1997)

Hendricks, B, *International Encyclopedia of Horse Breeds* (Norman, 1995)

Hillier, B, 'Horse Racing in China: Real, Surreal, or Virtual? ', *Thoroughbred Racing Commentary*, 2014 <https://www.thoroughbredracing.com/articles/horse-racing-china-real-surreal-or-virtual-pt-i/?tid=Racing>

Hong Lee, L, and A Stefanowska, *Biographical Dictionary of Chinese Women* (Hong Kong, 2003)

'Horse Talk in the Year of the Horse' <http://themiddleland.com/cultural/eastern/item/789-horse-talk-in-the-year-of-the-horse>

Jackson, R, 'Xi Jinping Imposes Austerity Measures on China's Elite', *New York Times*, 27 March 2013

Jacobs, A, 'Once-Prized Tibetan Mastiffs Are Discarded as Fad Ends in China', *New York Times*, 17 April 2015

Jowett, P, *Chinese Civil War Armies 1911–49* (Oxford, 1997)

Kelekna, P, *The Horse in Human History* (Cambridge, 2009)

Lees, J, 'Quarantine Issues Delay Fixture in China', *Racing Post*, 25 October 2013

Levine, M, *Prehistoric Steppe Adaptation and the Horse* (Oxford, 2003)

Li, P, 'Stop Cruelty to Animals in the Chinese Entertainment Industry', *Humane Society International*, 2009 <http://www.hsi.org/news/news/2009/08/china_movie_cruelty_081309.html>

Liu, J, 'Polo and Cultural Change: From T'ang to Sung China', *Harvard Journal of Asiatic Studies*, 45/1 (1985), 203–24

———, *The Chinese Knight Errant* (Chicago, 1967)

Lopez, L, 'China's President Just Declared War on Global Gambling', *Business Insider UK*, 6 February 2015

Man, J, *Kublai Khan* (London, 2007)

Mathieson, A, 'Chinese Riders Aim for 2016 Olympics in Rio', *Horse and Hound*, 15 June 2012

Miller, A, 'The Woman Who Married a Horse: Five Ways of Looking at a Chinese Folktale', *Asian Folklore Studies*, 54 (1995), 275–305

Minford, J, and J Lau, *Classical Chinese Literature: An Anthology of Translations* (New York, 2002)

van Moorsel, L, *An Overview of China's Equestrian Industry* (Shanghai, 2010)

Olsen, S, 'The Horse in Ancient China and Its Cultural Influence in Some Other Areas', *Proceedings of the Academy of Natural Sciences of Philadelphia*, 140/2 (1988)

Osburg, J, *Anxious Wealth: Money and Morality Among China's New Rich* (Stanford, 2013)

Osnos, E, 'Is Corruption Souring China on Gold Medals? ', *New Yorker*, 29 January 2015

Paludan, A, *Chinese Emperors: The Reign-by-Reign Record of the Rulers of Imperial China* (London, 2009)

Perdue, P, 'Military Mobilization in Seventeenth and Eighteenth Century China, Russia and Mongolia', *Modern Asian Studies*, 30/4 (1996)

Ramzy, A, 'China Cracks Down on Golf, the "Sport for Millionaires"', *New York Times*, 18 April 2015

Richburg, K, 'China's Xi Jinping to Party Officials: Simplify', *Wall Street Journal*, 5 December 2012

Robinson, D, *Martial Spectacles of the Ming Court* (Cambridge, MA and London, 2013)

Salvacion, M, 'Ancient Leather Balls Found in Xinjiang Show Polo as Sport in Early China', *Yibada*, 13 May 2015

Schafer, E, *The Golden Peaches of Samarkand: A Study of T'ang Exotics* (Oakland, 1985)

Shaughnessy, E, 'Historical Persepectives on the Introduction of the Chariot into China', *Harvard Journal of Asiatic Studies*, 48/1 (1988), 189–237

Smith, P, *Taxing Heaven's Storehouse: Horses, Bureaucrats and the Destruction of the Sichuan Tea Industry, 1074–1224* (Cambridge, MA, 1991)

Snow, E, *Red Star Over China* (New York, 1994)

Spring, M, 'Fabulous Horses and Worthy Scholars in Ninth-Century China', *T'oung Pao*, 74/4/5 (1988), 173–210

Thomas, N, 'China's Equestrian Industry Faces Big Hurdles', *Reuters*, 19 March 2014

'Tough, but Bright – Dongguan's World Champion on Horse-Riding in China', *Dongguan Today*, 2 April 2013 <http://www.dongguantoday.com/news/dongguan/201302/t20130204_1748193.shtml>

Tze-wei, N, 'Land Row Leaves Racehorses Starving', *South China Morning Post*, 12 November 2009

Tzu, C, and J Legge, *The Writings of Chuang Tzu* (Oxford, 1891)

Tzu, L, and A Waley, *The Way and Its Power: A Study of the Tao Te Ching and Its Place in Chinese Thought* (London, 1934)

'US Has 9.5 Million Horses, Most in World, Report Says', *Veterinary News*, 2007 <http://veterinarynews.dvm360.com/us-has-95-million-horses-most-world-report-says?rel=canonical>

Waley, A, 'The Heavenly Horses of Ferghana: A New View', *History Today*, 1955, 95–103

———, *The Real Tripitaka and Other Stories* (London, 1952)

Wong, E, 'Survey in China Shows a Wide Gap in Income', *New York Times*, 19 July 2013

Wood, F, *The Silk Road: Two Thousand Years in the Heart of Asia* (Oakland, 2004)

Xiaoyan, W, 'The Continuation and Abolishment of Official Tea-Horse Trade during Qing Dynasty', *China's Borderland History and Geography Studies*, 4 (2007)

Xiuqin, Z, 'Emperor Taizong and His Six Horses', *Orientations*, 32/2 (2001)

WAR

Baum, D, 'The Price of Valor', *New Yorker*, 12 July 2004

Bazay, C, 'Rider and Handler Effect on Horse Behavior', *TheHorse.com*, 2011 <http://www.thehorse.com/articles/28267/rider-and-handler-effect-on-horse-behavior>

Beamish, H, *Cavaliers of Portugal* (New York, 1969)

Beckman, M, and E Painter, 'Riding Rehab: Veterans with Limb Loss Benefit', *North American Riding for the Handicapped Association*, 2009 <http://www.pathintl.org/images/pdf/resources/horses-heroes/RidingRehab.pdf>

Beetz, A, K Uvnäs-Moberg, H Julius, and K Kotrschal, 'Psychosocial and Psychophysiological Effects of Human-Animal Interactions: The Possible Role of Oxytocin', *Frontiers in Psychology*, 3 (2012), 234

Boiselière, E, *Eperonnerie et Parure Du Cheval de l'Antiquité À Nos Jours* (Brussels, 2005)

Brown, P, 'Rehabilitation of the Combat-Wounded Amputee', in *Orthapedic Surgery in Vietnam*, ed. by W Burkhalter (Washington DC, 1968)

Crane, S, 'Chivalry and the Pre/Postmodern', *Postmedieval: A Journal of Medieval Cultural Studies*, 2/1 (2011), 69–87

Davidson, A, 'The Soldier, the Telephone, and the Rose', *New Yorker*, 2010 <http://www.newyorker.com/news/amy-davidson/the-soldier-the-telephone-and-the-rose>

Davis, N, 'Horse Genome Sequence and Analysis Published in Science', *Broad Institute of MIT and Harvard*, 2009 <https://www.broadinstitute.org/news/1373>

DiMarco, L, *War Horse: A History of the Military Horse and Rider* (Yardley, PA, 2008)

DiNardo, R, *Mechanized Juggernaut or Military Anachronism?: Horses and the German Army of World War II* (New York, 1991)

Ellis, J, *Cavalry, the History of Mounted Warfare*, (Barnsley, 2004)

van Emden, R, *Tommy's Ark: Soldiers and Their Animals in the Great War* (London, New York, 2010)

Epona TV, *Dominance as Culture*, 2015 <http://epona.tv/dominance-as-culture>

———, *Horse Culture*, 2015 <http://epona.tv/horse-culture>

———, *The Ideal Stallion*, 2015 <http://epona.tv/the-ideal-stallion>

Fallows, J, 'The Tragedy of the American Military', *The Atlantic*, January/February 2015

Finkel, D, *Thank You for Your Service* (New York, 2013)

Goody, J, *Metals, Culture and Capitalism: An Essay on the Origins of the Modern World* (Cambridge, 2012)

Graeber, D, *Debt: The First 5,000 Years* (Brooklyn NY, 2011)

Gray, J, *Straw Dogs: Thoughts on Humans and Other Animals* (London, 2003)

Gronow, R, *Captain Gronow's Last Recollections: Being the Fourth and Final Series of His Reminiscences and Anecdotes* (London, 1866)

Groopman, J, 'The Grief Industry', *New Yorker*, 26 January 2004

Halpern, S, 'Virtual Iraq', *New Yorker*, 19 May 2008

Hoexter, M Q, G Fadel, A C Felício, M B Calzavara, I R Batista, M A Reis, and others, 'Higher Striatal Dopamine Transporter Density in PTSD: An in Vivo SPECT Study with [(99m)Tc]TRODAT-1', *Psychopharmacology*, 224/2 (2012), 337–45

Hyland, A, *The Medieval Warhorse from Byzantium to the Crusades* (Far Thrupp, Stroud, Gloucestershire, Dover NH, 1994)

———, *The War Horse in the Modern Era: Breeder to Battlefield, 1600 to 1865* (Stockton-on-Tees UK, 2009)

———, *The Warhorse, 1250–1600* (Stroud, 1998)

Keeling, L J, L Jonare, and L Lanneborn, 'Investigating Horse-Human Interactions: The Effect of a Nervous Human', *Veterinary Journal*, 181/1 (2009), 70–71

Kropotkin, P, *Mutual Aid, a Factor of Evolution* (Boston, 1955)

Law, R, *The Horse in West African History: The Role of the Horse in the Societies of Pre-Colonial West Africa* (Oxford, New York, 1980)

Lesté-Lasserre, C, 'Dopamine and Horses: Learning, Stereotypies, and More', *TheHorse.com*, 2015 <http://www.thehorse.com/articles/36130/dopamine-and-horses-learning-stereotypies-and-more>

———, 'Study: Horses More Relaxed Around Nervous Humans', *TheHorse.com*, 2012 <http://www.thehorse.com/articles/29455/study-horses-more-relaxed-around-nervous-humans>

Loch, S, *The Royal Horse of Europe* (London, 1986)

Marvin, G, *Bullfight* (Oxford, UK and New York, USA, 1988)

Masters, N, 'Equine Assisted Psychotherapy for Combat Veterans with PTSD' (Washington State University Vancouver College of Nursing, 2010) <https://research.wsulibs.wsu.edu/xmlui/bitstream/handle/2376/3434/N_Masters_011005659.pdf?sequence=1>

'Memorial to Fallen Horses Unveiled', *HorseTalk*, 12 August 2010 <http://horsetalk.co.nz/news/2010/08/124.shtml#axzz3toWiQkWM>

Mills, D, *The Domestic Horse: the Origins, Development and Management of Its Behaviour* (Cambridge, 2005)

Mitchell, P, Horse Nations: The Worldwide Impact of the Horse on Indigenous Societies Post-1492 (Oxford, 2015)

de Montaigne, M, and C Cotton, 'Of the War-Horses, Called Destriers', in The Essays of Michael Seigneur de Montaigne, ed. by E Coste (London, 1776)

Morris, D, 'How Much Does Culture Really Matter to PTSD?' New Yorker, 16 July 2013

Natterson-Horowitz, B, Zoobiquity: The Astonishing Connection between Human and Animal Health, 2013

Rees, L, The Horse's Mind (New York, 1985)

Samatar, S, 'Somalia's Horse That Feeds His Master', African Languages and Cultures, Supplement No. 3, Voice and Power: The Culture of Language in North-East Africa. Essays in Honour of B. W. Andrzejewski, 1996, 155–70

Segman, R, R Cooper-Kazaz, F Macciardi, T Goltser, Y Halfon, T Dobroborski, and others, 'Association between the Dopamine Transporter Gene and Posttraumatic Stress Disorder', Molecular Psychiatry, 7/8 (2002), 903–7

Singleton, J, 'Britain's Military Use of Horses 1914–1918', Past & Present, 139, 1993, 178–203

Stanton, D, Horse Soldiers: The Extraordinary Story of a Band of Special Forces Who Rode to Victory in Afghanistan (London, 2010)

Steele, R, Mediaeval Lore from Bartholomew Anglicus (London, 1905)

Taylor, L, Mourning Dress: a Costume and Social History (London, 2009)

Thompson, K, 'Binaries, Boundaries and Bullfighting: Multiple and Alternative Human–Animal Relations in the Spanish Mounted Bullfight', Anthrozoos: A Multidisciplinary Journal of The Interactions of People & Animals, 23/4 (2010), 317–36

Thompson, K, 'Le Voyage Du Centaur: La Monte à La Lance En Espagne (XVIe–XXIe)', in A Cheval: Ecuyers, Amazones et Cavaliers, ed. by D Roch and D Reytier (Versailles, 2007)

'Two Military Horses Saluted for Exemplary Service to US Military', Horsetalk, 26 May 2014 <http://horsetalk.co.nz/2014/05/26/two-military-horses-saluted-exemplary-service-us-military/#axzz3toWiQkWM>

'War Horses Conference', School of Oriental and African Studies (London, 2014) <http://www.soas.ac.uk/history/conferences/war-horses-conference-2014/>

Wathan, J, A M Burrows, B M Waller, and K McComb, 'EquiFACS: The Equine Facial Action Coding System', PLoS One, 10/8 (2015)

Webbe Dasent, G, The Story of Burnt Njal (New York, 1900)

West, M, Indo-European Poetry and Myth (Oxford, 2007)

For further information on sources, please visit susannaforrest.com

Index